U0179265

Introduction to Civil Engineering

土木工程导论

·第3版·

汉英对照

主　编：刘荣桂　胡白香
　　　　唐小卫

副主编：韩　豫　陆春华
　　　　戴　丽　朱　炯

主　审：吕志涛

江苏大学出版社
JIANGSU UNIVERSITY PRESS

镇江

图书在版编目(CIP)数据

土木工程导论：汉英对照 / 刘荣桂，胡白香，唐小
卫主编. — 3 版. — 镇江：江苏大学出版社，2022.8
ISBN 978-7-5684-1678-8

Ⅰ.①土… Ⅱ.①刘… ②胡… ③唐… Ⅲ.①土木工
程－高等学校－教材－汉、英 Ⅳ.①TU

中国版本图书馆 CIP 数据核字(2022)第 089491 号

土木工程导论(第 3 版)

Tumu Gongcheng Daolun(Di-San Ban)

主　　编/	刘荣桂　　胡白香　　唐小卫
责任编辑/	李菊萍
出版发行/	江苏大学出版社
地　　址/	江苏省镇江市京口区学府路 301 号(邮编：212013)
电　　话/	0511-84446464(传真)
网　　址/	http://press.ujs.edu.cn
排　　版/	镇江市江东印刷有限责任公司
印　　刷/	镇江恒华彩印包装有限责任公司
开　　本/	787 mm×1 092 mm　1/16
印　　张/	20.5
字　　数/	573 千字
版　　次/	2013 年 5 月第 1 版　2016 年 7 月第 2 版　2022 年 8 月第 3 版
印　　次/	2022 年 8 月第 3 版第 1 次印刷　累计第 9 次印刷
书　　号/	ISBN 978-7-5684-1678-8
定　　价/	52.00 元

如有印装质量问题请与本社营销部联系(电话：0511-84440882)

序

几千年以来,土木工程历经了几个划时代的发展过程。从古埃及金字塔、古希腊雅典卫城、印度泰姬陵、中国长城,再到法国埃菲尔铁塔、澳大利亚悉尼歌剧院和中国的三峡工程、青藏铁路等,建成了无数伟大的工程。随着社会生产力的不断发展,人类对建筑功能的要求也日益复杂多样,大量不同类型的结构体系,与之相伴的设计、计算理论、实验与测试方法、施工技术与工程管理方法等都在变革、创新的驱动力作用下不断更新、发展与进步。如今土木工程已成为国民经济发展的主要支柱产业之一。为了让本专业及对本专业感兴趣的中外学生了解土木工程的概况,作者决定编写双语版《土木工程导论》一书。

《土木工程导论》是以全国土木工程专业建设指导委员会最新专业指导意见(2012年)为指南,为土木工程及相关专业的中外大学生了解土木工程学科及其分支领域历史沿革与发展状况而编写的概论性教材,采用中英文双语编写。

本教材以“引领专业学习,激发专业兴趣”为主旨,立足于反映近20年土木工程领域的主要理论与应用成果以及未来发展趋势,通过系统地介绍土木工程(含部分工程管理知识)学科的基本框架及土木工程专业的基本概念、主要特点、基础理论与技术方法等,帮助学生在进入专业学习之前了解土木工程行业及学科的基本情况与发展态势、执业资格认证体系及就业导向等,最终激发学生对本专业学习、研究的兴趣,为今后的专业学习奠定基础;同时,通过“知识拓展”“相关链接”和“小贴士”等形式,使用网络等技术,扩展知识面,提供学生自学通道,为学生的学业生涯设计、就业核心竞争力的提升提供专业支撑。

作者及编写组的多数成员已从事土木工程专业教学20年以上,并一直活跃在土木工程科研和社会服务一线,理论与实践经验比较丰富。本书尽作者所能,力图使编写内容从整体上反映土木工程学科的综合性、理论性、技术性和实用性,重点展现我国土木工程大建设、大发展、大提升的最新成果与未来发展趋势,并与国际接轨。

　　本教材充分借鉴国内外土木工程概论类教材和课程教学的先进经验,如美国爱荷华州立大学工程学院开设的"土木工程的伟大成就(The great achievements of civil engineering)"课程等,贯穿式地介绍了土木工程在推动人类社会进步过程中的巨大作用、土木工程建设与可持续发展的关系、工程全寿命期管理等。本书内容丰富,观点独特,写作严谨认真,力求文字精准,图文并茂。本书的出版将弥补目前有关土木工程导论(或概论)类教程编写的不足,也可为专科生、研究生等不同层次读者了解土木工程学科提供参考。

中国工程院院士　吕志涛

2013年4月

前　言

　　《土木工程导论》(双语)是以全国土木工程专业指导委员会最新(2012年)专业指导意见为指南,为土木工程及相关专业的中外大学生编写的,旨在介绍土木工程学科及其分支领域历史沿革与发展状况的概论性教材,采用中英文双语编写。

　　本教材以"引领专业学习,激发专业兴趣"为主旨,立足于反映近20年土木工程领域的基本概况、国内外主要工程应用成果以及未来发展趋势,力图通过系统地介绍土木工程(含部分工程管理内容)学科的基本框架和关键知识,让学生形成对本学科和相关专业的宏观了解,激发学生对本专业学习、研究的兴趣,为今后的专业学习奠定基础。教材通过二维码、网站链接等形式,打开学生自学的通道,为学生的学业生涯及职业生涯设计、就业核心竞争力的提升提供专业支撑。

　　编者力求从整体上反映土木工程学科的综合性、理论性、技术性和前沿性,重点展现我国土木工程大建设、大发展、大提升的最新成果与未来发展趋势。本书的部分内容融合了编者及其科研团队最新的研究成果,写作风格力求文字精准,图文并茂。

　　本教材由绪论统领全书,以土木工程各子学科和分支体系,以及土木工程实施中的关键管理任务为基本架构展开介绍,最终通过系统和全面的学科展望,力求提升学生的专业素养,建立起宏观的土木工程及相关知识体系框架,完成引领后续专业课程学习的任务。

　　本教材主编单位为南通理工学院,并得到南通市建筑结构重点实验室(编号为CP12015005)的资助。参加编写的单位有南通理工学院、江苏大学、苏中建设集团、扬州大学、南京建工集团和徐州工程学院等。参加编写本书人员的具体分工如下:

刘荣桂[①]、李琮琦[④]、孙进[①]编写第 1 章;戴丽[①]、陆春华[②]、钱红[③]、鲁开明[⑤]编写第 2 章;王海超[①]、高海建[①]、胡白香[②]编写第 3 章;陈好[②]、沈圆顺[②]编写第 4 章;延永东[②]、朱炯[⑥]、孙智鑫[①]编写第 5 章;徐荣进[②]、唐小卫[③]编写第 6 章;操礼林[②]、刘荣桂编写第 7 章;苏志忠[①]、谢桂华[②]编写第 8 章;殷杰[②]、朱炯编写第 9 章;韩豫[②]、吴旭[①]、孙莹[②]、杨峰[①]编写第 10 章。刘荣桂制定了编写大纲并对全书进行了最后统稿;翁煜[①]、姜佩弦[①]和张灵灵[①]对全书英文进行了校对。东南大学吕志涛院士对本书(第 3 版)进行了主审。在此表示一并感谢。

　　感谢省级科技服务平台培育项目(XQPT202102)对本书的资助,感谢苏中智慧产业学院为本书提供的案例库。

　　感谢东南大学、浙江大学、江苏省建筑科学研究院、扬州大学、徐州工程学院等兄弟单位的技术帮助与支持;感谢编者课题组的老师与研究生为本书编写所做的工作;感谢江苏大学出版社汪再非、李菊萍及其同仁为本书的顺利出版所做出的不懈努力。最后,要特别感谢东南大学的吕志涛院士对本书出版的指导与支持。

　　土木工程导论作为一本科普性、引导性教程,涉及的问题很多且较为复杂,尚有许多研究亟待完善。希望使用本书进行教学的各位同仁多提宝贵意见,以便再版时修改与完善。由于编者水平有限,书中难免存在不足之处,恳请读者批评指正。

刘荣桂

2022 年 6 月

---

① 南通理工学院;② 江苏大学;③ 苏中建设集团;④ 扬州大学;⑤ 南京建工集团;⑥ 徐州工程学院。

Preface

Introduction to Civil Engineering is a textbook written in both Chinese and English. It is compiled for Chinese and foreign students majoring in civil engineering or the related specialties, according to the latest professional guidance given by National Steering Committee of Civil Engineering Specialty in 2012. Students can learn about the historical evolution and development of the civil engineering and its branches from this book.

The aim of this textbook is to lead students to professional fields and arouse their professional interest. Therefore, this book is designed to report the general situation of civil engineering over the past 20 years, the main achievements in engineering application both at home and abroad, and the development trend in the future. The framework and the key knowledge of civil engineering (including some project management knowledge) are introduced systematically to help students get the basic information of civil engineering and the related fields, arouse students' interest in professional learning and research, and lay the foundation for professional learning in the future. The book provides students with the way of studying professional knowledge by themselves, and making plans for their academic and professional career and improving the core competitiveness in getting jobs through the form of QR code and website links, etc.

As a whole, the content reflects the comprehensive, theoretical, technical and frontier properties of civil engineering discipline, while the latest achievements of great constructions in China, the improvement, and the development trend in the future in civil engineering field are emphasized. Part of the book is the latest research results of the author and his team members. The author aims for accuracy in writing, and abundant pictures are included in the book.

This textbook is guided by the introduction and extended by the basic framework of civil engineering, its various sub-disciplines branch systems combined with the key management tasks in civil engineering implementation. Through systematic and comprehensive prospects for civil engineering, this textbook will enhance students' professional knowledge, help them set up a macroscopic knowledge system and framework for civil engineering as well as the related knowledge, and lead them to the following specialized courses.

This textbook is edited by Nantong Institute of Technology, and supported by Nantong Key Laboratory of Building Structures (No.CP12015005). Nantong Institute of Technology, Jiangsu University, Jiangsu Suzhong Construction Group, Yangzhou University, Nanjing Construction Engineering Group and Xuzhou University of Technology participated in the compilation. The specific division of labor of those involved in the preparation of this book is as follows: Liu Ronggui[1], Li Congqi[4], Sun Jin[1] wrote Chapter 1; Dai Li[1], Lu Chunhua[2], Qian Hong[3] and Lu Kaiming wrote Chapter 2; Chapter 3 was written by Wang Haichao[1], Gao Haijian[1] and Hu Baixiang[2]; Chen Yu[2] and Shen Yuanshun[2] wrote Chapter 4; Yan Yongdong[2], Zhu Jiong[6] and Sun Zhixin[1] wrote Chapter 5; Chapter 6 was written by Xu Rongjin[2] and Tang Xiaowei[3]; Cao Lilin[2] and Liu Ronggui wrote Chapter 7; Su Zhizhong[1] and Xie Guihua[2] wrote Chapter 8; Chapter 9 was prepared by Yin Jie[2] and Zhu Jiong; Han Yu[2], Wu Xu[1], Sun Ying[2] and Yang Feng[1] wrote Chapter 10. Liu Ronggui formulated the compilation outline and finished the final draft of the book. The book was proofread and revised in English by Weng Yu[1], Jiang Peixian[1] and Zhang Lingling[3]. Academician Lv Zhitao of Southeast University reviewed the book(3rd edition). Thank you all.

Thank to the Provincial Science and Technology Service Platform Cultivation Project (XQPT202102)for the funding of this book, and thanks to Suzhong Intelligent Architecture Industrial College for providing the case for this book.

Thanks for the technical assistance and support from Southeast University, Zhejiang University, Jiangsu Institute of Building Research, Yangzhou University, Xuzhou Institute of Technology and other fraternal units. Thanks to Wang Zaifei, Li Juping and his colleagues from Jiangsu University Press for their tireless efforts for the publication of this book. The graduates of the author made a great contribution to the book, to whom I owe many thanks. Finally, special thanks to Academician Lv Zhitao who guided and supported the publishment of this book.

As an educational and instructive textbook, *Introduction to Civil Engineering* involves a lot of complex problems to be resolved. Criticism and suggestions from the readers will be highly welcomed and appreciated. Due to the limited knowledge of the authors, defects in the textbook are unavoidable. We deeply appreciate your criticism and correction.

Liu Ronggui

June 2022

---

[1] Nantong Institute of Technology; [2] Jiangsu University; [3] Jiangsu Suzhou Construction Group; [4] Yangzhou university; [5] Nanjing construction Engineering Group; [6] Xuzhou university of Technology.

目 录

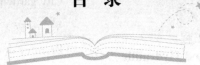

Contents

第1章 绪 论

Chapter 1　Introduction

　　本章主要介绍土木工程的基本概念及土木工程专业涉及的主要技术领域。通过本章的学习,可以加深对土木工程的含义、类型、发展历史的认识,了解土木工程未来的发展趋势,激发对土木工程专业知识的学习兴趣,明确土木工作者的责任。

1.1　土木工程的内涵

1.1.1　土木工程

　　国务院学位委员会在学科简介中对土木工程所下的定义如下:土木工程(civil engineering)是建造各类工程设施的科学技术的统称。它既指建设的对象,即建造在地上、地下、水中的工程设施,也指应用的材料与设备和进行的勘测、设计、施工、保养及维修等专业技术。土木工程是一个专业覆盖面极广的一级学科。

　　This chapter mainly introduces the basic concepts of civil engineering and its related technological fields. After studying this part, people will have a deeper understanding of the concepts, types, development history and the developing tendency of civil engineering. Besides, people can build their interests in civil engineering and realize civil workers' responsibilities.

1.1　Connotations of civil engineering

1.1.1　Civil engineering

　　Civil engineering is defined by the Academic Degrees Committee of the State Council in the introduction to subjects as follows: Civil Engineering is the general term of the construction technology of various projects. It not only refers to its objects such as the engineering facilities constructed on the ground, underground and underwater, but also includes the materials, equipment and the professional technology of investigation, construction and maintenance. Civil engineering is regarded as a first-level discipline with a wide professional coverage.

1828 年，英国土木工程师协会在其皇家宪章中给出这样的定义："土木工程是利用伟大的自然资源为人类造福的艺术"。其作为国内生产和交通的实现手段，促进了国内外贸易。例如，应用于道路、桥梁、渡槽、运河、内河航运和码头的建设，而这些建设又服务于内部流通和交换；应用于港口、码头、防波堤和灯塔的建设；应用于以商业为目的，以人为的力量进行的航海行为；应用于机械的建设和使用；应用于城市和乡镇的排水系统。

"Civil"一词来源于拉丁文"公民"。 1782 年，英国人 John Smeaton 为了把他的非军事工程工作区别于当时占优势地位的军事工程师的工作而采用该名词。自那时起，"土木工程"一词常被从事公共设施建设的工程师所使用，尽管其包含的领域更为广阔。"土木"在中国是一个古老的术语，意指建造房屋，而古代建房主要依靠"土"（包括岩石、沙、泥土、石灰及由土烧制成的砖、瓦等）和"木"（包括木材、茅草、竹子、藤条等），故将 civil engineering 译为"土木工程"。

随着时代的发展和技术的进步，土木工程被不断注入新的内容，显示出勃勃生机，其中工程材料的变革和力学理论的发展起着最为重要的推动作用。现代土木工程早已不再是传统意义上的砖、瓦、灰、砂石，而是由新理论、新材料、新技术、新方法武装起来的众多领域和行业不可缺少的大型综合性学科群，是一个古老而年轻的学科。

In 1828, in its Royal Charter, the Institution of Civil Engineers defined civil engineering as the art of benefiting human beings by using great natural resources. Civil engineering works as the means of production and transportation for external and internal trade. For example, it can be applied to the construction of roads, bridges, aqueducts, canals, river navigation and docks for internal intercourse and exchange; it can also be applied to the construction of ports, harbors, moles, breakwaters, lighthouses and the navigation by artificial power for the purposes of commerce, the construction and application of machinery, as well as the drainage system both in cities and towns.

The word "civil" derives from the Latin word "citizen". In 1782, Englishman John Smeaton used this term to differentiate his non-military engineering work from that of the military engineers which predominated at that time. Since then, the term civil engineering has often been used among engineers who build public facilities, although the field of civil engineering is supposed to be much broader. "Civil" is an ancient term in Chinese, which means construction. In ancient times, "soil" (including rock, sand, clay, lime, the brick and tile made of soil) and "wood" (including wood, thatch, bamboo and rattan) have been mainly used in the construction. Therefore, the meaning of "civil engineering" in Chinese includes the meaning of wood and soil.

With the development of technology and the times, civil engineering is developing quickly, in which the revolution of engineering material and the development of mechanical theory play the most important roles. The modern civil engineering is not just the brick and stone in the traditional sense, but an indispensable and comprehensive subject clot armed with the new theory, new materials, new technologies and new methods. It is a young subject with a long history of development.

1.1.2 土木工程的范畴

土木工程是工程学科分支之一,旨在为人们提供舒适而安全的生活。人们的生活离不开衣、食、住、行。其中"住"是人们最基本的生活需求之一,它与土木工程直接相关,而供水及灌溉工程的合理规划与设计可有效提高粮食产量。从古老的金字塔到今天的薄壳结构,所有的工程奇迹都是土木工程不断发展的结果,而公路、铁路、桥梁等运输线路也是土木工程师的工作成果。

随着近现代工程技术和科学技术的迅猛发展,土木工程逐渐被划分成一些专门分支学科,如结构工程、岩土工程、交通工程、环境工程、水利工程、建设工程、材料科学、测量学、城市工程等,其包含的内容和涉及的范围非常广泛。土木工程不仅为人类生存与发展建造了单体的建筑、桥梁、隧道、大坝等,也创造了城市、乡村、厂矿等综合的生态与环境。

若业主拥有充足的资金,那么建设项目的所有步骤可按图 1.1 所示的流程完成。

一个项目开工之初,土木工程师要对场地进行勘测、测绘,如地下水水位、下水道和电力线高程等。岩土工程师则进行土力学试验以确定土壤能否承受工程荷载。环境工程师研究工程对当地的影响,包括可能对空气和地下水产生的污染,对当地动植物的影响,以及如何让工程设计满足政府对环境保护的要求等。交通工程

1.1.2 The scope of civil engineering

Civil engineering is the branch of engineering, which aims to provide the comfortable and safe life for people. People cannot live without clothing, food, shelter and transportation. Shelter, one of the primary needs of mankind, directly relates to civil engineering. The efficient planning of water supply and irrigation system can increase the food production. The engineering marvels in the world, starting from the ancient pyramids to thin shell structures, are the results of the development in civil engineering. Besides, the transport routes like roads, railways, bridges are also the products of civil engineers.

With the rapid development of the modern engineering and the technology, civil engineering has been gradually subdivided into some specific branch disciplines, including structural engineering, geotechnical engineering, transportation engineering, environmental engineering, hydraulic engineering, construction engineering, materials science, surveying science and urban engineering, etc. The content and the scope of civil engineering are much broader. Civil engineering not only includes the monomer building for human survival and development, such as bridges, tunnels and dams, but also contains the ecology and environment in cities, villages, factories and so on.

Assuming that the house owner has enough money, all steps of the construction project for housing or industry can be implemented as follows in Fig.1.1.

When a project begins, the site should be surveyed and mapped by civil engineers, such as the stage of underwater, sewer and elevation of power lines. Geotechnical specialists will perform soil experiments to determine whether the earth can bear the weight of the project. Environmental specialists will study the impact of the project on the local area: the potential pollution for air and groundwater, the impact on local animals and plants, and how the project can be designed to meet government's requirements aimed at the protection of the environment. Transportation specialists will determine what

专家确定不同种类的必需设施,以减轻整个工程对当地交通网络的负担。同时,结构工程专家利用初步数据对工程作详细规划、设计和说明。

    从项目开始到结束,对这些土木工程师的工作进行监督和调配的则是施工管理专家(工程监理)。根据其他工程师所提供的信息,施工管理专家计算所需材料和人工的数量和开支,确定所有工作的进度表,订购工作所需要的材料和设备,雇佣承包商和分包商,还要做些额外的监督工作以确保工程能按时按质完成。

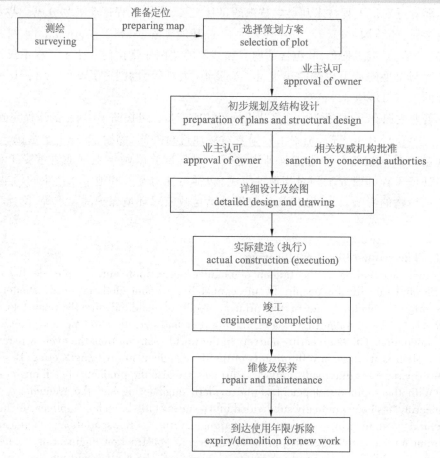

图1.1　工程建设概要

Fig. 1.1　Outline of the construction activity

kinds of facilities are needed to ease the burden on local transportation networks. Meanwhile, structural specialists will use preliminary data to make detailed designs, plans and specifications for the project.

    From the beginning to the end of the project, the construction management specialists will be in charge of the supervision and coordination of the civil engineering specialists (project supervison). Based on the information supplied by the other specialists, construction management specialists will estimate quantities and costs of materials and labor. They will also schedule all tasks, order materials and equipments, hire contractors and subcontractors, and perform other supervisory work to ensure the project will be completed on time and as specified.

1.2 土木工程发展简史

土木工程的发展可以分为 3 个阶段:古代土木工程、近代土木工程和现代土木工程。

1.2.1 古代土木工程

古代土木工程的历史跨度很长,大致从旧石器时代(约公元前 8 000 年)到 17世纪中叶。在这一时期内,人们修建各种设施时主要依靠经验,没有设计理论指导,所运用的材料也大多取自于自然,如石块、草筋、土坯等,大约在公元前 1 000 年才采用烧制的砖。这一时期,所用的工具也很简单,只有斧、锤、刀、铲和石夯等手工工具。尽管如此,古人还是以他们卓越的智慧建造了许多具有历史价值的建筑。

人类历史上最早的土木工程大约出现在公元前 4 000 年到公元前 2 000 年的古埃及和美索不达米亚,那时人类开始放弃游牧生活方式,因而需要有作为庇护所的建筑。

古巴比伦人和亚述人曾致力于解决包括水坝、大堤和运河在内的水利工程问题。他们通过考虑直角三角形的三边关系来解决问题,同时也能解一些简单的代数方程。他们能计算土地面积、砖石砌体体积及开挖运河所必需的土方量等。亚述帝国完成了第一次有组织的道路建设,而第一座具有技术含量的桥梁则于公元前 6 世纪建造在幼发拉底河上。

1.2 Brief history of civil engineering

The development history of civil engineering can be divided into three stages: ancient civil engineering, modern civil engineering and contemporary civil engineering.

1.2.1 Ancient civil engineering

The ancient civil engineering had a long time span, roughly from the Paleolithic Age (8 000 BC) to the mid-17th century. During this period, people built all kinds of facilities mainly depending on experience without any guidance. The materials are mostly taken from nature, such as rocks, grass and adobe. Fired brick was not adopted until 1 000 BC. During this period, the used tools were very simple, only include stone axes, hammers, knives, sickles, stone ram and other hand tools. Even without advanced tools, the outstanding ancients still built lots of buildings with historical value.

The earliest practice of civil engineering may appear between 4 000 BC and 2 000 BC in ancient Egypt and Mesopotamia, when human beings started to abandon their nomadic way of life. Instead, they needed one kind of construction as the shelter.

Babylonians and Assyrians struggled with problems of hydraulic engineering involving dams, levees and canals. They solved problems by considering the sides of right triangle, and they also solved simple algebraic equations. They calculated the areas of land, volumes of masonry, the necessary cubic contents of excavation for canals. The first organized road building was finished in the Assyrian Empire, and the first high-tech bridge was constructed over the Euphrates River in the 6th century BC.

　　古埃及人用最原始的力学原理和工具建造了众多的庙宇和金字塔,至今仍有许多不朽建筑屹立不倒,包括吉萨大金字塔和卡纳克阿蒙-拉神庙(见图1.2)。高146.6 m的大金字塔由225万块平均质量超过1.4 t的巨石建成。这种标志性建筑动用了大量的人力来修建。此外,埃及人也使用青铜作为切削工具,切割巨石建造方尖碑,其中一些巨石的质量超过900 t。

　　公元前6世纪到公元前3世纪,古希腊人在将理论引入工程方面取得了巨大的进展,并发展了线、角、面和体的抽象概念。古希腊建筑工程中的几何学基础包括正方形、矩形和三角形这样的图形。古希腊的工长通常既是设计者也是建造者,他们带领工匠、泥瓦匠和雕刻匠建造了很多杰出的工程。在古希腊时代,所有重要的房屋都由石灰石或大理石建成,如帕特农神庙就是由大理石建成的。这是一座公元前5世纪建造在雅典卫城之上的供奉雅典娜的神殿(见图1.3)。

大金字塔 The Great Pyramid

卡纳克阿蒙-拉神庙
The Temple of Amon-ra in Karnak

图 1.2　古埃及建筑
Fig. 1.2　Egyptian architecture

　　In ancient Egypt, the simplest mechanical principles and devices were used to construct many temples and pyramids that are still standing today, including the Great Pyramid at Giza and the Temple of Amon-ra in Karnak (Fig.1.2). The Great Pyramid, 146.6 meters high, was made of 2.25 million stone blocks with an average weight of more than 1.4 tons. A large number of labor were used during the construction of such monuments. The Egyptians also built obelisks by huge stone blocks with tools made of hard bronze. Some huge stones even weigh more than 900 tons.

　　The Greeks made great strides in introducing theory to engineering during the 6th century BC to the 3rd century BC. They developed an abstract knowledge of lines, angles, surfaces and solids rather than refer to specific objects. The geometric bases of Greek building construction includes figures such as the square, rectangle and triangle. The Greek architect was usually the designer, as well as the builder of the architectural and engineering masterpieces. Craftsmen, masons and sculptors worked under his supervision. In the classical period of Greece, all important buildings were built of limestone or marble. The Parthenon, for example, was built of marble. It was a temple of Athena, built in the 5th century BC on the Acropolis of Athens(Fig.1.3).

公元 2 世纪,罗马帝国进入全盛期,统治了从苏格兰至波斯的大片疆域。当罗马人征服其他民族时,那些战败国的工程理论和经验也为其所拥有和继承。其中,古埃及建筑对古罗马建筑影响最大。但是,古罗马的工程师发明了带有拱顶石的半圆拱结构,这表明他们当时已熟知砖石砌体受压的事实,尽管他们没有书面或正式的关于力平衡的知识。古罗马的建筑师是技术专家,他们设计并建造了桥梁、沟渠、公路及公共建筑。古罗马时期的筑路技术达到了古代土木工程的最高水平,除了著名的阿皮亚古道,古罗马人还修建了隧道、渡槽、石拱桥、海港、码头及灯塔等(见图1.4)。

图 1.3　帕特农神庙
Fig. 1.3　The Parthenon

渡槽 The aqueducts

阿皮亚古道 The Via Appia

图 1.4　古罗马工程
Fig. 1.4　The ancient Roman engineering

In its heyday of the 2nd century AD, Rome ruled the world from Scotland to Persia. As the Romans conquered other nations, they borrowed their captives' ideas and practised engineering. The influence of Egyptian building is especially noticeable. However, the Roman arch construction created a central keystone at the top, which indicated that Roman engineers were familiar with masonry under compression, although they had no written or formal knowledge about equilibrium of forces. The Roman technical experts designed and constructed bridges, aqueducts, roadways and buildings for public uses. The art of road building in ancient Rome reached its highest level in civil engineering filed. Besides the Via Appia, Romans also built tunnels for roadways, aqueducts, stone arch bridges, harbors, dock and lighthouse(Fig.1.4).

在我国黄河流域的仰韶文化遗址(公元前 5 000 年—公元前 3 000 年)和西安半坡遗址(公元前 4 800—前 3 600 年)发现了供居住的浅穴和直径为 5~6 m 的圆形房屋,这说明在新石器时代出现了原始的房屋建筑。进入封建社会后,我国开始了秦砖汉瓦的土木工程技术时代。战国时代建造的都江堰,秦朝开始修建的万里长城、大型宫殿、陵墓等,都是这个时期不朽工程的代表。中国古代的建筑多采用木结构,并逐渐形成与此相适应的风格(见图 1.5)。公元 14 世纪建造的北京故宫是世界上最大、最完整的古代木结构宫殿建筑群;建于辽代的应县木塔是世界上最高的木建筑,并成为世界木结构建筑的典范。

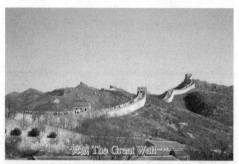 

图 1.5 中国古代建筑工程

Fig. 1.5 Chinese ancient architectural engineering

In China, the shallow holes for living and the round houses with the diameter of 5 to 6 meters have been found at Yangshao Culture Sites (5 000 BC—3 000 BC) of the Yellow River basin and Xi'an Banpo Ruins (4 800 BC—3 600 BC). It shows that the original construction appeared in the Neolithic Age. China started QinZhuanHanWa era of civil engineering after entering the feudal society. The Great Wall, large palaces and tombs built in Qin Dynasty as well as the Dujiang Dam built in the Warring States are immortal engineering representatives in the period of the feudal society. Most Chinese ancient architectures (Fig.1.5) were built with wooden structure, and gradually formed unique styles to match it. The Imperial Palace in Beijing, constructed in the 14th century, is the largest and most complete ancient wooden structure buildings in the world. The Yingxian Wooden Tower built in Liao Dynasty is the tallest wooden building in the world, and might be considered as the excellent example of wooden constructions.

值得一提的是,早在公元前 5 世纪,我国就已出现了以记述木工、金工等工艺为主且兼论城市、宫殿、房屋建筑规范的土木工程专著——《考工记》。

1.2.2 近代土木工程

近代土木工程历史阶段通常被认为是从 17 世纪中期到二战后超过 300 年的时间。在这段时期,土木工程成为一门独立学科,并进入了定量分析阶段。一些理论的发展,新材料的出现,新工具的发明,都使土木工程学科日渐完善和成熟。

在力学理论方面,1638 年意大利学者伽利略发表了《关于两门新科学的对话》一文,首次用公式表述了梁的设计理论。随后,在 1687 年牛顿总结出力学三大定律,为土木工程奠定了力学分析的基础。1825 年法国的维纳在材料力学、弹性力学和材料强度理论的基础上,建立了土木工程中结构设计的容许应力法。

在材料方面,1824 年波特兰水泥和 1867 年钢筋混凝土的发明成为土木工程历史上的重大事件。水泥和钢铁的大批量生产和应用,使得土木工程师可以利用这些材料实现大规模复杂工程设施的建设。在近代及现代建筑中,高耸、大跨、巨型、复杂的工程结构,绝大多数应用了钢结构或钢筋混凝土结构。

It is worth mentioning that, the earliest monographs about civil engineering had appeared in China as early as the 5th century B.C., in which the technology of carpenter, metalworking and the standards about city building were recorded.

1.2.2 Modern civil engineering

The history of modern civil engineering is generally regarded as the period from the middle of 17th century to the post-second world war. During this period, civil engineering has become an independent discipline, and stepped into a quantitative analysis stage. The development of some theories, the emergence of new materials and the invention of new tools made the civil engineering improved and mature gradually.

In mechanical theory, Italian scholar Galileo published *Dialogues Concerning Two New Sciences* in 1638, in which the design formula of the beam theory was put forward for the first time. In 1687, Newton summed up the three laws of mechanics, which laid the basis for the analysis of civil engineering. Then, the allowable stress method was established by the Frenchman Navier in 1825 for civil engineering structural design, based on material mechanics, theory of elasticity and strength of materials.

As for materials, the invention of Portland cement in 1824 and reinforced concrete in 1867 became the historic events in civil engineering. As the mass production of concrete and steel, civil engineers can make use of these materials to construct large and complex engineering facilities. In modern times, steel structures and reinforced concrete structures are applied to the vast majority of the tall, large span, huge and complicated engineering structure of modern buildings.

　　1825 年,英国采用盾构技术开凿泰晤士河底隧道;同年,英国人斯蒂芬森建成世界上第一条长达 21 km 的铁路;1863 年, 伦敦修建了世界上第一条地铁;1889 年,法国建成了高达 300 m 的埃菲尔铁塔(见图 1.6),该塔总重 8 500 t,现已成为巴黎乃至法国的标志性建筑;1890 年,英国在爱丁堡附近修建了福斯桥(见图 1.7),它是一座由众多钢管弦杆构件组成的双伸臂梁铁路桥, 被后人公认为桥梁史上的里程碑之一; 1931 年美国纽约建成 102 层的帝国大厦 (见图 1.8),378 m 高的钢骨架总重超过 50 000 t, 这一建筑高度保持世界纪录长达 40 年;1936 年美国旧金山建成了金门大桥(见图 1.9),该桥主跨 1 280 m,是世界上第一座主跨超过 1 000 m 的桥梁。

图 1.6　埃菲尔铁塔
Fig. 1.6　The Eiffel Tower

图 1.7　福斯桥
Fig. 1.7　The Forth Bridge

图 1.8　帝国大厦
Fig. 1.8　The Empire State Building

图 1.9　金门大桥
Fig. 1.9　The Golden Gate Bridge

　　In 1825, the British cut the Thames Tunnel using shield technology. The first railway in the world, 21 km long, was built by Stephenson in Britain in the same year. In 1863, the first underground railway was built in London. The Eiffel Tower (Fig.1.6), as high as 300 m, was built in 1889. The total weight of this building is 8 500 t, and it has become the landmark building for Paris even for France. In 1890, the British built the Forth Bridge (Fig.1.7) near Edinburgh, which is a railway bridge of double cantilever beam consisting of many steel chord members. It is recognized as one of the milestones in the history of bridge. The Empire State Building (Fig.1.8) which was built in 1931 in New York is 378 meters high, and the total weight of its steel skeleton is more than 50 000 t. The building height had kept the world record for 40 years. In 1936, the Golden Gate Bridge was built in San Francisco (Fig.1.9), with main span of 1 280 m, which is the first bridge of main span more than 1 000 m in the world.

　　在这一时期,我国由于近代历史原因,土木工程的发展处于落后状态,直到洋务运动后,才开始学习西方现代技术,并建造了一批有影响力的土木工程。例如,1909年詹天佑主持修建的京张铁路(见图 1.10),全长 200 km,它是首条由中国人自行建造、投入营运的铁路;1929 年建成中山陵(见图 1.11),它被誉为"中国近代建筑史上第一陵";1934 年建成的上海国际饭店一直是这座城市的中心(见图 1.12),作为上海最高楼的纪录保持了半个世纪之久;1937 年,茅以升主持建造了钱塘江大桥(见图 1.13),它是公路、铁路两用的双层钢结构桥,是我国近代土木工程的优秀成果。这些工程建设在中国近代土木工程史上都具有重要的历史意义。在建筑材料方面,1889 年在唐山建成了中国第一家水泥厂,1910 年开始生产机砖。

图 1.10　京张铁路
Fig. 1.10　Beijing-Zhangjiakou Railway

图 1.11　中山陵
Fig. 1.11　Sun Yat-sen Mausoleum

图 1.12　上海国际饭店
Fig. 1.12　The Park Hotel in Shanghai

图 1.13　钱塘江大桥
Fig. 1.13　The Qiantang River Bridge

　　During this period, due to the historical reasons in China, the development of civil engineering was backward for a long time. After the westernization movement, Chinese people began to learn western modern technologies, and a group of influential civil engineering was built. For example, Beijing–Zhangjiakou Railway (Fig. 1.10), with the total length of 200 km, was built by Zhan Tianyou in 1909. It is the first railway which is built and put into operation by the Chinese themselves. Sun yat-sen Mausoleum (Fig. 1.11), built in 1929, is known as "the first mausoleum in the architectural history of modern China". Since its opening in 1934, the Park Hotel in Shanghai (Fig. 1.12) has been regarded as the focus and the tallest building in this city for more than half a century. The Qiantang River Bridge(Fig. 1.13) built by Mao Yisheng in 1937 is a double layer steel structure bridge both for use of road and railway, and it is the outstanding achievement in the modern civil engineering in China. In terms of construction material, the first cement plant in China was built in Tangshan in 1889. In 1910, Chinese started to manufacture machine–made brick. These engineering constructions have important significance in the history of modern civil engineering in China.

1.2.3 现代土木工程

二战后,战后恢复的需要和现代科学的高速发展为土木工程提供了更为坚实的物质基础和理论指导,一个崭新的"现代土木工程"的辉煌时代到来了。这一时期的土木工程有以下几个特点。

(1)功能要求多样化、复杂化

现代土木工程的特征之一就是工程设施同它的使用功能或生产工艺紧密地结合在一起。公共建筑和住宅建筑要求建筑结构与水、暖、电、气的供应及室内温、湿度的自动控制等现代化设备相结合,并且与环境相协调。悉尼歌剧院因其独树一帜的帆状屋顶成为澳大利亚的象征,也被公认为是世上最与众不同的现代建筑(见图1.14)。工业建筑要求恒温、恒湿、防微振、防腐蚀、防辐射、防磁、除尘,并向大跨度、超重型、分隔灵活的方向发展。同时,具有特殊功能要求的各类特种工程结构,则对土木工程提出高标准的要求,如水利枢纽工程(见图1.15)、核电站中的安全壳要求具

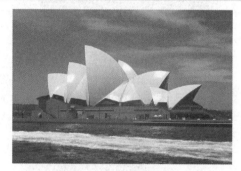

图 1.14 悉尼歌剧院
Fig. 1.14 Sydney Opera House

图 1.15 三峡水利枢纽工程
Fig. 1.15 Three Gorges Project

1.2.3 Contemporary civil engineering

After the Second World War, the needs of post-war recovery and the development of modern technology provided the material basis and theoretical guidance for the development of civil engineering. A new age for civil engineering has come. During that period, the features of civil engineering were as following.

(1) The diversification and complication of the function

One of the features of modern civil engineering is that the construction facilities are closely combined with their functions or manufacturing techniques. Public buildings and residential building require the combination of structures and the supply of water, heating, electricity, gas, as well as the automatic control of indoor temperature and humidity. Besides, they should be in harmony with the environment. The Sydney Opera House (Fig.1.14) is as representative of Australia as the pyramids of Egypt because of the distinguishing sails of the roof, and it is considered to be one of the most recognizable images of the modern world. Industrial constructions need to be constructed in constant temperature and humidity. They also ask for anti-corrosion, anti-slight shock, radiation protection, anti-magnetism, dedusting, to the large span, and flexible segmentation. Besides, all constructions of special structures ask for higher standards of civil engineering. For example, the water control project (Fig.1.15) and the containment of nuclear power station require high safety degree, while the ocean platform (Fig.1.16) fo-

有极高的安全度,而海洋平台(见图 1.16)则因其功能多样、使用环境恶劣、荷载复杂、施工困难而对工程结构提出了更高的要求。

(2)城市建设立体化

随着经济的发展和人口的膨胀,城市人口密度快速增长,造成城市用地紧张,交通拥挤,地价昂贵,迫使城市建设向立体化发展,表现为高层建筑的大量兴起,地下工程的高速发展和城市高架公路、立交桥的大量涌现。现代化城市建设在地面、空中、地下同时展开,形成了立体化发展的局面。位于阿拉伯联合酋长国迪拜的哈利法塔(见图 1.17),是目前世界上最高的人工建筑,共 169 层,总高度 828 m,历时 5 年建成,被称为"一座垂直而立的城市"。

(3)交通工程快速化

经济的繁荣和发展对运输系统提出了快速、高效的要求,同时现代技术的进步也为满足这种要求提供了条件。交通运输的高速化让世界变得越来越小,主要表现为高速公路的大规模修建、铁路电气化的形成和大量长距离海底隧道的出现。

图 1.16 海洋平台
Fig. 1.16 Ocean Platform

图 1.17 哈利法塔
Fig. 1.17 Burj Khalifa Tower

cuses on the construction structure based on its function diversity, bad operating environment, complicated load, and its difficulty in construction.

(2) Three-dimensional development in cities

With the development of economy and population, the population density in cities has grown quickly, which causes the tension of land use, traffic jam, high price of land and finally the three-dimensional development of city constructions. Many tall buildings, underground projects, urban elevated road and flyovers appear fast. Constructions are operated on the ground, underground and in the sky. The Burj Khalifa Tower (Fig.1.17) in Dubai, United Arab Emirates is now the highest artificial building in the world. It has 169 layers, 828 m in height, and it was finished within 5 years, and named as "a vertical city".

(3) Fast development of traffic engineering

The development of economy requires fast and efficient work of transportation, and the advanced technology also provides condition for such development. The world has become smaller with the fast development of transportation, such as the construction of highways, the electrification of railways and the long distance subsea tunnels.

　　现在,高速公路的里程数已成为衡量一个国家现代化程度的标志之一。我国从1988年建成第一条高速公路开始, 截止到2020年底通车里程已突破15万公里,跃居世界第二。铁路的电气化大大提高了其运输效率,日本的"新干线"铁路行车时速达210 km以上, 法国巴黎到里昂的高速铁路运行时速达260 km。截止到2020年底,我国高速铁路运营里程超过3.8万km,最高运行时速已达350 km。2006年建成通车的青藏铁路东起青海西宁,西至拉萨,全长1 956 km,是世界海拔最高、线路最长的高原铁路(见图1.18)。交通高速化直接促进桥梁、隧道技术的发展,不仅穿山越江的隧道日益增多,而且出现长距离的海底隧道。1993年,贯通英吉利海峡的英法海底隧道(见图1.19)实现通车,人们仅用35分钟就可以从欧洲大陆穿越英吉利海峡到达英国本土。

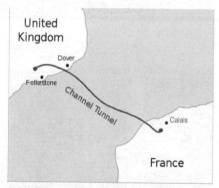

图 1.18　青藏铁路
Fig. 1.18　Qinghai-Tibet railway

图 1.19　英法海底隧道
Fig. 1.19　Channel Tunnel

　　Now the mileage of highway has become the symbol of a country's modernization degree. The first highway in China was built in 1988. By the end of 2020 the mileage of highway had been more than 150 000 km, which ranking the second in the world. Electrification of railways greatly improves the efficiency of the transport. The operation speed of Japanese Shinkansen is 210 km/h or more, and the French high speed railway operates from Paris to Lyon with the speed of 260 km/h. By the end of 2020, there has been more than 38 000 km of high speed railway in China, and the highest running speed has reached 350 km/h. The Qinghai-Tibet railway(Fig.1.18), completed in 2006 with the total length of 1 956 km, is from Xining in Qinghai to Lhasa. It is the highest and longest plateau railway in the world. High speed traffic directly promotes the development of the bridge and the tunnel technology. Some long distance subsea tunnels have been put into operation. In 1993, the Channel Tunnel (Fig.1.19) went into trial operation. People can travel from the continental Europe to England in only 35 minutes.

（4）工程材料轻质高强化

工程材料中的混凝土材料发展迅速，由普通混凝土向轻骨料混凝土、加气混凝土和高性能混凝土的方向发展，既降低了混凝土的重量，又提高了其强度，同时其他性能也得到了很大改善。此外，诸如低合金、高强度钢材的发展，铝合金、建筑塑料等轻质高强材料的应用为建筑材料的轻质高强化创造了条件。纤维增强材料的发展与应用，也是 21 世纪新材料在土木工程中应用的重要标志，如碳纤维（CFRP）、芳纶纤维（AFRP）、玄武岩纤维（BF）、金属纤维及混杂纤维等，它们既可以作为增强材料或智能材料直接掺入混凝土中，也可以制成片材或棒材作为结构构件的补强或加筋材料。

（5）施工过程工业化、装配化

在工厂成批生产房屋、桥梁的各种构配件、组合体，在现场进行拼装的生产方式极大地加快了施工速度，减少了户外工作时间和施工污染。此外，各种先进的施工手段也得到了很大发展，为复杂、大型、高耸的建（构）筑物的工程施工提供了条件。

（6）设计理论精确化、科学化

结构理论的发展与完善也是现代土木工程快速发展的重要基础和标志。现代力学和分析方法及计算机技术的发展，使得土木工程学科的理论基础得到了迅速发展，结构设计方法实现了从经验方法、安全系数法到可靠度设计方法的过渡。进入21 世纪，基于性能设计理论、抗连续倒塌设计理论、结构耐久性理论、结构的振动控制理论、结构实验技术等又有了重大发展，为建（构）筑物的安全、可靠、合理、经济提供了设计保证。

(4) Lighter and more qualified engineering materials

In engineering materials, the concrete material developed quickly. Normal concrete now tends to be lightweight aggregate concrete, aerated concrete and high performance concrete, which not only decreases the weight of concrete, but also advances its intensity and improves other properties greatly. The development of low content alloy, high strength steel and the application of aluminum alloys and building plastic make the engineering materials lighter and more qualified. The development of fiber reinforced material also reflects the development of new materials in the 21st century. Materials such as carbon fiber reinforced plastic (CFRP), aramid fibers reinforced plastic (AFRP), basalt fiber (BF), metal fiber and hybrid fiber can not only be mixed into concrete as strength materials, but also be made as slice and stick materials.

(5) The industrialization and assemblage of construction

It makes the construction faster and reduces outdoor working time and construction pollution at the same time, by producing the components of houses and bridges in factories while assembling in site. Besides, various advanced construction ways have been developed quickly, which makes the existence of complex tall buildings possible.

(6) The precision and scientization of design theories

The development of structure theories also helps modern civil engineering. Modern mechanics, analytical methods and the development of computer technologies accelerate the development of theoretical basis of civil engineering. Instead of experience method and safe coefficient method, reliability design method has been more accepted. From the 21st century, performance-based design theory, continuous collapse resistance design theory, durability theory of structure, the vibration control theory of structure and the structural experimental technology have been used widely, which guarantee the safety, reliability, rationality and economy of the buildings.

1.3 土木工程专业及其知识构成

1.3.1 土木工程专业简介

土木工程专业是为培养土木工程专门技术人才而设置的。早期的土木工程师很少受过正规教育。最早针对土木工程师的正规训练计划由法国国立路桥学院提供,该学院是为了路桥建设的科学发展而在 1716 年组建的,旨在为国家路桥兵团培养人才。

我国土木工程教育则始于 1895 年的天津大学。新中国成立前,我国工科学科的设置基本上是仿效英美的,土木工程没有明确的专业和统一的教学计划,更没有教学大纲,各校土木系开课很不一致,开设的课程很广泛。新中国成立后,特别是改革开放以来,我国土木工程专业教育有了很大的发展。针对专业划分过细、专业范围过窄、门类之间专业重复设置等问题,我国分别在 1982 年、1993 年、1997 年进行了三次专业目录的调整,坚持拓宽专业口径、增强适应性原则,主要按学科划分专业,使培养的人才具有较宽广的适应性。自 1997 年起建筑工程、交通土建、地下工程等近十个专业合并成为目前的"土木工程专业"。

按人才培养的层次划分,土木工程专业培养的人才有专科、本科(工学学士)、硕士(工学硕士)、博士(工学博士)等几个层次。在本科教育阶段,土木工程专业属于一级学科专业,下设建筑工程、铁道工程、道路与渡河工程等多个专业方向。进入硕士或博士教育阶段则具体分二级学科专业,如岩土工程、结构工程、防灾减灾与防护工

1.3 Civil engineering major and its knowledge composition

1.3.1 Brief introduction to civil engineering major

The civil engineering major is set up for training civil engineering specialists. Early civil engineers, as a general rule, had very little formal education. The earliest formal training program for civil engineers was offered by "the Ecole Nationale Des Ponts et Chaussees", which was formed in 1716 for the scientific advancement of bridge building and road building.

The education of civil engineering in China started in Tianjing University in 1895. Before the founding of New China, the setting-up of engineering subjects in China was learnt from Europe and America. There were no clear subjects, no formal teaching plan and no syllabus in civil engineering department. After 1949, civil engineering developed quickly, especially after the opening up to the outside world. China changed the catalogue of specialty in 1982, 1993 and 1997, aiming to solve the problems such as the too narrow professional range or repeated majors. The new plan has aimed at cultivating students' adaptability. Since 1997, the major civil engineering has combined nearly ten majors, such as constructional engineering, traffic engineering and underground construction.

Civil engineering major intends to cultivate students with three-year college, bachelor diploma, master degree and doctor degree. Civil engineering is regarded as first discipline major in the stage of undergraduate education. It comprises constructional engineering, railway engineering, road and river crossing engineering, etc. In the stage of post graduate and doctoral education, civil engineering is seen as second discipline

程、桥隧工程、市政工程等。

1.3.2 土木工程专业知识构成

土木工程专业的工作对象及业务范围非常广，每个工作对象中的工作内容又分勘察、设计、施工、监理、管理等多个方面和环节。因此，专业培养中应贯彻"大土木"的人才培养理念。在这一理念下，应坚持"宽口径、厚基础"的培养原则，这样才能培养出既符合现实工程需要，又符合土木工程未来发展要求的适应能力强的专门人才。

土木工程专业的知识体系由 4 部分组成，分别为工具性知识、人文社会科学知识、自然科学知识和专业知识。每个知识体系所涵盖的知识领域见表 1.1。

表 1.1　土木工程专业知识体系和知识领域
Tab. 1.1　Knowledge hierarchy and knowledge areas of major civil engineering

序号 No.	知识体系 Knowledge hierarchy	知识领域 Knowledge area
1	工具性知识 instrumental knowledge	外国语、信息科学基础、计算机技术与应用 foreign languages, information science basis, computer technology and application
2	人文社会科学知识 humanities and social science knowledge	政治、历史、伦理学与法律、心理学、管理学、体育运动 politics, history, ethics and law, psychology, management, sports
3	自然科学知识 natural science knowledge	工程数学、普通物理学、普通化学、环境科学基础 engineering mathematics, general physics, general chemistry, environmental science basis
4	专业知识 professional knowledge	力学原理和方法、材料科学基础、专业技术相关基础、工程项目经济与管理、结构基本原理和方法、施工原理和方法、计算机应用技术 mechanics principle and method, material science basis, professional technology, project economic and management, basic principle and method of structure, construction principle and method, computer application technology

subject. It comprises geotechnical engineering, structural engineering, disaster prevention, bridge and tunnel engineering, municipal engineering, etc.

1.3.2　The composition of knowledge in major civil engineering

Civil engineering has a wide range of project objects and businesses. Every task includes reconnaissance, design, construction, supervision and management. Therefore, people should regard this major as "vast civil engineering", learn more and set up a good foundation. By doing so, we will train students to be promising specialists with strong adaptability.

Knowledge in civil engineering can be divided into four parts: instrumental knowledge, humanities and social science knowledge, natural science knowledge and professional knowledge. The following chart is the knowledge hierarchy in each part (Tab. 1.1).

1.4 土木工程师的能力素质及职业发展

1.4.1 土木工程师的能力素质

21世纪是高科技时代,土木工程将引进更多的高新技术,不断提高、创新和发展,以满足人们日渐提高的社会生活和生产需求。为了适应新时期土木工程的发展,作为一名土木工程师必须具备扎实的专业基础、较强的实践能力和较高的综合素质。

(1) 土木工程师的专业技能

土木工程是一个应用性的学科,具有很强的个性和综合性,大量问题需要依靠工程师的经验和工程实例来解决。土木工程师想要把在学校里学到的专业基础知识、工程技术知识和实践技能应用到工程项目中去,就必须依靠他们自身的各种能力,包括工程能力、科技开发能力、组织管理能力、表达能力和公关能力等专业技能。

工程能力就是土木工程技术人员在从事土木工程工作时应用工程技术知识和技能的能力。科技开发能力就是在现有的设计方法和施工技术的基础上,对设计方法和施工技术提出改进设想并予以实施的能力。土木工程是一项群体性的工作,这要求土木工程师具有必要的组织管理能力,这里的管理包括人力资源管理、投资管理、进度管理、质量管理、安全管理、工程项目管理、各工种工作的协调管理等;土木工程师需要具有文字、图纸和口头的表达能力,以及社会活动、人际交往和公关的能力。更重要的是要有自我终身学习的能力。

1.4 The competence and career development of civil engineers

1.4.1 The competence of civil engineers

21st century is a high-tech age. Civil engineering will bring in more high technologies to keep development and innovation constantly, and fulfill people's increasing production needs. To adapt the development of civil engineering, a civil engineer should have solid foundation of professional knowledge, strong practical abilities and competence.

(1) The professional ability of civil engineers

Civil engineering is an applied science, which has its own characteristics and comprehensiveness. A lot of problems need to be solved by engineers' experience and project example. Civil engineers should combine knowledge and practice in actual constructional tasks which need their abilities, such as constructional ability, science development ability, management ability, expression ability and public relation ability.

The constructional ability is the ability to apply engineering technology when operating objects. Science development ability is the ability to improve the designed methods and construction skills. Civil engineering is a group work. Civil engineers should have the necessary management ability for the human resources, investment, quality control, schedule quality, safety control and coordination. Civil engineers should also have an excellent expression ability both literally and pictorially, as well as the communication abilities for social public. What is more important is to have the ability of lifelong learning.

（2）土木工程师的综合素质

土木工程师的综合素质一般包括4个方面的内容：个人修养、心理和体魄、自然科学知识、土木工程专业知识。土木工程师在个人修养方面，应具有良好的思想品德、社会公德和家国情怀，具有高尚的科学人文素养和精神；应具有健康的心理和良好体魄，能保持乐观、积极向上；应能了解当代科学技术发展的主要方面，学会科学思维的方法；在土木工程专业方面，土木工程师要有良好的职业道德、强烈的社会责任感与奉献精神，优秀的工匠精神与吃苦耐劳的品质。

（3）土木工程师的创新意识

土木工程师的创新能力和品质，是土木工程这个有"创造力的专业"永葆青春的基本保证。但创新意识不可能孤立地培养，土木工程师除了应有扎实的知识结构、良好的实践技能和完善的能力结构外，还必须结合各自工作自觉地培养创新意识。在大学学习期间，学生首先应结合自身特点，逐步明确今后的发展方向，做好学业规划。其次，在学习的不同阶段，循序渐进地培养自学能力。同时，接受艺术和美学方面的熏陶，对激发土木工程师的创新意识大有好处。

（4）土木工程师的法律意识

工程建设活动通常具有建设周期长、涉及面广、人员流动性大、技术要求高等特点。因此，工程建设活动应确保工程建设的质量与安全，应当符合国家的工程建设安全标准，应当遵守法律、法规的强制性规定，不得损害社会公共利益和他人的合法权益，而依法进行的建设活动也不得受到任何单位和个人的妨碍和阻挠。建设法规的作用主要表现为：规范与指导建设行为，保护合法建设行为，处罚违法建设行为。

（2）The competence of civil engineers

The competence of civil engineers normally includes personal cultivation, physical health and mental health, scientific knowledge and professional knowledge. Civil engineers should have good personalities, social morality (Family-Country emotion), healthy body and mind. They should be aware of the new technologies in civil engineering field. Civil engineers should be able to understand the development of contemporary science and technology, and the scientific thought. They also should have good professional ethics, social responsibility, spirit of utter devotion, excellent craftsman spirit and hard-working quality.

（3）The innovation consciousness of civil engineers

Innovation consciousness and creativity are the essence of civil engineering. Innovation consciousness cannot be cultivated in isolation. Civil engineers should not only have the solid foundation of academic knowledge, excellent practice skills and great knowledge structure, but also be aware of the coming tendency of civil engineering. What is more, civil engineers should study all the time and accept the nurture of art to inspire their innovation ability.

（4）The legal consciousness of civil engineers

Some construction tasks have the characteristics of long duration, as well as widely related, staff mobilized and high-tech requirement. Therefore construction progress should be safe and qualified. Civil engineers should follow the nation's law, regulations and disciplines. The construction law is aiming to rule and guide the constructional behaviors, to protect the legal constructional tasks and punish illegal engineering behaviors.

(5) 土木工程师的风险意识

建设工程项目在设计、施工和竣工验收等各个阶段中都大量存在未确定因素，这些未确定因素会不断变化，由此而造成的风险直接威胁工程项目的顺利实施和完工。为了降低风险，土木工程师在项目建设以前应进行风险评估，即在风险识别和风险估测的基础上把握风险发生的概率、损失严重程度，然后根据评估结果制订出完整的、切实可行的风险控制计划。

(6) 土木工程师的可持续发展意识

当前可持续发展已成为国际社会的共识。任何土木工程都要占据一定的自然空间并直接或间接地消耗大量的物质资源。为解决可持续发展过程中最基本的"资源有限"问题，土木工程师在价值观念、理论基础、方法原理和技术手段等方面应进行一系列的变革，以最大限度地提高自然资源的利用率，保护、恢复自然生态环境。

1.4.2 土木工程师的职业发展

由于自身的专业特点，土木工程专业的就业范围比较广泛，同时随着城市建设和交通工程建设的不断升温，土木工程专业的就业形势持续走好。随着我国执业资格认证制度的不断完善，土木工程师不但需要精通专业知识和技术，还需要取得必要的执业资格证书。总体来说，土木工程专业的主要就业方向可以概括为以下3个方面(见表1.2)。

(5) The risk consciousness of civil engineers

Because of many possible unsure factors, constructional projects are facing changes during the period of design, building and final check of construction. Any tiny problem may make the task unsuccessful. In order to reduce the risk, civil engineers need to estimate the risk before applying the project. Engineers need to be aware of the probability of the risk and work out a complete and feasible risk management plan.

(6) The sustainable development awareness of civil engineers

The sustainable development thought has been accepted by people all over the world. Any civil engineering project will occupy some natural space and consume lots of resources directly or indirectly. In order to save resources as many as possible, civil engineers should take actions to efficiently make use of the resources, which will be helpful in the environment protection.

1.4.2 The career development of civil engineering

In light of its specialties, the major civil engineering has a wide range of career choices, and as the urban and traffic engineering construction is heating up constantly, its career expectation has been on a good trend in recent years. With the improvement of the professional qualification certification system in China, civil engineers not only need to be excellent in professional knowledge and technique, but also should acquire some necessary professional qualification certificates. The career objectives of civil engineering, by and large, can be summarized to three aspects as follows (Tab. 1.2).

表 1.2 土木工程专业的主要就业方向
Tab. 1.2 Career objectives of civil engineering

就业方向 Career objectives	代表职位 Representative titles	代表行业 Representative industries	执业资格认证 Qualification certificates
工程技术 technical engineering	施工员、建筑工程师、结构工程师、岩土工程师、建造师、技术经理、项目经理等 construction crew, architectural engineer, construction engineer, geotechnical engineer, constructor, technical manager and project manager	建筑施工企业、房地产开发企业、路桥施工企业等 construction enterprises, real estate enterprises, road and bridge construction enterprises	全国一、二级注册建筑师,全国一、二级注册结构工程师,全国一、二级注册建造师,注册土木工程师(岩土)等 First or Second Class National Certified Architect, First or Second Class National Certified Construction Engineer, First or Second Class National Certified Constructor and Certified Civil Engineer (Geotechnical field)
设计、规划及预算 design, plan and budget	项目设计师、城市规划师、预算工程师等 project designer, city planner and budgetary engineer	工程勘察设计单位,房地产开发企业,交通或市政工程类,工程造价咨询机构等 project design, investigation and survey; real estate development; transportation and public work; project budget consulting	全国一、二级注册建筑师,全国一、二级注册结构工程师,注册土木工程师(岩土),注册造价工程师等 First or Second Class National Certified Architect, First or Second Class National Certified Construction Engineer, Certified Civil Engineer (Geotechnical field) and Certified Cost Engineer, etc.
质量监督及工程监理 quality control and engineering supervision	监理工程师、BIM 工程师 supervising engineer、BIM Engineer	建筑、路桥监理公司,工程质量检测监督部门等 building, road and bridge quality supervision company; construction quality detection and supervision sector	注册监理工程师、BIM 工程师 Certified Supervising Engineer、BIM Engineer

1.5 土木工程的发展趋势

1.5.1 工程材料向轻质、高强、多功能化发展

土木工程材料是新型结构出现与发展的基础,而新型结构的出现又是新材料产业的驱动力。在结构材料方面,高强、高性能混凝土已在工程中广泛应用。目前世界上研究的混凝土抗压强度可达 300 MPa,而且混凝土的各种性能,如工作性、耐久性等显著改善。但是,混凝土的生产消耗了大量自然资源。因此,从人类社会可持续发展的前景出发,混凝土也要坚持可持续发展的方向。近年来提出的发展"绿色混凝土"或"生态混凝土"正是上述要求的集中体现,这不仅贯穿从生产到使用的全过程,还包括材料的再循环使用问题等。近年来,钢材的性能与加工工艺得到了显著改善和提高,而耐腐蚀、耐高温、易焊接、易加工的新型超级钢铁材料的研究已成为多国科学家关注的热点。

为了满足工程结构保温、隔热、隔声、耐高温、耐高压、耐磨、耐火等方面的需求,化学合成高分子材料将会广泛应用于土木工程结构中,其应用范围将扩展至大面积围护结构以及抗力结构。同时,这类材料也可以实现材料的智能化,即材料本身能够自感知、自调节、自清洁、自修复,实现构筑物自我监控功能,以及可重复利用。

1.5 The development tendency of civil engineering

1.5.1 The engineering materials tend to be lighter, stronger and multi-functional

Civil engineering materials are the basis of the development of any new structures, which are the driving force of new material industry. Now more and more high strength concretes are used. At present, the compressive strength of concretes is 300 MPa, and the workability and durability are getting better. However, the production of concretes consumes a lot of resources. Therefore, concretes industry should develop with the sustainable consciousness. The "Green Concrete" and "Ecological Concrete" clearly define the core idea of the above mentioned opinion, which should be applied not only to every detail in the construction, but also to the material recycle. Nowadays, the performance and processing technology of the steel have been quickly developed. New types of steel that are corrosion/heat resisting and easy to be produced have become the focus of scientists' attention in many countries.

In order to meet the requirements of heat preservation, heat/noise insulation, resistance to high temperature and high pressure, abrasion resistance, fire resistance, etc., chemical synthetic polymer materials will be widely used in the structure of civil engineering. Its application scope will be extended to large areas, such as palisade structure and resistance structure. At the same time, the intelligent material can also be implemented, such as self-awareness, self-adjusting, self-cleaning, self-repairing, self-monitoring function, and reusing.

1.5.2　工程设计向工程全生命周期综合决策发展

所谓工程的"全生命周期"是指包括工程设计、建造、使用和老化的全过程。在不同阶段,工程的风险来源不完全相同。如建造阶段的风险来自对未完成结构及其支撑系统的分析不完全,以及对人为错误的失控;老化阶段的风险来自结构或材料长期在自然环境和使用环境中功能的逐渐退化。相对而言,工程使用阶段的平均风险率是最低的,其主要危险来自自然灾害和可能出现的人为灾害。以往的工程设计有的仅考虑使用阶段的工程安全,而现在除了要考虑使用阶段的安全,还要考虑安全以外的内容,如结构的功能能否得到保证以及耐久性问题等。对一些重大的高坝、水库、桥梁、高层建筑、海洋工程和港口工程,目前比较科学的做法是综合考虑建造、使用、老化三个阶段后,再做最后决策。

1.5.3　土木工程向海洋、太空、荒漠开拓

地球表面只有约 30% 的面积为陆地,而这其中又有大约 1/3 为沙漠或荒漠地区。随着地球上人口的不断增长,资源逐渐枯竭,随之带来的人类生存问题已迫在眉睫。因此,人类大力开发海洋、荒漠甚至太空资源已成为一种趋势。

现在世界各国已有许多这类成功案例,如日本关西国际机场是建在人工岛上的,可抵挡地震和台风的侵袭,中国香港大屿山国际机场劈山填海而建,而荷兰 1/5 的土地面积是 800 年来通过填海得到的。在中国西北部,利用兴修水利、种植固沙植物、改良土壤等方法,使一些沙漠变成了绿洲。这些都是成功造福人类的宏大工程。外太空星球上拥有丰富的资源,有些星球甚至有可能适宜人类居住,如果能够利用也将扩大人类的生存空间。

1.5.2　The engineering design tends to develop in the whole project life cycle

The project "Life Cycle" means the whole process that includes the construction, application and maturing. The project faces different risks in different stages. For example, in the construction process, designers may make a wrong judgement of the support system of an unfinished project. Materials will degenerate in the long operating and natural environment in the maturing stage. In comparison, there will be fewer risks in the application stage. Natural and man-made disasters are the main causes of the risks. Now people should be concerned about not only the safety in the application process, but also other problems such as the guarantee of the performances of the structures and the durability. As for those important projects such as high dams, reservoir, bridges, tall buildings, ocean engineering and harbor engineering, it is better to consider those three stages carefully and then make possible decisions.

1.5.3　Civil engineering has been exploited towards ocean, outer space and deserts

Only about 30% area of the earth is land, and about 1/3 of the land are deserts. With the growing population, people are facing serious living problems due to the lack of resources. It is a new tendency to open up more space from ocean, desert and outer space.

Now we have a lot of successful examples in the world, such as Japanese Kansai International Airport, which was built on man-made island and could resist the attack of earthquake and typhoon. The Hong Kong International Airport was built by moving mountains to the sea, and 1/5 area of Netherlands was obtained by filling the sea within 800 years. In northwestern China, people have made some deserts into oasis by water conservancy, planting sand binder and improving soil. Some outer spaces may be suitable for human habitation, which will be helpful for human to expand their living spaces.

1.5.4　工程信息化与智能技术的发展

信息技术、计算机技术、智能化技术在土木工程领域得到了愈来愈广泛的应用，并且是今后相当长一段时间内的重要发展方向。

智能化技术的应用主要体现在智能建筑与智能交通系统的发展与推广上。智能建筑的特征是所有设备都是用计算机的先进管理系统进行监测与控制的，并能通过优化控制来满足使用者对舒适、安全、能源利用率和可靠性的需求。智能交通系统将先进的信息技术、通信技术、传感技术、控制技术及计算机技术等有效地集成运用于整个交通运输管理体系，从而建立起实时、准确、高效、综合的运输和管理系统。

建筑信息建模(BIM)作为一种创新的工具与生产方式，是信息化技术在建筑业的直接应用，自 2002 年被提出后，已在欧美等发达国家引发了建筑业的巨大变革。BIM 技术通过建立数字化的 BIM 参数模型，涵盖与项目相关的大量信息，服务于建设项目的设计、建造、安装、运营等整个生命周期，在提高生产效率、保证生产质量、节约成本、缩短工期等方面发挥巨大的作用。虽然我国的 BIM 应用还处于初级阶段，但是认识并发展 BIM，实现行业的信息化转型已是势不可挡的趋势。

1.5.4　The development of engineering information and intelligent technology

Information technology, computer technology and intelligent technology have been widely used in civil engineering, and they will become an important development direction for quite a long time.

The application of intelligent technology is mainly reflected in the development of intelligent building and intelligent transportation system. By using highly controlled and supervised machines, the features of intelligent technology meet the needs of the users for comfort, safty, energy efficiency and reliability. Intelligent transportation system is to combine information technology with communication technology, sensor technology, control technology and computer technology, aiming to build up an accurate and efficient management system.

Building information modeling(BIM) is an innovative tool and production process, which is a typical application of information technology in construction industry. The rapid development of BIM has brought a gigantic revolution in construction industries in Europe, the United States and other developed countries since 2002. With BIM technology, a BIM parameter mode is constructed digitally, which contains the relevant data needed to support the design, construction, fabrication and procurement activities through the whole life cycle of the project. When properly implemented, the BIM users can gain a range of benefits that include improving productivity, enhancing quality, saving cost, shortening construction period and so on. Although the BIM application is still in its primary stage in China, recognizing and developing BIM to realize the reforming of Chinese construction industry has become an irresistible trend.

随着计算机技术的迅猛发展,计算机在土木工程中的应用已从早期的数值计算发展应用到了工程设计的各个阶段和许多环节。在土木工程领域中,计算机辅助设计(computer-aided design,CAD)已经被应用到规划、设计、施工、维修和加固等各个方面,CAD 技术朝着标准化、集成化、智能化等方向发展。通过土木工程的计算机仿真分析,可在计算机上模拟原形大小的工程结构在灾害荷载作用下从变形到倒塌的全过程,从而揭示结构不安全的部位和因素。地下工程开挖全过程的计算机模拟仿真可以预测和防止出现土体失稳或管涌、潜蚀、流沙、土洞等现象。

1.5.5 土木工程的可持续发展

随着人口增长,人类正面临生态失衡、资源枯竭等越来越严峻的生态环境恶化问题。世界环境与发展委员会(World Commission on Environment and Development,WCED)在 1987 年提出了"可持续发展"的概念。经过几十年的努力,可持续发展的观念已成为世界各国的共识,我国更是把可持续发展原则定为国策之一,并在这方面做了很大的努力。

土木工程要消耗大量资源(一般占总资源消耗的 25%)。为了得到土地,有时会毁林建设,围湖造地,导致环境失衡与恶化,可见土木工程师在贯彻可持续发展原则方面负有重大的责任。

With the increasing development of computer technology, computer has been used in many aspects in engineering design. The computer-aided design(CAD) has been used in design, construction, maintenance and many other aspects of civil engineering. The development of CAD technology tends to be standardized, integrated and intellectualized. The computer simulation analysis in civil engineering can simulate the whole process of the collapse of projects in disasters. This simulation will help people find the parts and factors of any dangerous structures. The computer simulation of underground projects will forecast and prevent disasters, such as soil instability or piping, shallow corrosion, sand drifting and soil cave.

1.5.5 Sustainable development of civil engineering

People are faced with more and more severe environmental degradation, such as ecological imbalance and resource depletion, etc. World Commission on Environment and Development(WCED) put forward the concept of "sustainable development" in 1987. After decades of effort, the concept of "sustainable development" has been accepted all over the world. China has even designed "sustainable development" as one of the national policies, and made great efforts in this aspect.

Civil engineering consumes a large amount of resources (typically account for 25% of the consumption resources). To occupy the land, sometimes people need to deforest and turn lake into land, which lead to environmental imbalance and deterioration, so civil engineers should be more responsible for implementing the principles of sustainable development.

　　在工程项目建设中,土木工程师应遵循"安全、经济、适用和可持续发展"的原则,综合考虑以下方面:尽量少占耕地,提高土地利用率;城市应加快地下空间的开发利用;尽量利用可再生资源,提高废物回收再利用的效率;建设过程中应采取措施减少对环境的影响;推广节能建筑;提倡循环用水等。

　　根据可持续发展原则,人类居住环境应保持与自然环境的和谐,为此专家提出了"绿色建筑"的理念。各国学者对"绿色建筑"的理解已普遍形成如下共识:在建筑物的全生命周期中,最低程度地占有和消耗地球资源,用量最小且效率最高地使用能源,最少量地产生废弃物及排放有害环境物质,使之成为与自然和谐共生、有利于生态系统与人居系统安全,健康且满足人类功能需求、心理需求、生理需求及舒适度需求的宜居可持续建筑物。可见,绿色建筑的目标是通过人类的建设行为,达到人与自然安全、健康、和谐共生,满足人类追求适宜生存居所的需求和愿望。

　　绿色建筑综合了大量的实践、技术和技能,以减少并最终消除建筑对环境和人类的影响。绿色建筑要根据地理条件,充分利用环境提供的天然可再生能源,如设置太阳能装置,利用植物建造绿色屋顶或雨水花园以减少雨水径流。为了补充地下水,可以使用砾石或透水混凝土来代替普通混凝土或沥青等传统路面材料。德国达姆施塔特工业大学设计的亚热带全生态减碳屋(见图1.20),将太阳能作为唯一的能源。

In construction projects, civil engineers should follow the principle of "safe, economy, applicable and sustainable development", which means they should take the following aspects into consideration: using the arable land as little as possible, improving land utilization, speeding up the development and utilization of underground, maximizing the use of renewable resources, improving the efficiency of waste recycling, taking measures to reduce the environmental impact during the construction process, promoting the energy-efficient buildings, water recycling and so on.

According to the principle of sustainable development, people's living environment should maintain harmony with the natural environment. The experts put forward the concept of "green building". As to the "green building", scholars have generally reached consensus as follows: minimizing the possession and consumption of the resources during the full life cycle of the buildings; using the energy in minimum amount and most efficiency; producing the least waste and discharging the least harmful substance to the environment; becoming a sustainable building which is in harmony with nature, conductive to the common safety and health for ecosystem and human settlements system, and satisfying the human functioning requirement, psychological needs, physical needs as well as comfort needs. Therefore, the purpose of green buildings is to meet the needs of safety, health and harmony between human and nature through the construction, and to satisfy people's demand of appropriate living condition.

Green buildings bring together a vast array of practice, techniques and skills to reduce and ultimately eliminate the impacts of buildings on the environment and human being. Taking advantage of renewable resources is emphasized, e.g., using sunlight through passive solar, active solar, and photovoltaic equipments, using plants and trees through green roofs, rain gardens to reduce rainwater run-off. Many other techniques are used, such as low-impact building materials, packed gravel or permeable concrete instead of conventional concrete or asphalt to enhance replenishment of ground water. So-

绿色建筑的标准在不同国家和地区都存在差异,但绿色建筑的设计都必须包含以下方面:选址和结构的能效、能耗、水资源的能效、材料能效、室内环境改善、运行和维护优化、废弃物减排等。绿色建筑的本质是以上一个或多个原则的优化。通过适当的协同设计,各个绿色建筑技术可以共同工作以发挥更大的累积效应。

中国地域广大,人口众多,环境条件多样,人均资源匮乏且存在分布不合理的现象,经济与社会条件差异性突出。因此,发展绿色建筑具有特殊的科学地位、政治作用、社会意义和经济价值。

图 1.20 达姆施塔特工业大学设计的亚热带全生态减碳屋
Fig. 1.20 Subtropical ecosystem carbon reduction house designed by Darmstadt University of Technology

lar energy is the only energy in the subtropical ecosystem carbon reduction house designed by Darmstadt University of Technology in Germany(Fig.1.20).

While the standard for green buildings is constantly evolving and may differ from region to region, the fundamental principles are as follows: siting and structure design efficiency, energy efficiency, water efficiency, materials efficiency, indoor environmental quality enhancement, operations and maintenance optimization, waste and toxics reduction. The essence of green building is an optimization of one or more of these principles. With the proper synergistic design, individual green building technologies may work together to produce a greater cumulative effect.

China has a vast area with a large population, complex environmental condition, insufficient resource and irrational distribution. The differences of economy and social condition are highlighted. Therefore, the development of green building has the special scientific status, political role, social significance and economic value.

注:本章插图 1.2~1.20 来源于网络。
Note: the illustrations in this chapter Fig.1.2~Fig.1.20 are from the network.

知识拓展
Learning More

相关链接　Related Links

(1) 中国土木工程网

(2) 美国土木工程师协会官网

(3) 美国混凝土学会 ACI 官网

(4) 中国土木网

(5) 中国土木科技网

(6) 中国土木工程学会官网

(7) 美国发现工程网

思考题　Review Questions

(1) 什么是土木工程？思考土木工程在人类社会发展中的作用。

What is civil engineering? Think about the role of civil engineering in the development of human society.

(2) 了解和认识土木工程的历史、现状和未来。

Learn and understand the history, the present situation and the future of civil engineering.

(3) 土木工程专业的工程对象与业务范畴有哪些？

What are the project objects and scopes of civil engineering?

参考文献
References

[1] 高等学校土木工程学科专业指导委员会.本科教育培养目标和培养方案及课程教学大纲[M].北京:中国建筑工业出版社,2002.

[2] 高等学校土木工程学科专业指导委员会. 高等学校土木工程本科指导性专业规范[M].北京:中国建筑工业出版社,2011.

[3] 叶列平.土木工程科学前沿[M].北京:清华大学出版社,2006.

[4] 赵鸿佐,胡鹤钧.中国土木建筑百科辞典[M].北京:中国建筑工业出版社,1999.

[5] 叶志明.土木工程概论[M].3 版.北京:高等教育出版社,2009.

[6] 周新刚.土木工程概论[M].北京:中国建筑工业出版社,2011.

[7] Bhavikatti S S. Basic Civil Engineering[M].New Age International Ltd.,2010.

[8] 丁大钧,蒋永生.土木工程概论[M].2 版.北京:中国建筑工业出版社,2010.

[9] 帕拉理查米.土木工程概论[M].3 版.北京:机械工业出版社,2005.

[10] Nisture S P, Pawar A D. Basic Civil Engineering[M].Technical Publications Pune,2006.

[11] 《绿色建筑》教材编写组.绿色建筑[M].北京:中国计划出版社,2008.

[12] 崔京浩.新编土木工程概论:伟大的土木工程[M].北京:清华大学出版社,2013.

[13] 张志国.土木工程概论[M].武汉:武汉大学出版社,2014.

第2章 土木工程材料

Chapter 2　Civil Engineering Material

先导案例
Guide Case

在土木工程中,材料是产生新结构体系的原动力。人类社会初期,土木工程主要采用土、木、石材建造,结构体系相对简单。工业革命后,混凝土和钢材逐渐成为建造土木工程的主要结构材料,并由此推动了现代土木工程结构体系的发展。然而,由于混凝土与钢材的生产需要消耗大量的自然资源并产生一定的环境污染,人们希望找到可回收和可循环利用的、寿命长及节能的结构材料来建造可持续的土木工程。纤维增强复合材料(FRP)因其良好的力学和耐久性能,逐渐成为继钢材和混凝土材料之后的第三类结构材料。常见的FRP材料主要有碳纤维增强复合材料(CFRP)、玻璃纤维增强复合材料(GFRP)、芳纶纤维增强复合材料(AFRP)和玄武岩纤维增强复合材料(BFRP)等几种。

In civil engineering(CE) field, materials are usually regarded as the motivity to produce new structural system. In early human society, CE was mainly built by soil, timber or stone, and the CE structural systems were simple. After the industrial revolution, concrete and steel gradually became the main structural materials to build CE, which promoted the modern structural systems appeared in CE. During the process of producing concrete and steel, however, a great deal of natural resources will be consumed and the environment will be also polluted. People hope to find some structural materials, which are recyclable, long-life and energy-saving, to build sustainable development CE. Due to its good mechanical and durable performance, fiber reinforced polymer (FRP) has gradually become the third kind structural material except for concrete and steel. Currently, the FRP materials mainly include carbon fiber reinforced polymer (CFRP), glass fiber reinforced polymer (GFRP), aramid fiber reinforced polymer (AFRP), basalt fiber reinforced polymer (BFRP), and so on.

　　下面将通过两个实际工程案例来介绍 CFRP 材料在现代土木工程结构中的应用。第一个案例是国内首座碳纤维 CFRP 索斜拉桥(见图 2.1)。该桥位于江苏大学校内,是一座连接学生宿舍与食堂的人行桥,由东南大学、江苏大学和北京特希达公司共同研究与开发,并于 2004 年建成。该斜拉桥有两跨,跨径为 30 m+18.4 m,采用双柱式索塔,索塔两侧各布置 4 对斜拉索。该桥的最大亮点在于斜拉索采用 CFRP 索替代传统的钢索,从而充分发挥了 CFRP 材料轻质、高强、耐久、耐腐蚀等良好特性,为该类新材料结构体系的进一步推广应用起到了先导示范作用。

图2.1　国内首座碳纤维CFRP索斜拉桥
Fig. 2.1　First CFRP cable-stayed bridge in China

　　Then, two engineering cases will be used to introduce the application of CFRP materials in CE field. One case is the first CFRP cable-stayed bridge in China (see Fig.2.1). This footbridge locates in Jiangsu University, connecting students' dormitories and canteen. It was designed and built by Southeast University, Jiangsu University and Beijing Te Xi Da Corporation in 2004. This cable-stayed bridge has two spans with 30 meters and 18.4 meters respectively. Between the two spans, there is a two-column cable support tower, around which there are four tensile cables. The highlight of this cable-stayed bridge is the application of CFRP cables instead of steel cables. There are many advantageous characteristics of CFRP cable, such as low-weight, high-strength, durability and corrosion resistance etc. The application of CFRP cable in this bridge may become a leading case for the wider use of FRP materials in the new structural systems.

第二个案例是国内首座碳纤维 CFRP 索悬吊式建筑(见图 2.2)。该悬吊式建筑是一栋三层办公楼（底层为车库），位于南京诺尔泰复合材料设备制造有限公司内部。在该悬吊式建筑中，内部四根钢筋混凝土圆柱起到了承受和传递荷载的作用，即楼面荷载通过内外 CFRP 吊索传递到固定在圆柱上的水平悬吊梁，再通过圆柱传递到基础。用 CFRP 吊索替代传统的钢索，可充分发挥 CFRP 材料强度高的力学特性，能有效增大结构跨度并减少材料用量,使建筑的形式富于变化。

Another case is the first CFRP suspension type building in China (Fig.2.2). This suspension type building is a three-story office building, locating at Nanjing Loyalty Composite Equipment Company Ltd. In the center of this building, there are four main reinforced concrete (RC) circular columns, which can bear loads and also transmit them to the foundation. The loads on the second floor are transmitted through the inner and outer CFRP suspension cables to the horizontal suspended beams which were fixed on the circular columns. This alternative application of CFRP cables can give full play to the CFRP materials' mechanical properties of high strength. It can effectively increase the structure span and reduce the material consumption, also make the building form abundant in variation.

建筑外观
Architectural appearance

内部吊索
Internal suspension cable

外部吊索
External suspension cable

图2.2　国内首座碳纤维CFRP索悬吊式建筑
Fig. 2.2　First CFRP suspension type building in China

土木工程材料是土木工程建(构)筑物所使用的各种材料及制品的总称。它是一切土木工程的物质基础,决定了建筑结构形式及其施工方法。土木工程材料的发展与科技水平、工业化水平及环境保护要求密切相关。为了适应建筑工业化、提高工程质量、保护生态环境、实现可持续发展的要求,土木工程领域不断涌现出各种新型材料;新材料的出现,又促使建筑形式发生变化,结构设计和施工技术产生革新。本章就常见的土木工程材料进行简要的介绍。

2.1 土木工程材料分类概述

土木工程材料品种繁多,钢材、水泥、木材、混凝土、砖、砌块、沥青等都是常见的工程材料。土木工程材料的分类方法主要有如下 3 种。

① 按使用性能分类,可以分为结构材料(受力构件或结构所用的材料,如基础、梁、板、柱等所用的材料)、墙体材料(内外及隔墙墙体所用的材料,如砌墙砖、砌块、墙板、幕墙等所用的材料)、功能材料(具有专门功能的材料,如防水材料、保温隔热材料、装饰装修材料、地面材料及屋面材料等)。

② 按用途分类,可以分为建筑结构材料、桥梁结构材料、水工结构材料、路面结构材料等。

Civil engineering material (CEM) is a generic term of all kinds of materials and products used in civil engineering (CE). CEM is the material basis in the whole CE buildings. CEM determines the structural forms and construction methods. Its development is closely related with the current level of science and technology, industrialization and environmental protection. In order to adapt to building industrialization, improvement of engineering quality, protection of ecological environment and social sustainable development, new materials are constantly emerging in CE field. The appearance of new materials promotes the renovation of architectural form and the reformation of structural design and construction technique. The common CEMs will be introduced in this chapter.

2.1 Classification and overview of CEM

There are many kinds of CEM, such as steel, cement, wood, concrete, brick, building block and asphalt. Common classification methods mainly include:

① According to their usability, CEMs can be classified as structural materials (materials used for bearing members and structures, such as foundation, beam, slab, column, etc.), wall materials (materials used for external and internal walls, such as wall brick, building block, wallboard, curtain wall), and materials with specialized functions (such as waterproof materials, thermal insulation materials, decoration materials, roofing materials, etc.).

② According to their usages, CEMs can be classified as building structure materials, bridge structure materials, hydraulic structure materials and pavement structure materials, etc.

③ 按化学成分分类,土木工程材料又可以分为无机材料、有机材料及复合材料(见表 2.1)。

2.2　石材、砖、瓦和砌块

石材、砖、瓦和砌块等是最基本的建筑材料。无论是在古代还是现代的建筑领域中,石材、砖、瓦和砌块均处于不可替代的地位。

表2.1　土木工程材料按化学成分分类
Tab. 2.1　CEMs classified according to the chemical components

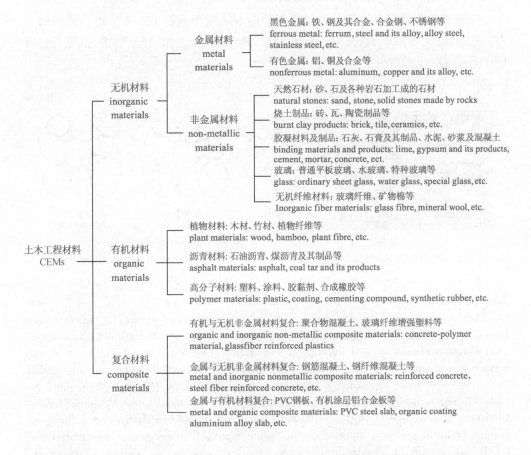

③ According to their chemical components, CEMs can be classified as inorganic materials, organic materials and composite materials, as shown in Tab.2.1.

2.2　Stone, brick, tile and building block

Stone, brick, tile and building block are basic building materials. They are irreplaceable in CE field, not only in ancient times, but also in modern times.

2.2.1 石材

凡采自天然岩石,经过加工或未经加工的石材,统称为天然石材(见图 2.3)。一般天然石材具有强度高、硬度大、耐磨性好、装饰性及耐久性好等优点。石材的使用有着悠久的历史,古埃及的金字塔、太阳神神庙,中国隋唐时期的石窟、石塔、赵州桥等,都是具有历史代表性的石材建筑。在现代建筑中,北京人民英雄纪念碑、毛主席纪念堂、人民大会堂等,都是使用石材的典范。石材被公认为是一种优良的土木工程材料,土木工程中常用的石料根据其加工程度分为毛石、片石、料石、装饰石材和石子等。在混凝土原材料中,石子常被称为粗骨料(见图 2.4)。

（1）毛石

岩石被爆破后直接获得的形状不规划的石块称为毛石。土木工程中使用的毛石,一般高度应不小于 150 mm。毛石可用于砌筑基础、堤坝、挡土墙等,乱毛石也可用作毛石混凝土的骨料。

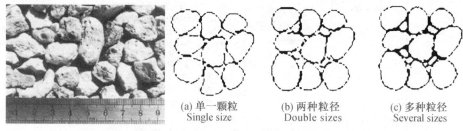

(a) 单一颗粒　　　　(b) 两种粒径　　　　(c) 多种粒径
Single size　　　　Double sizes　　　　Several sizes

图 2.3　天然石材　　　　　图 2.4　粗骨料颗粒级配图
Fig. 2.3　Natural stone　　　Fig. 2.4　Distribution diagram of coarse aggregates

2.2.1　Stone

Any stone collected from natural rock, whether processed or not, is called natural stone(Fig.2.3). Natural stone usually has the advantages of high strength, large hardness, good abrasion resistance and durability, etc. The use of stone in CE field has a long history. For example, the ancient Egyptian Pyramids, Sun God Temple, as well as rock caves, stone towers, Zhaozhou Bridge built in Sui and Tang Dynasties in China, are historic stone buildings. In modern architectures, the Monument to the People's Heroes, Chairman Mao Memorial Hall, and Great Hall of the People in Beijing are typical examples of buildings with stone material. Stone material is recognized as an excellent CEM, and it can be divided into rubble, flag, dressed stone, decorative stone and cobblestone according to the degree of processing. The cobblestone is usually called coarse aggregate in concrete material(Fig. 2.4).

（1）Rubble

Rubble is the out-of-shape stone which obtained from the blasted rocks. The average height of rubbles used in CE is not less than 150 mm. Rubble is available for laying foundation, dam, revetment, etc. Mess rubble can be also used as aggregate for rubble concrete.

（2）片石

片石也是由爆破而得的，形状不受限制，但薄片不得使用。一般片石的尺寸应不小于 150 mm，主要用来砌筑圬工工程、护坡、护岸等。

（3）料石

料石是由人工或机械开采出的较规则的六面体石块，再经人工加凿而成。料石一般由致密均匀的砂岩、石灰岩、花岗岩加工而成，用于土木工程结构物的基础、勒脚、墙体等部位。

（4）装饰石材

建筑物内外墙面、柱面、地面、栏杆、台阶等处装修用的石材称为装饰石材。装饰石材一般采用大理石或花岗岩制成，大理石板材可用于室内装饰，而花岗岩板材主要用于土木工程的室外饰面。

（5）石子

在混凝土的组成材料中，砂为细骨料，石子为粗骨料（见图 2.4）。石子除用作混凝土粗骨料外，也常用于路桥工程、铁道工程的路基等。石子又分为碎石和卵石：由天然岩石或卵石经破碎、筛分而得到的粒径大于 5 mm 的颗粒，称为碎石或碎卵石；岩石由于自然条件作用而形成的粒径大于 5 mm 的颗粒，称为卵石。

（2）Flag

Flag is also obtained from the blasted rocks, but its shape is not restricted, and the slice can't be used. Generally, the size of flag should be no less than 150 mm. Flag can be used for masonry engineering, slope protection, revetment, etc.

（3）Dressed Stone

Dressed stone is gained from hexahedral stones, which is exploited by artificial or mechanical method. Generally, dressed stone is made by dense homogeneous sandstone, limestone, granite. It can be used for foundation, plinth, and wall of CE structures.

（4）Decorative stone

Decorative stone is used for wall, cylindrical surface, ground, railing, and step surface of the building. Decorative stone is usually made of marble or granite. The former is used for indoor decoration, and the latter is mainly for outside decoration.

（5）Cobblestone

In concrete ingredients, sand is fine aggregate, and cobblestone is coarse aggregate (Fig.2.4). Besides, cobblestone also can be used for the roadbed in bridge and road engineering, railway engineering, etc. Cobblestone is classified as gravel and pebble. Gravel is gained from broken natural rock or pebble, and its diameter is greater than 5 mm. The rock, which is formed by natural conditions, is called pebble, and its diameter is also greater than 5 mm.

2.2.2 砖

砖是一种常用的砌筑材料。砖有多种分类方法:① 按生产工艺可分为两类,一类是通过焙烧工艺制成的,称为烧结砖;另一类是通过蒸压工艺制成的,称为蒸压砖,也称非烧结(免烧)砖。② 按所用原材料可分为黏土砖、页岩砖、煤矸石砖、粉煤灰砖、炉渣砖和灰砂砖等。③ 按有无孔洞砖又可分为实心砖、多孔砖和空心砖。其中,孔洞率大于等于 25%,且孔的尺寸小而数量多的砖为多孔砖(见图 2.5),常用于承重部位;孔洞率大于等于 40%,且孔的尺寸大而数量少的砖为空心砖(见图2.6),常用于非承重部位。

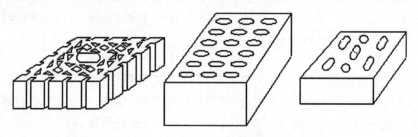

图 2.5　烧结多孔砖
Fig. 2.5　Fired perforated brick

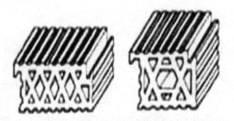

图 2.6　烧结空心砖
Fig. 2.6　Fired air brick

2.2.2 Brick

Brick is a kind of masonry materials. Usually, bricks can be classified as follows: ① Classified by manufacturing technique, brick can be sorted into fired brick and autoclaved brick. The former is made by roasting process, and the latter is made by autoclaved process. ② Classified by raw material, brick can be sorted into clay brick, shale brick, colliery wastes brick, fly ash brick, cinder brick, sand-lime brick, etc. ③ Classified by the hole ratio inside, brick can be sorted into solid brick, perforated brick and air brick. Perforated brick is the brick with void ratio not less than 25%. The void size is small and the number of hole is large (Fig. 2.5). Therefore, it is commonly used in load-bearing areas. Air brick is the brick with void ratio not less than 40%. The void size is big but the number of hole is small (Fig.2.6). It is commonly used in non load-bearing areas.

砖的标准尺寸为 240 mm×115 mm×53 mm,通常将 240 mm×115 mm 的面称为大面,240 mm×53 mm 的面称为条面,115 mm×53 mm 的面称为顶面,如图 2.7 所示。

由于生产烧结普通砖(以黏土砖为主)的过程中要大量占用耕地,能耗高、污染环境,施工生产中劳动强度高、工效低,因此我国绝大多数城市现已全面禁止使用烧结普通砖。目前应用较广的是蒸压砖,主要有蒸压灰砂砖、蒸压粉煤灰砖等。其他一些非烧结砖正在研发中,如石灰非烧结砖,水泥、石灰黏土非烧结空心砖等。

2.2.3 瓦

瓦,一般指黏土瓦,属于屋面材料(见图 2.8)。它以黏土为主要原料,经泥料处理、成型、干燥和焙烧等环节制成。由于黏土瓦材质脆、自重大、片小、施工效率低及破坏与污染环境等缺点,与黏土砖一样,目前已经禁止使用。

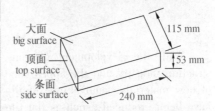

图 2.7 砖及其各部分名称
Fig. 2.7 Brick and its faces

The standard size of brick is 240 mm×115 mm ×53 mm. The face with size of 240 mm×115 mm is commonly called big face. The face with size of 240 mm×53 mm is called side face, and the face with size of 115 mm×53 mm is called top face, as shown in Fig.2.7.

Owing to the use of a large number of cultivated land, high energy consumption, pollution, high labor intensity and low work efficiency, the application of fired common brick is prohibited in most cities in China. Autoclaved brick is widely used at present, which includes autoclaved sand-lime brick, autoclaved flyash-lime brick, etc. Many non-clinker bricks are researched, such as lime non-vitrified brick, which will become a promising material.

图 2.8 黏土瓦
Fig. 2.8 Clay tile

2.2.3 Tile

Tile generally referred to the clay tile in the past. It is a kind of roof material (Fig. 2.8). The raw material of tile is clay, and it is made by several processes such as sludge processing, forming, drying, and roasting. Since clay tile has many shortcomings, such as brittle, heavy, pollution and low work efficiency, it is also prohibited at present.

随着建筑工业的发展,新型建筑材料涌现,目前我国生产的瓦的种类很多:按形状,可分为平瓦和波形瓦(见图 2.9)两类;按所用材料,可分为陶土烧结瓦、混凝土瓦、石棉瓦、钢丝网水泥瓦、聚氯乙烯瓦、玻璃钢瓦、沥青瓦等。

2.2.4 砌块

砌块是人造板材,外形多为直角六面体(见图 2.10)。砌块建筑在我国始于 20 世纪 20 年代,近十年来发展较快。砌块可以充分利用地方资源和工业废渣,节省黏土资源和改善环境,实现可持续发展,且其具有生产工艺简单,原料来源广,适应性强,制作及使用方便灵活等优点。砌块除用于砌筑墙体外,还可用于砌筑挡土墙、高速公路隔音屏障及其他构筑物。

图 2.9 波形瓦
Fig. 2.9 Pantile

With the development of construction industry and the appearance of new materials, many kinds of tile have been produced, such as flat tile and pantile (Fig.2.9), which are sorted by shape. Classified by materials, there are clay sintered tile, concrete tile, asbestos tile, steel mesh cement tile, polyvinyl chloride tile, glass steel tile, asphalt tile, etc.

2.2.4 Building block

Building block is a kind of artificial panel, and its shape is right angle hexahedron(Fig. 2.10). In China, block buildings can be dated back to 1920s, and were developed rapidly in the recent decade. Blocks can make use of local resources and industrial wastes. It is a good way to save clay resources and improve environments, and then to achieve sustainable development. Block has lots of advantages, such as simple production process, wide raw material source, good adaption, and flexible usage. Block can be used not only in the external wall, but also in the retaining wall, the highway noise barriers and other constructions.

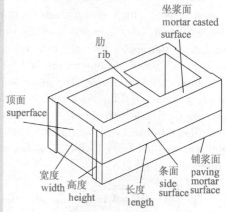

坐浆面
mortar casted surface

肋
rib

顶面
superface

宽度
width

高度
height

长度
length

条面
side surface

铺浆面
paving mortar surface

图 2.10 砌块及部位名称
Fig. 2.10 Building block and its parts

2.3　胶凝材料

土木工程中,凡是经过一系列物理、化学作用,能将散粒材料(如沙子、石子等)或块状材料(如砖、石块和砌块等)黏结成具有一定强度的整体的材料,称为胶凝材料。胶凝材料的分类见表 2.2。

气硬性胶凝材料在水中不能硬化,只能在空气中硬化,保持并发展其强度,因而不能用于潮湿环境和水中。水硬性胶凝材料不仅能在空气中硬化,还能更好地在水中硬化,保持并继续发展其强度,因此它既适用于地上,也适用于潮湿环境或水中。下面主要介绍土木工程中常见的胶凝材料及其拌合物。

2.3.1　水泥

1824 年,英国工程师约瑟夫·阿斯帕丁发明了"波特兰水泥"(即 Portland 水泥,我国称硅酸盐水泥),并取得了生产专利,这标志着水泥的诞生。水泥是一种粉状矿物材料,它与水拌和后形成塑性浆体,能在空气和水中凝结硬化,并能把砂、石等材料胶结成整体,形成坚硬石状体的水硬性胶凝材料。普通水泥的主要成分包括硅酸三钙($3CaO \cdot SiO_2$)、硅酸二钙($2CaO \cdot SiO_2$)和铝酸三钙($3CaO \cdot Al_2O_3$)。

表 2.2　胶凝材料的分类
Tab. 2.2　Classification of cementitious material

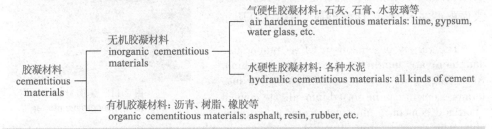

2.3　Cementitious materials

The material, which can bond granular materials (such as sand, stone, etc.) and massive materials (such as brick, stone, block, etc.) together as a whole unit by a series of physical and chemical effects, is called cementitious (binding) material. Its classification is shown in Tab.2.2.

Air hardening cementitious material can be hardened and develop its strength only in air rather than in water, so it can not be used in damp environment. Hydraulic cementitious material can be hardened not only in air but also in water, and it can be applied on the ground, damp environment and under water. Common cementitious materials and its mixtures used in CE will be introduced briefly.

2.3.1　Cement

Joseph Aspdin, a British engineer, discovered the "Portland cement" (which is called silicate cement in China) in 1824, and obtained production patent. This event marked the birth of cement. Cement is a kind of powdery mineral material. It becomes plastic slurry after mixed with water, and can condense and harden in air and water. This plastic slurry can combine the materials such as sand and stone into a whole, and then form a hard body like stone. The main compositions of Portland cement include $3CaO \cdot SiO_2$, $2CaO \cdot SiO_2$ and $3CaO \cdot Al_2O_3$.

土木工程中应用的水泥品种众多,在我国就有上百个品种。按水泥的主要水硬化物质可分为硅酸盐系水泥、铝酸盐系水泥、硫铝酸盐系水泥、铁铝酸盐系水泥、磷酸盐系水泥、氟铝酸盐系水泥等系列;按水泥的用途和性能可分为通用水泥、专用水泥和特性水泥三大类。

2.3.2 砂浆

砂浆是由胶凝材料、细骨料、水,有时也加入适量掺和料和外加剂混合,按适当比例配制而成的土木工程材料,在工程中起黏结、衬垫和传递应力的作用。在结构工程中,砂浆可以把砖、砌块和石材等黏结为砌体;在装饰工程中,墙面、地面及混凝土梁、柱等需要用砂浆抹面,起到保护结构和装饰的作用。

按胶凝材料不同,砂浆可以分为水泥砂浆、水泥混合砂浆、石灰砂浆、石膏砂浆和聚合物砂浆等。其中,水泥混合砂浆是在水泥砂浆中加入一定量的掺和料(如石灰膏、黏土膏、电石膏等)制成的,以此来改善砂浆的和易性,降低水泥用量。

按用途不同,砂浆又可以分为砌筑砂浆、抹面砂浆和特种砂浆等。将砖、石、砌块等黏结成砌体的砂浆称为砌筑砂浆。它起着黏结砌块、传递荷载、均匀应力、协调变形的作用,是砌体的重要组成部分。凡粉刷于建筑物或建筑构件表面的砂浆,统称为抹面砂浆,如图 2.11 所示。

图 2.11　抹面砂浆
Fig. 2.11　Decorative mortar

The variety of cement in CE is numerous, and there are hundreds of varieties in China. According to the main water hardening materials, cement can be sorted into silicate cement (Portland cement), aluminate cement, sulphur aluminate cement, iron aluminate cement, phosphate, fluorine aluminate cement and other series. According to its usage and performance, cement can also be divided into general cement, special cement and characteristic cement.

2.3.2　Mortar

Mortar is made up of cementitious materials, fine aggregate and water with appropriate proportion. Sometimes, the suitable amount of admixture and additive will be added. Mortar plays functions as bond, pad and transfers stress in CE. In structural engineering, mortar can bond brick, block and stone to build masonry. In decorative engineering, mortar can protect and decorate the surface of structural element, such as wall, ground, concrete beam and column, etc.

According to cementitious materials, mortar can be sorted into cement mortar, mixed cement mortar, lime mortar, plaster mortar and polymer mortar, etc. Mixed cement mortar is formed by adding a certain amount of admixture into cement mortar, such as lime paste, clay, and gypsum, which can improve the workability of mortar and reduce the amount of cement.

According to its application, mortar can also be divided into masonry mortar, decorative mortar and special mortar. Masonry mortar is used to mix brick, stone and building blocks. It can bond blocks, transfer load, make stress distribution more uniform and coordinate deformation, so it is an important part of the masonry. The mortar which is painted on the surface of the building or its component is called decorative mortar, as shown in Fig.2.11. Decorative mortar can protect the base materials to meet the service-

抹面砂浆具有保护基层材料,满足使用要求和装饰的作用。特种砂浆是指具有某些特殊功能的抹面砂浆,主要有绝热砂浆、吸声砂浆、耐酸砂浆和防辐射砂浆等。

2.3.3　沥青

沥青是一种褐色或黑褐色的有机胶凝材料,在房屋建筑、道路、桥梁等工程中有着广泛的应用,采用沥青作为胶凝材料的沥青拌和料是公路路面、机场跑道面的一种主要材料(见图 2.12)。由于沥青属于憎水材料,因此也广泛应用于水利工程以及其他防水、防渗工程中。

2.4　建筑钢材

2.4.1　钢材分类

钢是由生铁冶炼而成的。在理论上凡碳质量分数在 2.06%以下,含有害杂质较少的铁碳合金均可称为钢。钢的品种繁多,分类方法也很多(见表 2.3),通常有按化学成分、质量、用途等几种分类方法。土木工程常用钢材可划分为钢结构用钢和混凝土结构用钢两大类, 两者所用的钢种基本上都是碳素结构钢和低合金高强度结构钢。

图 2.12　沥青路面
Fig. 2.12　Bituminous pavement

ability and achieve adornment effect. Special mortar has some special functions, mainly contain sound absorption, insulation, acid-proof and radiation protection, etc.

2.3.3　Asphalt

Asphalt is a kind of brown or dark brown organic cementitious material, which has been widely used in house buildings, roads, bridges and so on. Asphalt cementitious material is mainly used in highway pavement and airport runway, as shown in Fig.2.12. Considering hydrophobic nature, asphalt is widely used in hydraulic projects and other waterproof, anti-seepage projects.

2.4　Construction steel

2.4.1　Classification of steel

Steel is made from pig iron. It is the iron-carbon alloy, in which the carbon mass fraction is less than 2.06%, and few harmful impurities exist. There are many varieties of steel, which are usually classified by chemical composition, quality, application, etc., as shown in Tab.2.3. In CE, steel can be divided into two kinds, namely, steel for steel structure and steel for the reinforcement of concrete. All steels are basically carbon steels and low-alloy high strength structural steels.

表 2.3　钢的分类
Tab. 2. 3　Classification of steel

分类 Classification	类别 Category		特性 Character
按化学成分 分类 classified by chemical composition	碳素钢 carbon steel	低碳钢 low-carbon steel	碳质量分数 < 0.25% carbon mass fraction <0.25%
		中碳钢 medium-carbon steel	碳质量分数 0.25%～0.60% carbon mass fraction 0.25%～0.60%
		高碳钢 high-carbon steel	碳质量分数 > 0.60% carbon mass fraction > 0.60%
	合金钢 alloy steel	低合金钢 low-alloy steel	合金元素质量分数 < 5% mass fraction of alloy elements < 5%
		中合金钢 medium-alloy steel	合金元素质量分数 5%～10% mass fraction of alloy elements 5%～10%
		高合金钢 high-alloy steel	合金元素质量分数 > 10% mass fraction of alloy elements > 10%
按脱氧程度 分类 classified by deoxidization degree	沸腾钢 rimmed steel		脱氧不完全，硫、磷等杂质偏析较严重，代号为"F" deoxidization is incomplete; segregation of impurities such as sulfur and phosphorus is serious, which is code named "F"
	镇静钢 killed steel		脱氧完全，同时去硫，代号为"Z" deoxidization is complete; sulfur is abandon, which is code named "Z"
	半镇静钢 balanced steel		脱氧程度介于沸腾钢和镇静钢之间，代号为"b" deoxidization degree is between the rimmed steel and killed steel, which is code named "b"
	特殊镇静钢 special killed steel		比镇静钢脱氧程度还要充分彻底，代号为"TZ" deoxidization degree fully completely, which is code named "TZ"
按质量分类 classified by quality	普通钢 ordinary steel		硫质量分数≤0.055%，磷质量分数≤0.045% sulfur mass fraction ≤0.055%, phosphorus mass fraction ≤0.045%
	优质钢 high quality steel		硫质量分数≤0.03%，磷质量分数≤0.035% sulfur mass fraction ≤0.03% phosphorus mass fraction ≤0.035%
	高级优质钢 high-grade fine steel		硫质量分数≤0.02%，磷质量分数≤0.027% sulfur mass fraction ≤0.02% phosphorus mass fraction ≤0.027%

续表　continued

分类 Classification	类别 Category	特性 Character
按用途分类 classified by application	结构钢 structural steel	工程结构构件用钢、机械制造用钢 engineering structural and machinery steel
	工具钢 tool steel	各种刀具、量具及模具用钢 all kinds of cutter, gauge and mould
	特殊钢 special steel	具有特殊物理、化学或机械性能的钢,如不锈钢、耐热钢、耐酸钢、耐磨钢、磁性钢等 special physical, chemical or mechanical performance, such as stainless steel, heat-resistant steel, acid-resistant steel, wear-resistant steel, magnetic steel, etc.

2.4.2　钢结构用钢材

钢结构用钢主要有型钢、钢板和钢管。型钢有热轧及冷弯成形两种;钢板有热轧(厚度为 0.35~200 mm)和冷轧(厚度为 0.2~5 mm)两种;钢管有热轧无缝钢管和焊接钢管两大类。钢结构的连接方法有焊接、螺栓连接和铆接 3 种,如图 2.13 所示。

(1) 型钢

型钢主要有热轧型钢和冷弯薄壁型钢两大类。其中,常用的热轧型钢有角钢(等边的和不等边的)、工字钢、槽钢、T 型钢、H 型钢、Z 型钢等,如图 2.14 所示。冷弯薄壁型钢通常由 1~6 mm 薄钢板冷弯或模压而成,有角钢、槽钢等开口薄壁型钢及方形、矩形等空心薄壁型钢,主要用于轻型钢结构。

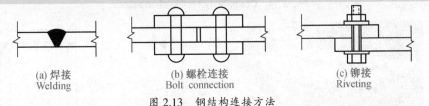

(a) 焊接　　　　　　(b) 螺栓连接　　　　　　(c) 铆接
Welding　　　　Bolt connection　　　　Riveting

图 2.13　钢结构连接方法
Fig. 2.13　Connection method of steel structure

2.4.2　Steel for steel structure

In steel structure, the shape steel, steel plate and steel pipe are mainly used. Shape steel can be formed by hot rolling or cold bending. There are two kinds of steel plate: the hot rolled steel (thickness: 0.35~200 mm) and cold rolled steel (thickness: 0.2~5 mm). There are two kinds of steel pipe: the hot rolled seamless pipe and the welded pipe. Connection methods of steel structure are welding, bolt connection and riveting, as shown in Fig.2.13.

(1) Shape steel

Shape steel mainly includes two kinds: hot rolled shape steel and cold-formed thin-walled steel. The common hot rolled shape steels contain angle steel, I-shaped steel, U-shaped steel, T-shaped steel, H-shaped steel, and Z-shaped steel, as shown in Fig. 2.14. Cold-formed thin-walled steel is made by 1~6 mm thin steel plate, which includes angle steel, U-shaped steel, square or rectangular hollow thin-walled steel, etc. It is mainly used in thin-walled steel structure.

（2）钢板

用光面压辊轧制而成的扁平钢材，且以平板状态供货的称为钢板(见图2.15)。钢板有热轧钢板和冷轧钢板两种，热轧钢板按厚度分为厚板(厚度大于4 mm)和薄板(厚度为0.35~4 mm)两种，冷轧钢板只有薄板(厚度为0.2~4 mm)。

一般厚板用于焊接结构，薄板主要用于屋面板、墙板和楼板等。在钢结构中，单块板不能独立工作，必须用几块板通过连接组合成工字形、箱形截面等构件来承受荷载。图2.16所示为用钢板焊接组成的工字形截面和箱形截面。

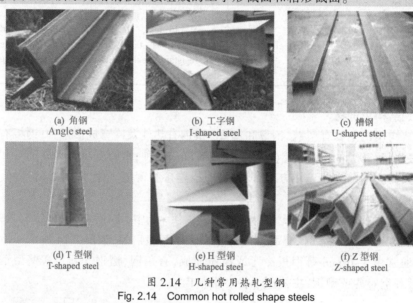

(a) 角钢
Angle steel

(b) 工字钢
I-shaped steel

(c) 槽钢
U-shaped steel

(d) T型钢
T-shaped steel

(e) H型钢
H-shaped steel

(f) Z型钢
Z-shaped steel

图 2.14　几种常用热轧型钢
Fig. 2.14　Common hot rolled shape steels

图 2.15　钢板
Fig. 2.15　Steel plate

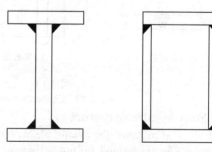

图 2.16　焊接组成截面
Fig. 2.16　Welded cross section

（2）Steel plate

Steel plate is a kind of flat steel which is rolled with a smooth roller, as shown in Fig.2.15. There are two kinds of steel plate, namely, hot rolled plate and cold rolled plate. According to its thickness, the hot rolled plate can be divided into thick plate (thickness>4 mm) and thin plate (thickness 0.35~4 mm). The thickness of cold rolled plate is 0.2~4 mm.

Generally, thick plate is used for welding structure, while thin plate is mainly for roof panel, wall panel and floor, etc. In steel structure, single plate cannot work independently. The element formed by connecting several single plates into I-section or box section could bear loads, as shown in Fig.2.16.

（3）钢管

按生产工艺不同,钢结构所用钢管(见图 2.17)分为热轧无缝钢管和焊接钢管两大类。在土木工程中,钢管多用于制作桁架、塔桅、钢管混凝土等,广泛应用于高层建筑、厂房柱、塔柱、压力管道等工程中。

2.4.3　混凝土结构用钢材

混凝土结构主要包括钢筋混凝土结构和预应力混凝土结构,其所用钢材主要有普通钢筋和预应力筋。

（1）普通钢筋

普通钢筋指用于钢筋混凝土结构中的钢筋和预应力混凝土结构中的非预应力钢筋,其材质包括普通碳素钢和普通低合金钢两大类。普通钢筋按生产工艺性能和用途的不同可分为以下几类。

① 热轧钢筋。用加热钢坯轧成的条形成品,称为热轧钢筋。按轧制外形,可分为热轧光圆钢筋和热轧带肋钢筋（见图 2.18）两类,其中肋的形式有等高肋和月牙肋(见图 2.19)。按热轧工艺,热轧带肋钢筋又可分为普通热轧钢筋和细晶粒热轧钢筋。

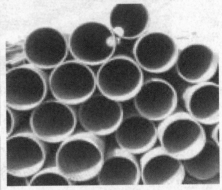

图 2.17　钢管
Fig. 2.17　Steel pipe

（3）Steel Pipe

According to the production process, steel pipes are divided into two categories, namely, the hot rolled seamless pipe and the welded pipe, as shown in Fig.2.17. In CE, steel pipes are used to make truss, tower mast and concrete filled steel tubes, etc., which are widely used in high-rise buildings, columns, pillar, pressure pipe, etc.

2.4.3　Steel used in concrete structure

Concrete structure mainly includes the reinforced concrete(RC) structure and prestressed concrete(PC) structures. The steels used in concrete structure are mainly ordinary reinforcement and prestressing tendon.

（1）Ordinary reinforcement

Ordinary reinforcement is the steel used in RC and the non-prestressing steel in PC. The materials include two types: plain carbon steel and ordinary low-alloy steel. According to production process and usage, ordinary reinforcement can be divided into the following categories.

① Hot rolled bars. Hot rolled bars are rolled by heating billet. According to rolled appearance, they can be divided into hot rolled plain bars (HPB) and hot rolled ribbed bars (HRB, as shown in Fig.2.18). There are two types of ribbed bars: contour ribs and crescent ribs(as shown in Fig.2.19). According to hot rolled process, HRB can be divided into the ordinary hot rolled bars and fine grain hot rolled bars.

② 冷拉钢筋。为了提高强度以节约钢筋，工程中常按施工规程对钢筋进行冷拉。冷拉后钢筋的强度提高，但塑性、韧性变差。因此，冷拉钢筋不宜用于受冲击或重复荷载作用的结构。

③ 冷轧带肋钢筋。冷轧带肋钢筋是采用普通低碳钢或低合金钢热轧的圆盘条，经冷轧在其表面形成两面或三面有肋的钢筋。

④ 热处理钢筋。热处理钢筋是用热轧螺纹钢筋经淬火和回火的调质处理而成的，公称直径主要有 6,8,10,12,14 mm 5 个规格。热处理钢筋具有高强度、高韧性和高黏结力及塑性降低少等优点。

图 2.18　热轧钢筋
Fig. 2.18　Hot rolled bars

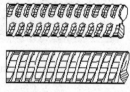 

等高肋 Contour ribs　　　　　月牙肋 Crescent ribs

图 2.19　带肋钢筋
Fig. 2.19　Ribbed steel bars

② Cold-drawn bars. In order to improve strength and save steel, cold drawing process is usually done according to the construction regulations. The strength of cold-drawn steel bars will be improved, but ductility and toughness will be deteriorated. Therefore, cold-drawn steel should not be used in structures with impact loads or repeated loads.

③ Cold rolled ribbed bars. Using the hot rolled rods of low-carbon or low-alloy steel, the cold rolled ribbed bars with two or three ribbed sides are formed by cold rolling technique.

④ Heat-treated bars. Heat-treated bars are made with hot rolled ribbed bars by quenching and tempering treatment. The nominal diameters of this bar are 6, 8, 10, 12 and 14 mm. Heat-treated bars have some advantages such as high strength, high toughness, high bond strength and less plasticity decreases.

（2）预应力筋

预应力筋主要包括预应力钢丝、钢绞线和螺纹钢筋。

① 预应力钢丝。预应力钢丝是采用优质碳素钢或其他性能相当的钢种，经冷加工及时效处理或热处理而制得的高强度钢丝（见图 2.20 a）。根据《预应力混凝土用钢丝》（GB/T 5223—2002），钢丝分为冷拉钢丝和消除应力钢丝（包括光圆钢丝、刻痕钢丝和螺旋肋钢丝）两类。它的强度比普通热轧钢筋高许多，可节省钢材、减少截面、节省混凝土，主要用于桥梁、吊车梁、大跨度屋架、管桩等预应力钢筋混凝土构件中。

② 钢绞线。预应力混凝土用钢绞线由冷拔钢丝制造而成。钢绞线的规格有 2 股、3 股、7 股、19 股等，其中 7 股钢绞线（见图 2.20 b）由于面积较大、柔软、施工操作方便，目前已成为国内外应用最广的一种预应力钢筋。钢绞线具有强度高、柔性好、质量稳定、成盘供应无需接头等优点，适用于大型结构、薄腹梁、大跨度桥梁等负荷大、跨度大的预应力混凝土结构。

(a) 钢丝　　　　　　　　　　(b) 钢绞线(7 股)
　　Wires　　　　　　　　　　　Strand(7 wires)

图 2.20　钢丝和钢绞线
Fig. 2.20　Wires and strand

（2）Prestressing tendons

Prestressing tendons mainly contain prestressing wire, strand and deformed bar.

① Prestressing wire. Prestressing wire is made of high-quality carbon steel or other similar performance steel by cold working and aging treatment or heat treatment (Fig. 2.20 a).According to *Steel Wires for the Prestressing of Concrete* (GB/T 5223—2002), the wires can be divided into cold drawn steel wire and stress relief wire (including light round wire, nicked wire and spiral wire). The strength of prestressing wire is much higher than that of HRB, so the amount of steel bars and concrete can be reduced. Prestressing wires are mainly used in many PC elements, such as bridges, crane beams, large-span roof trusses and pipe piles.

② Strand. The strands used in PC are made with cold drawn steel wires. Specifications of strands include 2, 3, 7 and 19 steel wires and so on. Due to large area, soft and operability, the 7 steel wires strand (Fig.2.20 b) has become the most widely used prestressing strand at home and abroad. Strand has many advantages, such as high strength, good flexibility, stable quality and no joints. Therefore, it is suitable for large structures, thin abdominal beams, large span bridges which are typical PC structure with large load and span PC structures.

③ 预应力螺纹钢筋。预应力混凝土用螺纹钢筋又名精轧螺纹钢筋,按屈服强度分为 PSB785,PSB835,PSB930,PSB1080 四个级别。具有连接锚固简便、张拉锚固安全可靠、黏着力强等特点,主要用于制造高强度、大跨度的混凝土制品,如核电站、水电站、桥梁、隧道、高速铁路等重点工程。

2.5 混凝土及其构件

2.5.1 混凝土

混凝土是指由胶凝材料、骨料(或称集料)、水按一定比例配制(也常掺入适量的外加剂和掺合料),经搅拌振捣成型,在一定条件下养护而成的人造石材。混凝土常简写为"砼(tóng)",它是现代土木工程中用途最广、用量最大的建筑材料之一。

(1) 混凝土的分类

混凝土品种众多,主要有以下几种分类方法:

① 按胶凝材料,可分为无机胶凝材料混凝土(如水泥混凝土、石膏混凝土、硅酸盐混凝土、水玻璃混凝土等)和有机胶凝材料混凝土(如沥青混凝土、聚合物混凝土、树脂混凝土等)。

② 按表观密度,可分为重混凝土(表观密度>2 800 kg/m³)、普通混凝土(表观密度为 2 000 kg/m³~2 800 kg/m³,一般在 2 400 kg/m³ 左右)和轻混凝土(表观密度<2 000 kg/m³)。

③ Prestressing screw-thread bar. Deformed bar used in PC are also named as refined rolled deformed bar. According to yield strength, prestressing screw-thread bars can be divided into four levels: PSB785, PSB835, PSB930 and PSB1080. The straight bars have many advantages including easy connecting anchorage, reliable tensioned anchorage and strong adhesion. It is mainly used in the manufacture of high-strength, large span RC products, such as nuclear power plants, hydroelectric plants, bridges, tunnels, high-speed railway and other key projects.

2.5 Concrete and its components

2.5.1 Concrete

Concrete is a kind of artificial stone block made with cementitious materials, aggregate, water with certain percentage by vibrating and curing, in which some admixtures are usually added. Concrete is one of the widely-used building materials.

(1) Classification of concrete

Concrete can be classified as follows:

① Divided by cementitious materials, there are inorganic cementitious materials concrete(such as cement concrete, gypsum concrete, silicate concrete, water glass concrete, etc.) and organic cementitious materials (such as asphalt concrete, polymer concrete, resin concrete, etc.) .

② Divided by apparent density, there are heavy concrete (apparent density > 2 800 kg/m³), ordinary concrete (apparent density is between 2 000 kg/m³ and 2 800 kg/m³, generally is about 2 400 kg/m³) and lightweight concrete(apparent density < 2 000 kg/m³).

③ 按使用功能,可分为结构混凝土、保温混凝土、装饰混凝土、防水混凝土、耐火混凝土、水工混凝土、海工混凝土、道路混凝土、防辐射混凝土等。

④ 按生产和施工工艺,可分为离心混凝土、真空混凝土、灌浆混凝土、喷射混凝土、碾压混凝土、挤压混凝土、泵送混凝土等。

⑤ 按掺和料种类,可分为粉煤灰混凝土、硅灰混凝土、矿渣混凝土和纤维混凝土等。

⑥ 按混凝土抗压强度等级,可分为低强度混凝土(抗压强度 f_{cu}<30 MPa)、中强度混凝土(f_{cu} 为 30~60 MPa)、高强度混凝土($f_{cu} \geqslant$60 MPa)、超高强混凝土($f_{cu} \geqslant$100 MPa)。

此外,随着混凝土的发展和工程的需要,还出现了膨胀混凝土,加气混凝土等各种特殊功能的混凝土。商品混凝土以及新的施工工艺给混凝土施工带来方便。

（2）水泥混凝土

水泥混凝土是指以水泥为胶凝材料,以砂、石为骨料,以水为稀释剂,并掺入适量的外加剂和掺和料拌制成的混凝土,也称普通混凝土。沙子和石子在混凝土中起骨架作用,故称为骨料(或称集料),沙子称为细骨料,石子(碎石或卵石)称为粗骨料;水泥和水形成水泥浆,包裹在砂粒表面并填充砂粒间的空隙而形成水泥砂浆,水泥砂浆又包裹在石子表面并填充石子间的空隙而形成混凝土,其结构如图 2.21 所示。

③ Divided by its functions, there are structural concrete, insulation concrete, decorative concrete, waterproof concrete, fire-resistant concrete, hydraulic concrete, marine concrete, road concrete and radiation concrete.

④ Divided by production and construction process, there are centrifugal concrete, vacuum concrete, grout concrete, roller compacted concrete, extruded concrete and pumping concrete, etc.

⑤ Divided by the type of admixtures, there are fly ash concrete, silica fume concrete, slag concrete and fiber concrete.

⑥ Divided by compressive strength of concrete, there are low strength concrete (compressive strength f_{cu} < 30 MPa), medium strength concrete (f_{cu} is from 30 to 60 MPa), high-strength concrete ($f_{cu} \geqslant$60 MPa), ultra high strength concrete ($f_{cu} \geqslant$ 100 MPa).

In addition, with the development of concrete and engineering needs, some special concretes, such as expansion concrete and aerated concrete are produced. Commercial concrete and new construction techniques make concrete construction more convenient.

（2）Cement concrete

Cement concrete, also known as ordinary concrete, is made of cement, sand, stone aggregate, water, and usually incorporated with the right amount of admixtures and additives to concrete mixture. Sand and gravel form the concrete skeleton, and they are called aggregate. Sand is known as fine aggregate, and gravel (crushed stone or gravel) is known as coarse aggregate. Cement and water form water slurry, parcel the sand surface and fill the voids among the sand grains to form cement mortar. Cement mortar wraps the surface of stones and fills the voids among the stones and then concrete is formed(structure is shown in Fig.2.21).

适量地掺入外加剂(如减水剂、引气剂、缓凝剂、早强剂等)和掺和料(如粉煤灰、硅灰、矿渣等)是为了改善混凝土的某些性能以及降低成本。

砂浆与水泥混凝土的区别在于砂浆不含粗骨料,因此可以认为砂浆是混凝土的一种特例,也可称为细骨料混凝土。

2.5.2 钢筋混凝土

不配筋的混凝土称为素混凝土,其主要缺陷是抗拉强度很低(只有一般抗压强度的1/20~1/10),也就是说混凝土受拉、受弯时易产生裂缝,并发生脆性破坏。为了克服混凝土抗拉强度低的弱点,充分利用其较高的抗压强度,一般在受拉一侧加设抗拉强度很高的(受力)钢筋,即形成钢筋混凝土 (reinforced concrete,RC)。图2.22为某简支梁破坏示意图。

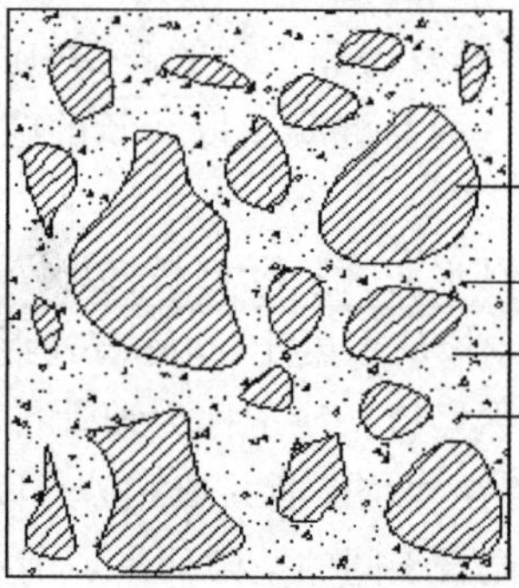

1. 石子 gravel

2. 砂子 sand

3. 水泥浆 cement paste

4. 气孔 pore

图 2.21 混凝土的结构

Fig. 2.21 The structure of concrete

In order to improve the performance of concrete and reduce costs, the right amount of additives (such as water reducer, air-entraining agent, retarder, and early strength agent) and admixtures(such as fly ash, silica fume, and slag) are added.

The difference between cement mortar and concrete is that no coarse aggregate is added in mortar. Thus cement mortar can be considered as a special case of concrete, which is called fine aggregate concrete.

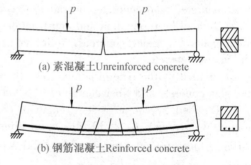

(a) 素混凝土 Unreinforced conerete

(b) 钢筋混凝土 Reinforced concrete

图 2.22 简支梁受力破坏示意图

Fig. 2.22 Damage of simply supported beam

2.5.2 Reinforced concrete

The main flaw of unreinforced concrete (referred to plain concrete) is its low tensile strength (only 1/20~1/10 of compressive strength). That is to say, when the concrete is in tension or bending, cracks are easy to occur, which results in brittle failure. In order to overcome the shortcoming of low tensile strength and make full use of its high compressive strength, the steels with high tensile strength are set in the tensile zone of concrete, that is the reinforced concrete (RC). Fig.2.22 is a schematic diagram of damage for simply supported beam.

在混凝土中合理地配置钢筋,可以充分发挥混凝土抗压强度高和钢筋抗拉强度高的特点,使两者共同承受荷载并满足工程结构的需要。如对混凝土梁(受弯构件)来说,除了在受拉一侧配置纵向受力钢筋外,一般还要加设箍筋及弯起钢筋,以防止它沿斜裂缝发生破坏。同时,在梁的上部另加直径较小的钢筋作为架立钢筋,它与受力钢筋、箍筋和弯起钢筋一起扎结成钢筋架,如图 2.23 所示。目前,钢筋混凝土是使用最多的一种结构材料。

2.5.3 预应力混凝土

钢筋混凝土虽然可以充分发挥混凝土抗压强度高和钢筋抗拉强度高的特性,但其在使用阶段往往是带裂缝工作的,这对某些结构如储液池等来说是不容许的。为了控制混凝土构件受荷后的应力状态,在构件受荷之前(制作阶段),人为给受拉区混凝土施加预压应力,使其减小或抵消荷载(使用阶段)引起的拉应力,将构件受到的拉应力控制在较小范围内,甚至处于受压状态,即可控制构件在使用阶段不产生裂缝,这样的混凝土称为预应力混凝土(prestressed concrete,PC)。

按照施加预应力的方法(施工工艺),预应力混凝土可分为先张法预应力混凝土(简称"先张法")和后张法预应力混凝土(简称"后张法")两大类。先张法是先将预应力筋张拉到设计控制应力,用夹具将其临时固定在台座或钢模上,进行绑扎钢筋,支设模板,然后浇筑混凝土;待混凝土达到规定的强度后,切断预应力筋,借助于它

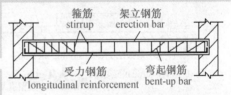

图 2.23　混凝土梁内钢筋
Fig. 2.23　Steel bars in concrete beam

Reasonable configuration of reinforcement in concrete can make use of the high compressive strength of concrete and high tensile strength of reinforcement to bear loads together and meet the needs of the engineering structure. As for concrete beam (flexural member), in addition to the longitudinal reinforcement set in tensile zone, the stirrup and bent-up bar will also be configured to prevent the diagonal crack damage in beam. At the same time, some small diameter steel bars are set in the upper portion of the beam as erection bars. All the steels in concrete beam are shown in Fig. 2.23. At present, RC is one of the most frequently used structural materials.

2.5.3　Prestressed concrete

RC can give full play to the compressive strength of concrete and high tensile strength of steel. However, cracks often occur in the service phase, which is not allowed for some structures such as the liquid storage tank. In order to control the stress state in concrete component, the compressive stress is applied to the concrete in tensile zone before loads are added, which can reduce or offset the tensile stress in concrete in the stage of service. Furthermore, the concrete in tensile zone may be in compressive state, which can control the appearance of cracks in the service stage. Such concrete is called prestressed concrete (PC).

According to the method of applying prestress, PC can be divided into two categories: the pre-tensioned PC (pre-tensioning method for short) and post-tensioned PC (post-tensioning method for short). The former can be expressed as follows:

们之间的黏结力,在预应力筋弹性回缩时,使混凝土获得预压应力,其工序如图 2.24 a 所示。后张法是先浇筑混凝土构件,并在预应力筋的位置预留出相应孔道,待混凝土强度达到设计规定的数值后(一般不低于混凝土设计强度标准值的 75%),穿入预应力筋进行张拉,并利用锚具把预应力筋锚固,最后进行孔道灌浆,使砼产生预压应力,其工序如图 2.24 b 所示。

2.6 木 材

木材是人类使用最早的工程材料之一。我国使用木材的历史不仅悠久,而且在技术上还有独到之处,如保存至今已达千年之久的山西佛光寺正殿、山西应县木塔等都集中反映了我国古代土木工程中应用木材的水平。

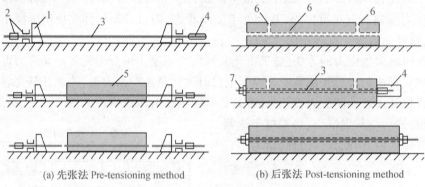

(a) 先张法 Pre-tensioning method　　(b) 后张法 Post-tensioning method

1—台座 stretching bed；2—夹具 grip；3—预应力筋 prestressing tendon；4—千斤顶 Jack；
5—构件 element；6—预留孔道 prepared hole；7—锚具 anchorage

图 2.24　先张法和后张法施工工艺
Fig. 2.24　Construction processes of pre-tensioning method and post-tensioning method

first, draw the prestressing tendons to the designed control stress and fix them in the stretching bed temporarily; second, assemble reinforcements and place concrete; finally, cut off the prestressing tendons when the strength of concrete reaches to design strength, and the compressive stress can be obtained in concrete through the adhesive force between concrete and tendon. All processes are shown in Fig.2.24 a. The construction processes of the latter can be expressed as follows: first, prepare concrete element with some prepared hole where the prestressing tendons are placed; second, draw the prestressing tendons after the concrete strength reaches to the design value (usually not less than 75% of the standard concrete design strength value) and fix the two ends with anchorage; at last, grout the hole with cement mortar. All processes are shown in Fig.2.24 b.

2.6 Timber

Timber is one of the earliest engineering materials used by human beings. In China, the use of timber not only has a long history, but also has unique technologies. For example, the main hall of Shanxi Foguang Temple and Shanxi Yingxian Wood Tower, which have existed for more than one thousand years, reflecting the application level of timber in ancient CE.

2.6.1　木材的分类

木材是由树木加工而成的,树木的种类很多,一般按树种分为针叶树和阔叶树两大类。针叶树树叶细长呈针状,树干直而高,易得大材,纹理平顺,材质均匀,木质较软而易于加工,故又称软木材。建筑上针叶树多用于承重结构构件和门窗、地面材及装饰材,常用树种有松树、杉树、柏树等。阔叶树树叶宽大呈片状,多为落叶数,树干通直部分较短,材质较硬,较难加工,故又称硬木材。建筑上阔叶树常用于制作尺寸较小的构件,常用树种有榆树、水曲柳、桦树等。

2.6.2　木材的主要性质

木材的构造决定着木材的性能,其宏观构造如图 2.25 所示。木材的性质包括物理性质和力学性质。物理性质主要有密度、含水率、热胀干缩等,力学性质主要有抗拉、抗压、抗弯和抗剪 4 种强度。

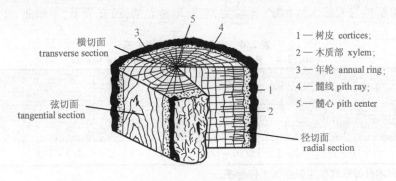

1 — 树皮 cortices;
2 — 木质部 xylem;
3 — 年轮 annual ring;
4 — 髓线 pith ray;
5 — 髓心 pith center

横切面 transverse section
弦切面 tangential section
径切面 radial section

图 2.25　木材的 3 个切面
Fig. 2.25　Three sections of timber

2.6.1　Classification of timber

Timber is made of wood. There are many kinds of trees, which can be generally divided into two major categories: conifer and broad-leaved tree. Conifer has long leaf, straight and tall trunk. With the features of smooth texture and uniform material quality, the wood of conifer is soft and easy to process, so it is also called soft wood. It is usually used for load-bearing member, door and window, the ground material and decorative material. Common tree species of conifer are pine, cedar, cypress and so on. Broad-leaved trees have large and flaky leaves, and their trunk is straight and short. As the material is hard and difficult to process, it is also known as hard wood. It is usually used as small size component. Common species of broad-leaved trees are elm, ashtree, birch and so on.

2.6.2　Main properties of timber

The structure of timber determines the performance of timber, and its macro-structure is shown in Fig.2.25. The properties of timber include the physical and mechanical properties. Its physical natures include density, moisture content, thermal expansion and shrinkage. It's mechanical properties include tensile strength, compressive strength, flexural strength, and shear strength.

木材有很好的力学性质,但木材是有机各向异性材料,顺纹方向与横纹方向的力学性质有很大差别,见表2.4。木材的顺纹抗拉和抗压强度均较高,但横纹抗拉和抗压强度较低。木材强度还因树种而异,并受木材缺陷、荷载作用时间、含水率及温度等因素的影响,其中以木材缺陷及荷载作用时间两者的影响最大。

2.6.3 木材的应用

在工程中,除直接使用原木外,木材一般都加工成锯材(板材、方材等)或各种人造板材使用。原木可直接用作屋架、檩、椽、木桩等。

为减小使用中发生的变形和开裂,锯材须经干燥处理。干燥能减轻自重,防止腐朽、开裂及弯曲,从而提高木材的强度和耐久性。锯材的干燥方法可分为自然干燥和人工干燥两种。

木材经加工成型和制作构件时,会留下大量的碎块废屑,以这些废料或含有一定纤维量的其他作物做原料,采用一般物理和化学方法加工而成的产品即为人造板材。这类板材与天然木材相比,板面宽,表面平整光洁,没有节子,不翘曲、开裂,经加

表2.4 木材各强度之间关系
Tab. 2.4 Relationship among strength of woods

抗压强度 Compression strength		抗拉强度 Tensile strength		抗弯强度 Flexural strength	抗剪强度 Shear strength	
顺纹 along grain	横纹 cross grain	顺纹 along grain	横纹 cross grain		顺纹 along grain	横纹 cross grain
1	1/10～1/3	2～3	1/20～1/3	3/2～2	1/7～1/3	1/2～1

注:以木材的顺纹抗压强度为1作标准。
Note:The standard compression strength along grain is 1.

Timber has excellent mechanical properties. However, timber is an organic anisotropic material, and there are great differences in mechanical properties between the longitudinal direction and transverse direction, as shown in Tab.2.4. The tensile and compressive strength along grain are higher. The strength of wood also varies among species, and it will be affected by timber defects, loading time, moisture content and temperature, among which the timber defects and loading time are the greatest ones.

2.6.3 Application of timber

Besides the direct usage, timber is generally processed into lumber (sheet,etc.) or artificial plate in engineering. Logs can be directly used as roof frame, purlin, rafter, wood, etc.

In order to reduce the deformation and cracking in usage, timber drying process must be approved. Drying process can reduce the weight, prevent decaying, cracking and bending, therefore the strength and durability of timber will be improved. Timber drying methods can be divided into natural and artificial drying methods.

A large number of pieces of scrap will be left during the process of making models and component. The waste or other crops that contain a certain amount of fiber can be taken as raw materials. Artifical plate can be made from these raw materials through general physical methods and chemical methods. Compared with natural timber, artifical plate has the characteristics of wide and smooth surface, without knot, warping and cracking. Besides, artifical plate has the feature of waterproof, fireproof, anticorrosive,

工处理后还具有防水、防火、防腐、防酸等性能。常用人造板材有胶合板、纤维板、刨花板、木屑板等,如图 2.26 所示。

2.7 土木工程材料的发展前景

土木工程材料是土木工程的重要组成部分,它和工程设计、工程施工及工程经济之间有着密切的关系。自古以来,工程材料和工程建(构)筑物之间就存在着相互依赖、相互制约和相互推动的关系。一种新材料的出现必将推动建筑设计方法、施工程序或形式的变化,而新的结构设计和施工方法必然要求提供新的、更优良的材料。

近几十年来,随着科学技术的进步和土木工程发展的需要,一大批新型土木工程材料应运而生,出现了仿生智能混凝土(自感知混凝土、自愈合混凝土、透光混凝土等)、高强钢材、新型建筑陶瓷和玻璃、纳米技术材料、新型复合材料(纤维增强材料、夹层材料)等。随着社会的进步,环境保护和节能减排的新形势,对土木工程材料提出了更多、更高的要求。因而,今后一段时间内,土木工程材料将向以下几个方向发展。

  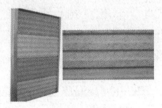

(a) 胶合板　　　　　　　　(b) 刨花板　　　　　　　　(c) 木屑板
　　Plywood　　　　　　　　　Chipboard　　　　　　　　Sawdust board

图 2.26　人造板材
Fig. 2.26　Artificial plate

and anti-acid performance after a series of processing. Common artifical plates are plywood, fiberboard, chipboard, sawdustboard, etc., as shown in Fig.2.26.

2.7 Developing prospects for CEM

The CEMs are the important ingredients of civil engineering. There are close relationships among CEM, the construction and costs of buildings. Since ancient times, the engineering materials and buildings have inseparably interconnected and developed together. The appearance of a new material will improve the way of structural design and construction. Conversely, the advanced structural type and construction method will inevitably induce the innovation of excellent materials.

In recent decades, with the development of science and technology in CE, a great deal of new materials has been invented, such as bionic intelligent concrete (including self-sensing concrete, self-healing concrete and light transmission concrete etc.), high strength steel, new-style building ceramics and glass, nanotechnology material, new-style composite material (including fiber reinforced material, sandwich material). It is predictable that more and more performance-advanced materials are required to meet the needs of environmental protection, energy-saving and emission-reduction of our society. Therefore, the developing prospect for CEMs will be listed as follows.

2.7.1 高强高性能材料

（1）高性能混凝土

在 20 世纪，混凝土的强度得到了较大幅度的提高，但高强度混凝土的延性、抗火性能均较差，严重影响了混凝土结构的抗灾性能。近十年来的研究表明，在混凝土的组分中引入纳米材料、短切复合材料或有机聚合物，可以更好地改善混凝土，并不断研究开发出纳米混凝土、高延性纤维混凝土、高耐久性混凝土、良好抗疲劳和耐磨混凝土材料。此外，将混凝土材料与其他聚合物复合，可以增加混凝土材料的阻尼特性，发展高阻尼混凝土材料，提高结构的抗震性能。

（2）高强钢材

采用高强钢材可显著减小钢结构构件的尺寸和结构重量，相应的减少焊接工作量和焊接材料用量，减少各种涂料(防锈、防火等)的用量及其施工工作量，所取得的经济效益可使整个工程总造价降低。同时，在建筑物使用方面，减小构件尺寸能够带来更大的使用空间。目前工程应用的钢材强度已经达到 460 MPa 以上，甚至开始推荐使用屈服强度为 500,590,620,690 MPa 的更高强度的结构钢。我国国家体育场"鸟巢"采用了 700 多吨板厚达到 110 mm 的 Q460E/Z35 高强度高性能钢材(见图2.27)。

图 2.27　中国国家体育场"鸟巢"实景图
Fig. 2.27　National Stadium "Bird's Nest"

2.7.1 High strength and performance materials

（1）High performance concrete（HPC）

In the 20th century, the strength of concrete was greatly improved. However, the ductility and fire resis-tance of high strength concrete were still bad, which greatly affected the anti-disaster ability of concrete structure. It can be found in recent studies that the concrete performance can be greatly improved when some additives, such as nanophase material, short composite material, and organic polymer are added into concrete mixture. Besides, many high performance concretes including nanometer concrete, high ductility fiber concrete, good durability concrete, anti-fatigue and anti-abrasion concrete are developed and used. The combination of concrete materials with organic polymer can effectively improve the seismic performance of concrete structures.

（2）High strength steel

High strength steels can evidently diminish the sizes and weights of steel structural elements. Accordingly, it can reduce the welding works, the amounts of welding materials and coating materials. All related construction work will be decreased obviously. As a result, the total cost of building will be reduced effectively. Besides, from the viewpoint of building's serviceability, it is advantageous to use smaller elements to get larger space. Nowadays, the strength of steel used in existed buildings has reached over 460 MPa, which is the yield strength of steel. Furthermore, some higher strength structural steels with yield strengths of 500, 590, 620 and 690 MPa begin to appear in practical engineering. For example, the National Stadium "Bird's Nest" (Fig.2.27) in China consumed more than 700 t high strength steels of the type of Q460E/Z35 with the thickness of 110 mm.

中央电视台新址采用了 2 300 多吨 Q460E/Z35 高强度高性能钢材。Q460 高强度型钢已经应用于输电塔架。

欧洲已将 S460~S690 级高强度结构钢材列入规范；美国已在桥梁建设中应用屈服强度 485 MPa 级和 690 MPa 级的高性能钢材；澳大利亚在高层和大跨度建筑中成功应用了屈服强度 690 MPa 级钢材。可见，发展高强度和高性能钢材符合世界各国的技术发展规划，以及可持续发展和环境保护的基本理念，是土木工程材料的重要发展方向之一。

(3) 纤维增强复合材料

在本章先导案例中，我们曾介绍了用碳纤维 CFRP 索替代钢索建造斜拉桥和悬吊式建筑的两个工程实例，分别见图 2.1 和图 2.2。

纤维增强复合材料(FRP)具有轻质、高强、耐久、高阻尼等特性，已成为一种重要的土木工程结构材料。随着研究的深入，利用 FRP 复合材料替代传统的土木工程材料受到了愈来愈多的关注，如 FRP 套管替代钢管约束混凝土结构、FRP 筋代替传统的钢筋以及 FRP 索代替钢索等。实践表明，纤维增强复合材料 FRP 因其良好的力学性能和耐久性，将成为继钢材和混凝土材料之后的第三类结构材料。

Another unique new building is China Central Television (CCTV), which adopted the same high strength steels with the amount of 2 300 t. Besides, the Q460 steels have also been used in transmission towers.

The high level structural steels with the yield strengths of S460 to S690 have been listed in European codes. Some high performance steels with yield strengths of 485 MPa to 690 MPa have been adopted in bridge structures in America. In Australia, the steel with yield strength of 690 MPa has been successfully used in high-rise and large-span buildings. It can be clearly seen that the action to develop high-strength or high-performance steels is in accordance with sustainable development and environmental protection all over the world. High strength steel has become one of the important developing fields in CEM.

(3) Fiber reinforced composite materials

In the section of introductory case, we have introduced two typical engineering cases about the use of CFRP cable instead of steel cable in cable-stayed bridge and suspension type building, as shown in Fig. 2.1 and Fig. 2.2 respectively.

Fiber Reinforced Polymer/Plastics(FRP) has the characteristics of low-weight, high-strength, durability and high-damping, and it has become an important structural material in civil engineering. With the deep research on FRP, it has become a hot issue on how to adopt FRP materials to replace the traditional CEMs all over the world. For example, how to use FRP tube to replace the steel tube in steel-concrete composite structures, as well as how to use FRP bars to replace steel bars in concrete structures and to replace prestressing cables in prestressed structures. It can be concluded from engineering practice that with the excellent mechanical properties and durability, FRP material will become the third structural material following concrete and steel.

2.7.2　绿色节能材料

（1）节能减排材料

土木工程材料的生产能耗和建筑物使用能耗，一般占国家总能耗的20%~35%左右，因此研制和生产低能耗的新型节能土木工程材料是构建节约型社会的需要。另外,充分利用工业废渣(如粉煤灰、矿渣等)、生活废渣以及建筑垃圾等生产土木工程材料,将各种废渣尽可能地资源化,以保护环境、节约自然资源,是人类社会实现可持续发展的需要。生态混凝土是近几年研究开发出来的一种有利于改善生态环境和自然景观(植生绿化功能)的新型混凝土,在我国已有一定的应用(见图2.28),这是一种很有发展潜力的混凝土材料。

（2）绿色建材

绿色建材就是指采用清洁的生产技术,少用天然资源、大量使用工业或城市固体废弃物和农植物秸秆生产的无毒、无污染、无放射性,有利于环保与人体健康的建材。发展绿色建材,改变长期以来存在的粗放型生产方式,选择资源节约型、污染最低型、质量效益型、科技先导型的生产方式是21世纪建材工业的必然出路。

图2.28　国内某生态混凝土江堤工程

Fig. 2.28　One Eco-concrete river embamkment engineering in China

2.7.2　Green and energy-saving material

（1）Energy-saving and emission reduction materials

Normally, the total energy consumption during the processes of producing materials and building using accounts for nearly 20%~35% of the national energy consumption. Therefore, it is a strong requirement for conservation-oriented society to develop and produce the energy-saving materials in CE. Besides, the emission reduction of buildings is another hot issue nowadays. In order to protect environment and save natural resources, it is a wise way to use industrial and construction waste to reproduce CEMs. It is also an inevitable demand to achieve sustainable development of society. Currently, the eco-concrete (or environmentally friendly concrete) is one kind of the new-style concretes, which can bring benefits to ecological system and improve environment with plant growth. It has already been used in China (Fig.2.28) and will have a good future.

（2）Green building materials

Green building materials are favorable to environmental protection and human's health. With clean technologies, the green building materials are usually produced by industrial and construction waste or plant straw instead of natural resources. In order to change the current extensive mode of production and create a new one with resource-saving, low pollution and leading technology, it is an inevitable way for building material industry to develop green building materials.

当前,我国墙体材料的"绿色化"进程已取得了一定的成果。

2.7.3　纳米智能材料

（1）智能混凝土

通过掺入功能性材料，使传统材料在保持原有基本力学性能不变的情况下,获得一些特殊功能,是材料科学发展的一个主要趋势。20 世纪 90 年代,这一概念也得到了混凝土研究学者的认同,并提出了"智能混凝土"的概念。目前,混凝土的智能化主要通过以下 3 个途径来实现:①在混凝土内复合某些导电或半导体纳米材料,使混凝土具备自感知的功能,制备出自感知混凝土;②将混凝土材料与压电材料、磁致伸缩材料或形状记忆材料等"智能材料"相复合,制备出自集能混凝土制品;③在混凝土内埋设一些传感器或感知骨料,使混凝土具有相应的传感功能。

（2）智能工程材料

对于智能材料的研究始于 20 世纪 80 年代的航空航天领域,目前已经在包括土木工程、机械工程、生物医学工程等各个领域中得到了广泛的研究和应用。过去的十多年里,以压电陶瓷、电/磁致伸缩材料、电磁流变液以及形状记忆材料等为代表的智能材料在土木工程领域得到了长足的发展:足尺的磁流变阻尼器已经应用于桥梁、海洋平台以及多层建筑的振动控制;形状记忆合金在古建筑的加固以及隔震座限位器等方面得到了应用;智能型压电摩擦阻尼器和磁致阻尼器也逐步走向了示范工程。进入 21 世纪,智能材料与智能土木工程结构是土木工程领域最具创新、最有活力的研究方向之一,也是发展高性能土木工程结构和可持续土木工程结构的重要途径。

Currently, some achievements have been obtained in green wall material's reformation in China.

2.7.3　Smart materials and nanophase materials

（1）Smart concrete

By adding some functional materials, the traditional materials will hold some special functions without changing their basic mechanical properties, which is the main trend in the development of material science. From the 1990s, this idea was accepted in concrete research, and the conception of smart concrete was put forward. At present, the characteristics of smart concrete have usually been achieved through the following ways: ① self-sensing concrete, in which some electric and semiconductor nanophase materials are added; ② self-energy collection concrete, which is a composite material by compounding concrete and "smart material", such as piezoelectric material, magnetostrictive material, shape memory material; ③ sensing concrete, which has certain sensing functions by setting in some sensors or sensing aggregates in concrete.

（2）Intelligent engineering materials

Intelligent materials were firstly researched in aerospace field in the 1980s. Today, it has been widely studied and applied in many fields including CE, mechanical engineering, biomedical engineering and so on. In the last decade, many representative smart materials, such as piezoelectric ceramics, electrostrictive or magnetostrictive material, magnetorheological fluid and shape memory materials, have been greatly developed and used in CE field. The intelligent materials and their related structures have become one of the most creative fields in CE field since the 21st century.

（3）纳米技术材料

纳米为细微的长度单位,等于十亿分之一米,记作 nm。今后建材的主导方向是绿色、环保及高性能,而这些建材的制备主要依靠纳米技术来实现。纳米材料对颜料、陶瓷、水泥等制品的改性有很大贡献,如把纳米氧化铝加入陶瓷中,则陶瓷强度、韧性的增加将非常显著。纳米无机涂料,可以解决混凝土表面腐蚀、老化及渗水等问题。这种涂料在混凝土或水泥浆表面形成玻璃态或离子化胶态,注入微裂或孔隙中与水泥反应形成新的硅酸盐复合体,不仅可以使弯曲强度提高 2~3 倍,还可起到防水作用。纳米建材现在还仅仅处于起步阶段,其进一步的发展应用还有很长的路要走。

总而言之,随着研究工作的深入和新材料的不断涌现,以及与材料有关的基础学科的日益发展,人类对材料内在规律有了进一步的了解,对各种材料的共性知识有了初步科学的抽象认识,从而形成了"材料科学"这一新的学科领域。材料科学(更准确地说应该是材料科学与工程)是介于基础科学与应用科学之间的一门应用基础科学,其主要任务在于研究材料的组分、结构、界面与性能之间的关系及其变化规律,从而达到按使用要求设计材料、研制材料及预测使用寿命的目的。土木工程材料也属于材料科学的研究对象,随着人们逐渐将土木工程材料的研究纳入材料科学的轨道,在不久的将来土木工程材料的发展必将有重大突破,土木工程也将出现翻天覆地的变化。

（3）Nanotechnology materials

Nanometre(nm) is a kind of length units, which equals to 10^{-10} m. Many high performance materials with the direction of green, environmental protection, and high efficiency, can be invented by nanotechnology. Nanotechnology can bring great contributions to improving material performance, such as paint, ceramics and cement-based products. For example, the ceramic strength and toughness can be remarkably improved by adding nanophase alumina into ceramics. Nanophase inorganic coating can solve some problems to concrete with surface erosion, aging and water percolation. However, nanotechnology materials used in CE field are still in the initial stage, and further efforts are needed for their development.

To sum up, with the fast development of material-based scientific research and the appearance of new materials, people can get more information about materials inherent properties and make some conclusions about common knowledge of all kinds of materials. Gradually, a new discipline, Material Science, comes into being. Material Science, which can be exactly described as Material Science and Engineering, is a basic application science between basic science and application science. Its main tasks are as follows: ① study the relationships between material's constituents, structures and interfaces with its performance; ② study their change rules to design materials, invent new materials and predict their service lives. The materials used in CE are also the research objects in Material Science. In the future, the great development will be obtained in CEM, and the enormous changes will happen in CE.

 **知识拓展**
Learning More

相关链接　Related Links

(1) 建设材料设备网

(2) 水泥网

(3) 中国工程机械工业协会钢筋及预应力机械分会

小贴士　Tips

(1) 水泥的分类、组分与材料可参见现行标准《通用硅酸盐水泥》(GB 175—2007)，主要有硅酸盐水泥、普通硅酸盐水泥、矿渣硅酸盐水泥、火山灰硅酸盐水泥、粉煤灰硅酸盐水泥和复合硅酸盐水泥。

More information about the classification, ingredient and stuff of cement can be obtained from the code *Common Portland Cement* (GB 175—2007). There are Portland cement, ordinary Portland cement, slag portland cement, portland pozzolan cement, fly-ash portland cement and composite portland cement.

(2) 钢筋混凝土结构用的热轧带肋钢筋可参见线性标准《钢筋混凝土用钢——第 2 部分：热轧带肋钢筋》(GB 1499.2—2007)，主要包括普通热轧钢筋(牌号：HRB335、HRB400 和 HRB500)和细晶粒热轧钢筋(牌号：HRBF335、HRBF400 和 HRBF500)。其中，HRB 是 Hot Rolled Ribbed Bars 的缩写，后面的数字是屈服强度；HRBF 是在 HRB 后面加"细"的英文(Fine)首位字母。

The information about hot rolled ribbed bars used in RC structures can be obtained from the code *Steel for the Reinforcement of Concrete—Part 2: Hot Rolled Ribbed Bars* (GB 1499.2—2007). There are two kinds of bars, ordinary hot rolled ribbed bars including HRB335, HRB400 and HRB 500, where the number is the yield strength of steel bar, and fine hot rolled ribbed bars including HRBF335, HRBF400 and HRBF 500.

思考题　Review Questions

(1) 土木工程中常用的天然石料有哪些？它们各有何特点？

What kinds of natural stone are used in CE? What characteristics do they have?

(2) 气硬性胶凝材料与水硬性胶凝材料有何区别？请举例说明。

What are the differences between air hardening cementitious materials and hydraulic cementitious materials? Please show some examples.

(3) 钢材的分类方法有哪些？土木工程中常用什么钢材？

How many classification methods for steel? Which steels are commonly used in CE?

(4) 水泥混凝土的组成材料有哪些？与砂浆有什么不同？

What are the components of cement concrete? What are the differences between mortar and concrete?

参考文献
References

［1］郑晓燕,胡白香.新编土木工程概论[M].2版.北京:中国建材工业出版社,2012.

［2］柯国军.土木工程材料[M].2版.北京:北京大学出版社,2012.

［3］刘正武.土木工程材料[M].上海:同济大学出版社,2005.

［4］柳俊哲.土木工程材料[M].3版.北京:科学出版社,2014.

［5］江见鲸,叶志明.土木工程概论[M].北京:高等教育出版社,2001.

［6］刘瑛.土木工程概论[M].北京:化学工业出版社,2005.

［7］丁大钧,蒋永生.土木工程概论[M].2版.中国建筑工业出版社,2010.

［8］Palanicharmy M S. Basic Civil Engineering[M].北京:机械工业出版社,2005.

［9］Shen Zuyan. Introduction of Civil Engineering[M].北京:中国建筑工业出版社,2005.

［10］Shan Somayaji,Yan Peiyu. Civil Engineering Materials [M].2nd ed. Englewood: Prentice Hall,2001.

［11］茹继平,刘加平,曲久辉,等.建筑、环境与土木工程[M].北京:中国建筑工业出版社,2011.

第3章　结构工程

Chapter 3　Structural Engineering

先导案例
Guide Case

　　各类建筑物应以结构工程可以实现为前提条件,建筑结构是形成建筑空间的骨架,属于土木工程的重要组成部分。随着人类文明及科学技术的发展,呈现建筑外形的各种建筑结构形式不断被推出,有拱结构、网壳结构、连续梁结构、框架结构、框架剪力墙结构、筒体结构等,它们分别由钢材、混凝土等制作而成。

　　2008年北京奥运会主场馆——国家体育馆(鸟巢,见图3.1)即为大跨度空间钢结构,总跨度294 m,结构形式为巨型门式钢架,钢材强度等级为Q460,用钢量达4.2万吨,可容纳10万观众。

　　位于阿联酋的哈利法塔(见图3.2)是当今世界最高的建筑(Burj Khalifa Tower),原名迪

图 3.1　中国国家体育场
Fig. 3.1　National Stadium of China

国家体育场

All kinds of buildings should be based on the premise that structural engineering can be realized. Building structure is the skeleton of building space, which belongs to an important part of civil engineering. With the development of human civilization and science and technology, various forms of architectural structures have been introduced, such as arch struture, reticulated shell structure, continuous beam structure, frame structure, frame shear wall structure, cylinder structure, etc., which are made of steel, concrete and other materials.

The National Stadium (Bird's Nest, as shown in Fig. 3.1), the main stadium of the 2008 Olympic Games, is a large-span steel structure with a span of 294 meters. The structure is in the form of a giant portal frame, with a steel strength grade of Q460, 42 000 tons of steel, and the total number of spectators can reach 100 000.

Burj Khalifa (Fig.3.2) in the United Arab Emirates

图 3.2　哈利法塔
Fig. 3.2　Burj Khalifa Tower

哈利法塔

拜塔,又称迪拜大厦或比斯迪拜塔。该塔总高828.4 m,计162层,造价15亿美金,由韩国三星公司建造。哈利法塔的下部为钢筋混凝土剪力墙结构,上部为钢框架支撑结构,塔内提供居住和工作场所,可容纳1.2万人。

本章首先介绍结构设计的基本思想、基本构件及其应用,在此基础上进一步介绍根据不同结构体系建造的各种形式的建筑结构物。

3.1 荷 载

外部环境等施加于结构体系上给结构带来的影响统称为作用。直接加载在建筑结构上的作用称为直接作用,如荷载即为直接作用,包括建筑结构构件、建筑装饰等的自重等。由于温度的改变、地基的沉陷、地震等外部环境的影响导致结构产生相应变形或位移,称为间接作用。

荷载按照作用时间的长短、大小和位置随时间变化等不同情况分为永久荷载和可变荷载;荷载按作用的性质可以分为静载和动载,如吊车荷载就是动荷载。

荷载的取值是按照数理统计的方法确定的,把具有规定可靠度保证的荷载取值

(UAE) is the world's tallest building, formerly known as Burj Dubai. With a total height of 828.4 meters and 162 storeys, it was built by the South Korean company Samsung at a cost of 1.5 billion dollars. The tower has a reinforced concrete shear wall structure at the bottom and a steel frame support structure at the top, providing housing and work spaces for 12 000 people.

This chapter first introduces the basic idea of structural design, the basic components and their applications, on this basis, it further introduces the various forms of building structures constructed by different structural systems.

3.1 Loads

The influence of the external environment on the structural system to the structure is collectively referred to as the effect, and the direct effect of the direct load on the building structure is referred to as the direct effect. For example, the load is the direct effect, including the self-weight of building structural components, building decoration and so on. Due to the change of temperature, the subsidence of the foundation, earthquake and other external environment, resulting in the corresponding deformation and replacement, known as the indirect effect.

Load is divided into permanent load and variable load according to the duration of action, its size and position change with time. According to the nature of load action, it can be divided into static load and dynamic load, such as crane load is dynamic load.

The value of load is determined according to the method of mathematical statistics. The value of load with guaranteed reliability is called the standard value of load. The standard value is the basic representative value for structural design. The design value of the load is the progressive adjustment on the basis of the standard value, according to the different requirements of the structure function to take the standard value or quasi

叫作荷载的标准值,标准值是用于结构设计的基本代表值。荷载的设计值是在标准值的基础上的进一步调整,按照不同的结构功能要求取标准值或取准永久值。具体的计算方法可查阅《建筑结构荷载规范》(GB 50009—2019)及《建筑抗震设计规范》(GB 50223—2019)。

3.2　基本构件

建筑结构是建筑物的骨架,撑起建筑空间并承担各种作用,实现建筑结构功能。建筑结构的破坏甚至倒塌会导致人类生命和财产的重大损失,因此,建筑结构功能的基本要求就是确保建筑物的安全性。和人体结构一样,建筑结构也是由最基本的构件在空间上通过节点的合理连接形成的。建筑结构的基本构件主要承受拉、压、弯、剪、扭等作用,其具体形式为杆、梁、柱、索、板、墙、拱等。

3.2.1　杆

截面尺寸远小于其长度的直线形构件称为杆。杆的受力形式比较简单,通常只承受拉力和压力,也称二力杆。杆承受与长度方向一致的轴向力作用,根据作用力的方向不同,分为轴心受拉杆和轴心受压杆(见图 3.3)。杆的横截面积沿杆的长度方向

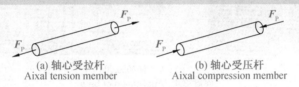

(a) 轴心受拉杆　　　　　　　　　　　　(b) 轴心受压杆
Aixal tension member　　　　　　　　Aixal compression member

图3.3　轴心受力杆件
Fig. 3.3　Axial load member

permanent value. For specific calculation methods, please refer to *Code for Load of Building Structures* (GB 50009—2019) and *Code for Seismic Design of Buildings* (GB 50223—2019).

3.2　Basic structural members

The building structure is the skeleton of the building, which supports the building space and assumes various functions to realize the structural functions of the building. The destruction or even collapse of the building structure will lead to great loss of human life and property. So, the basic requirement of structural function is to ensure the safety of the building. Like the human body structure, the architectural structure is formed by the most basic components in space through the reasonable connection of nodes. The basic components of the building structure are used to withstand the action of pulling, pressing, bending, shearing and torsion, and their specific forms are bars, columns, cables, plates, walls, arches, etc.

3.2.1　Bar

The section size is far less than the length of the linear member, known as the rod, generally speaking, the rod force form is relatively simple, usually only bears the tension and pressure, also known as the two force bar. Under the action of axial force consistent with the length direction, due to the different directions of the force, it is divided into axial tie bar and axial pressure bar (Fig. 3.3). The cross-sectional area of the rod does

不发生变化的称为等截面直杆,建筑结构中主要应用的是等截面直杆;杆的横截面积沿杆的长度方向发生变化的称为变截面直杆。等截面直杆常见于桁架、网架和支撑体系等结构中(见图 3.4)。

3.2.2 梁

在建筑结构中,汇集并承担竖向载荷的构件称为梁,又称为横力弯曲构件,即受弯构件(见图3.5)。受弯构件是建筑架构中十分重要的结构构件,有着广泛的应用。梁支撑于墙、柱上面,承受由屋面、楼面系统传下来的荷载,或直接承受荷载作用,通常水平搁置,也有斜梁,如楼梯梁(见图3.6)等。

图3.4　杆的应用
Fig. 3.4　Application of bars

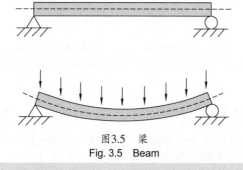

图3.5　梁
Fig. 3.5　Beam

图3.6　楼梯梁
Fig. 3.6　Stair beam

not change along the length direction of the rod, which is called the equal section straight rod. A change in the cross-sectional area along the length of the bar is called a variable cross-sectional straight bar. Straight bars of equal section are commonly found in trusses, grids and structural support systems (Fig. 3.4).

3.2.2　Beam

In the building structure, the member that gathers and bears the vertical load is beam, usually called the transverse force bending member, namely the bending member (Fig. 3.5). It is a very important structural member in the building architecture and has a wide range of applications. Beam supports above the wall, column, bear the load that passes down by the roof, floor system, or bear the load action directly, it's placed horizontally, there is inclined beam, like the beam of stair(Fig. 3.6).

　　梁有多种截面形式,常用的有矩形截面梁、T形截面梁、倒T形截面梁、花篮梁、工形截面梁(见图3.7)等,还有槽型截面梁、箱型截面梁、组合截面梁和叠合梁等(见图3.8)。

　　根据用途不同,梁有单跨(见图3.9)、多跨梁(见图3.10),以及静定梁、超静定梁之分。常用的单跨梁有简支梁、悬臂梁、外伸梁,属于静定梁,桥梁工程中有多跨静定梁。一般情况下,建筑结构大多是超静定的梁。

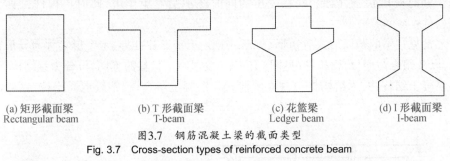

(a) 矩形截面梁　　　(b) T 形截面梁　　　(c) 花篮梁　　　(d) I 形截面梁
Rectangular beam　　　T-beam　　　Ledger beam　　　I-beam

图3.7　钢筋混凝土梁的截面类型
Fig. 3.7　Cross-section types of reinforced concrete beam

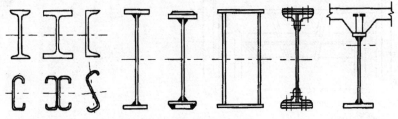

图3.8　钢梁的截面类型
Fig. 3.8　Cross-section types of steel beam

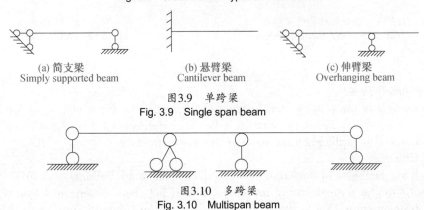

(a) 简支梁　　　　　(b) 悬臂梁　　　　　(c) 伸臂梁
Simply supported beam　　　Cantilever beam　　　Overhanging beam

图3.9　单跨梁
Fig. 3.9　Single span beam

图3.10　多跨梁
Fig. 3.10　Multispan beam

　　Beam has a variety of section forms, commonly used rectangular section beam, T section beam, inverted T section beam, flower basket beam, I section beam (Fig. 3.7), and groove section beam, box section beam, composite section beam and composite beam(Fig. 3.8).

　　According to different uses, beams have single span(Fig. 3.9), multi-span beam(Fig. 3.10), statically indeterminate beam, statically indeterminate beam. Commonly used single-span beams are simply supported beams, cantilever beams and overhanging beams, which belong to statically indeterminate beams. There are multi-span statically indeterminate beams in bridge engineering. In general, building structures are mostly statically

根据梁的用途和所在位置的不同,可把梁分成主梁、次梁,或者圈梁、过梁、联系梁等。图3.11表示框架结构的梁格布置。

3.2.3 柱

柱是建筑结构中最重要的结构构件,不仅承担其上部结构的全部竖向荷载,还承担水平荷载的作用。柱子主要承受压力,由于轴心受压的柱子很少,大部分柱子处于偏心受压状态,因此,柱在受压的同时还承受弯矩作用,此时该构件也称为压弯构件。

常见柱子的截面形式有矩形、方形、圆形、工字形柱;L形、十字形等称为异形截面柱,以及组合截面柱、格构柱(见图3.12)等。根据制作材料的不同,可分为钢结构柱、钢筋混凝土结构柱、木结构柱、石柱、砖柱、钢–混凝土组合柱、钢管混凝土柱、钢骨混凝

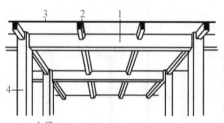

1—主梁 Primary beam;
2—次梁 Secondary beam;
3—楼板 Floor;
4—柱 Column

图3.11 某框架结构的梁格布置
Fig. 3.11 Floor framing system

图3.12 格构柱
Fig. 3.12 Lattice column

indeterminate beams.

According to the purpose and location of the beam, the beam can be divided into main beam, secondary beam; there are ring beams, lintel, contact beams and so on. Figure 3.11 shows the grillage arrangement of the frame structure.

3.2.3 Column

Column is the most important structural component in building structure, which bears not only the vertical load of its superstructure, but also the horizontal load. Pillars mainly bear pressure. Due to the axial compression of the column is rare, most columns have eccentric compression state. The column in compression at the sametime bears the bending moment, so it is also called flexural member.

The section form of the common column is rectangular, square, round, I-shaped column; L-shaped, cross-shaped column is called special-shaped section; there are combined section columns, lattice columns (Fig.3.12), and so on. According to the different materials, it can be divided into steel structure column, reinforced concrete structure column, wood structure column, stone column, brick column, steel-concrete composite column, concrete filled steel tube column, steel reinforced concrete column, etc. Walls and foundation walls are also vertical load-bearing components, which have the same

土柱等(见图3.13)。墙和基础墙等也是竖向承力构件,具有与柱子相同的力学性能。

3.2.4 索

索是一种柔性结构构件,用于悬挂荷载,而自身承受拉力。索可把竖向荷载转化为沿索方向的拉力,其垂度是按照优化结构设计确定的。索通常由高强度钢丝束制作而成,直径可达1.2~1.5 m,也有采用FRP等新材料制作的。索结构具有非线性特性,常见于悬索桥、张弦结构和薄膜结构中,可形成较大的建筑空间(见图3.14)。

石柱 Pillar

砖柱 Brick column

木柱 Wooden column

钢柱 Steel column

钢筋混凝土柱 Concrete column

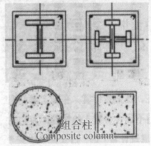
组合柱 Composite column

图 3.13 柱的种类
Fig. 3.13 Column types

镇江悬索桥

图 3.14 索的应用
Fig. 3.14 Application of cables

mechanical properties as columns. (Fig. 3.13)

3.2.4 Cable

A cable is a flexible structural member that is used to suspend loads while itself is subjected to tension. The vertical load is converted into the tension along the cable direction, and its sag is determined according to the optimal structural design. It is usually made of high strength steel wire bundles with a diameter of 1.2~1.5 m. There are also cables made of new materials such as FRP. Cable structure has nonlinear characteristics, it often can be found in suspension bridge, tension-string structure and membrane structure, which can form a large building space (Fig. 3.14).

3.2.5 板

板是指平面尺寸远大于其厚度的受弯构件,通常水平搁置,有时也斜放(如板式楼梯等)。板常见于楼板、屋面板、基础底板、桥面板结构等。

板可制作成实心板、空心板,也可制作成槽形板、T形板、密肋板等(见图3.15);按所用材料可分为木板、钢板、钢筋混凝土板等;按照工艺可分为预制板、现浇板和预应力板等。

3.2.6 墙

墙是指高和宽方向的尺寸比较大,而厚度相对较小的竖向承力构件,其力学性能与柱子相仿。

图 3.15　空心板与槽形板
Fig. 3.15　Hollow-core slab and channel slab

3.2.5　Plate

Board refers to its plane scale is far greater than the thickness of the bending components, usually horizontal use, sometimes inclined　(such as board stairs, etc.). Plates are commonly found in floor slabs, roof slabs, foundation plates, bridge panels structures and so on.

Plate can be made into solid plate, hollow plate, it can also be made into groove plate, T-shaped plate, ribbed plate, etc.(Fig. 3.15); according to the material, plate can be divided into wood plate, steel plate, reinforced concrete plate, etc.; according to the process, plate can be divided into prefabricated plate, cast-in-place plate and prestressed plate, etc.

3.2.6　Wall

A wall is a vertical load-bearing member whose size is relatively large in the direction of height and width, but whose thickness is relatively small. Its mechanical properties are similar to those of a column.

　　墙体按照使用功能不同,可分为承重墙和非承重墙。对于承重墙,主要承受竖向荷载,在多高层建筑体系中,用于承接并传递水平风荷载。剪力墙结构体系中的剪力墙,用于承担全部水平荷载和竖向荷载作用。

　　根据墙所处位置的不同,可将墙体分为横墙、纵墙、山墙、外墙和内墙等。

　　根据不同的施工工艺,墙可以分为预制墙、现浇墙和砌筑墙。

3.2.7　拱

　　拱是古老的结构形式,属于曲线型结构构件,构件的截面以承受轴向压力作用为主。拱结构可广泛应用于桥梁结构等构筑物中,以古罗马和哥特式的结构为典型代表(见图3.16)。

赵州桥

赵州桥 Zhaozhou Bridge

凯旋门 The arc de triomphe

圣路易斯拱门 Gateway

南京奥林匹克体育场
Nanjing Olympic Stadium

图 3.16　拱的应用
Fig. 3.16　Application of arch

　　According to the use function of wall, it can be divided into load-bearing walls and non load-bearing walls. For a load-bearing wall, it mainly bears vertical load. In multi-high-rise building system, it is used to undertake and transfer horizontal wind load. Shear walls belong to load-bearing walls, it is used to bear all horizontal load and vertical load.

　　According to the different locations of the wall, the wall can be divided into transverse wall, longitudinal wall, gable, external wall and internal wall.

　　According to different construction techniques, walls can be divided into prefabricated walls, cast-in-place walls and masonry walls.

3.2.7　Arch

　　Arch is an ancient structural form, which belongs to the curved structural member, and the section of the member is mainly subjected to axial pressure. Arch structure can be widely used in bridge structure and others, taking ancient Rome and Gothic structure as the typical representative(Fig.3.16)

拱在竖向荷载作用下,支座不仅产生竖向反力,还产生水平推力,这使得拱截面的弯矩比较小,横截面应力比较均匀,因此更能发挥材料的作用。拱结构以承受轴向压力为主,因此适合采用抗压性能好、抗拉性能差的材料,如砖、石、混凝土等。

3.3 建筑物分类

任何建筑物的骨架——建筑结构都是由基本构件组成的,因此建筑物可以有多种分类形式。

3.3.1 按使用功能分类

建筑物根据其使用功能的不同,分为工业建筑、民用建筑和农业建筑。民用建筑主要包括宅类建筑(如民用住宅宿舍、公寓等)和公共建筑(如教学楼、体育馆等)。工业建筑主要指各类厂房和生产辅助用房(如生产车间和仓库等)。农业建筑主要包括饲养牲畜、存放农具和农产品的用房(如蔬菜大棚、养鸡场等)。

3.3.2 按照使用的建筑材料分类

建筑物根据其所用建筑材料的不同,可分为木结构建筑、钢筋混凝土结构建筑、

Under the action of vertical load, the support not only produces vertical reaction force, but also produces horizontal thrust. The bending moment of the arch section is relatively small, and the cross-section stress is more uniform, so it plays a more important role of materials. The arch structure mainly bears axial pressure, so it is suitable for materials with good compressive performance and poor tensile performance, such as brick, stone, concrete and so on.

3.3 Classification of building

The skeleton of any building—The building structure is made up of the basic components, so the building can be classified in a variety of forms.

3.3.1 Classification by function

Buildings are divided into industrial buildings, civil buildings and agricultural buildings according to their different use functions. Civil buildings mainly include residential buildings (such as residential buildings, dormitories, apartments, etc.) and public buildings (such as teaching buildings, gymnasiums, etc.). Industrial buildings mainly refer to all kinds of factories and auxiliary production rooms (production workshops and warehouses, etc.). Agricultural buildings mainly include houses for raising livestock, storing farm tools and products (such as vegetable greenhouses, chicken farms, etc.).

3.3.2 Classification by material

Buildings can be divided into timber structure, reinforced concrete structure, steel

钢结构建筑、钢-混凝土组合结构建筑、砌体结构建筑和其他结构形式建筑（见图3.17）。

3.3.3 按照建筑层数分类

建筑按照其建造的层数可分为单层建筑、多层建筑、高层建筑等。下面将主要按该分类方式对建筑工程的常用结构形式进行介绍。

3.4 单层建筑

单层建筑通常指层数为一层的建筑，可分为一般单层建筑和大跨度建筑。

图3.17 不同材料建成的结构
Fig. 3.17 Building structures constructed by various materials

structure, steel-concrete composite structure, masonry structure and other structural forms according to the different building materials used (Fig. 3.17).

3.3.3 Classification by storey number

Buildings can be divided into single-storey buildings, multi-storey buildings and high-rise buildings according to the number of floors they are built. The following will mainly introduce the commonly used structural forms of construction engineering according to the classification of methods.

3.4 Single-storey building

Single-storey building usually refers to the number of floors for a building, can be divided into single-storey building and large span building.

3.4.1 一般单层建筑

一般单层建筑按照房屋的使用功能可分为民用单层建筑和单层工业厂房。

单层民用建筑一般是指砖木结构、砖混结构的房屋(墙体采用砖砌体,屋盖采用木屋架、钢屋或采用钢筋混凝土的平顶屋盖形式),多用于单层住宅、公共建筑等(见图3.18)。

单层工业厂房通常由钢屋架(或钢木组合屋架)、天窗架、吊车梁、柱子、维护墙体结构和支撑系统等构成(见图3.19)。

图 3.18 民用单层建筑
Fig. 3.18 Civil single-storey buildings

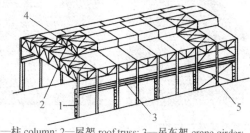
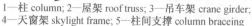

1—柱 column; 2—屋架 roof truss; 3—吊车架 crane girder;
4—天窗架 skylight frame; 5—柱间支撑 column braceing

图 3.19 单层工业厂房
Fig. 3.19 Industrial single-storey buildings

3.4.1 General single-storey building

General single-storey buildings can be divided into single-storey civil buildings and single-storey industrial factories according to the use function of the house.

Single-story civil buildings generally refer to the buildings with brick-wood structure and brick-concrete structure (walls with brick masonry, roof with wooden frame, steel roof truss or flat roof with reinforced concrete), which are mostly used in single-story residences and public buildings (Fig. 3.18).

Single-storey industrial buildings are usually composed of steel roof truss (or steel and wood combined roof truss), skylight truss, crane beam, column, maintenance wall structure and support system, etc. (Fig. 3.19).

单层工业厂房按结构形式可分为排架结构和刚架结构体系。排架结构体系中柱子与基础是刚性连接,屋架与柱子顶端是铰连接;刚架结构的梁与柱的连接均为刚性连接。

3.4.2 大跨度建筑

跨度在 60 m 以上的建筑统称为大跨度建筑,在民用建筑中主要应用于影剧院、体育场馆、展览馆和航空港等,如图 3.20 所示。在工业建筑中主要应用于生产大体量产品的厂房建筑,如飞机装配车间、飞机库和其他大跨度厂房,如图 3.21 所示。

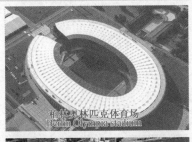
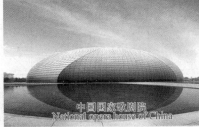

国家大剧院

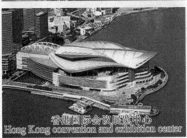

大兴国际机场

图 3.20 民用大跨度建筑
Fig. 3.20 Civilian long-span structures

图 3.21 工业大跨度建筑
Fig. 3.21 Industrial long-span structures

According to the structure form, the single-storey industrial workshop can be divided into frame structure and rigid frame structure system. In the frame structure system, the column and the foundation are rigidly connected, and the roof truss and the top of the column are hinged. The connection of beam and column of rigid frame structure is rigid connection.

3.4.2 Long-span structures

The long-span structure refers to the one whose span is over 60 m. Long-span structures are widely used in civil buildings (such as cinema, stadium, exhibition hall, aviation port, as shown in Fig.3.20). In industrial buildings it is mainly used in factory buildings producing large volumes of products (such as aircraft assembly workshops, hangars and other long-span factories, as shown in Fig.3.21).

大跨度建筑结构体系有很多种,如桁架结构、网架结构、网壳结构、悬索结构、膜结构和薄壳结构等。

(1)桁架结构

桁架结构是由直杆铰接而成的平面和空间结构。桁架通常由上弦杆、下弦杆、腹杆(竖腹杆和斜腹杆)组成,如图 3.22 所示。它将屋面荷载简化为作用在桁架节点上的节点荷载,通过节点传递作用力,使桁架杆受压或者受拉,充分发挥材料的力学性能,其制作方便,形状规则,广泛应用于一般大跨度建筑,如图 3.23 所示。

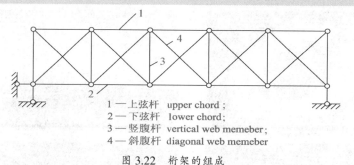

1—上弦杆 upper chord;
2—下弦杆 lower chord;
3—竖腹杆 vertical web memeber;
4—斜腹杆 diagonal web memeber

图 3.22 桁架的组成
Fig. 3.22 Composition of truss

图 3.23 桁架结构的应用
Fig. 3.23 Application of truss

The long-span structures can be classified into many types, such as truss, space grid structure, latticed shell structure, suspended structure, membrane structure, thin-shell structure, etc.

(1)Truss

The truss structure is a plane and space structure hinged by straight rods. The truss usually consists of upper chord, lower chord, and belly (vertical and diagonal)(Fig. 3.22). The roof load is simplified as the joint load acting on the truss joint, and the force is transferred through the joint to make the truss rod under compression or tension, giving full play to the mechanical properties of the material, which is convenient to manufacture and has regular shape, and is widely used in general large-span buildings(Fig. 3.23).

桁架结构根据所用的材料不同,可分为木桁架、钢-木组合桁架及钢筋混凝土桁架;按照桁架的平面形状,可分为三角形桁架、梯形桁架、抛物线形桁架等;按照其承力体系,可分为平面桁架和空间桁架。

(2) 网架结构和网壳结构

网架结构是空间结构体系,由许多空间位置不同的直杆通过球节点,按照一定的规律连接形成网状空间结构系统。网架可分为平板网架(见图 3.24 a)和曲面网架(也称网壳,见图 3.24 b),网架的应用十分广泛。网架结构在节点处通常采用焊接球(见图 3.25 a)或螺栓球(见图 3.25 b)进行连接。

(a) 平板网架
Flat grid

(b) 曲面网架（网壳）
Surface grid (latticed shell structure)

图 3.24　网架结构
Fig. 3.24　The grid structure

(a) 焊接球连接节点
Welded ball joint

(b) 螺栓球连接节点
Bolt ball joint

图 3.25　网架结构连接节点
Fig. 3.25　Joints of grid structure

According to the different materials used, the truss structure can be divided into wood truss, steel truss, steel-wood combination truss and reinforced concrete truss. According to the plane shape of the truss, it can be divided into triangle truss, trapezoidal truss, parabolic truss and so on. According to its bearing system, it can be divided into plane truss and space truss.

(2) Space grid structure and latticed shell structure

The grid structure is a spatial structure system, which is connected by straight rods at different spatial positions through the ball nodes in accordance with certain rules connecting the formation of the network space structure system. It can be divided into flat grid (Fig 3.24 a) and curved grid (also known as reticulated shell, Fig 3.24 b). Grid is widely used. The grid structure adopts welded ball joints　(Fig 3.25 a) and bolted ball joints (Fig 3.25 b) to connect at the joints.

（3）悬索结构

悬索结构是以索作为主要受力构件抵抗外荷载作用的一种结构体系（见图3.26）。悬索结构的优点是自重轻、跨度大、节省材料,根据悬索设计可实现不同的大空间功能要求。索的材料通常采用抗拉性能良好的钢丝束、钢丝绳、钢绞线等线材。

（4）膜结构

膜结构是由高强薄膜材料(PVC或Teflon)依托于加强构件(钢架、钢柱或钢索)形成的一种空间结构形式。膜结构分为张拉膜结构和充气膜结构两大类。张拉膜结构通过支撑柱、刚架或索张拉成型(见图3.27 a);充气膜结构通过向膜空腔内充入一定压力的空气后成型(见图3.27 b)。膜结构适应变形能力强,自重轻,具有良好的抗震性能,制作方便、施工速度快、造价低。

图 3.26　悬索结构(代代木体育馆)
Fig. 3.26　Suspended structure (Yoyogi Stadium)

(a) 张拉膜结构
Tensioned membrane structure

(b) 充气膜结构
Air-supported membrane structure

图 3.27　膜结构
Fig. 3.27　Membrane structure

（3）Suspended structure

Suspension cable structure is a structural system in which cables are the main load-bearing members to bear external loads (Fig. 3.26). The advantages of the suspension cable structure are light weight, large span and material saving. According to the design of the suspension cable, different functional requirements of large space can be realized. The material of cable usually adopts wire bundle, wire rope, steel strand with good tensile performance.

（4）Membrane structure

The membrane structure is a form of spatial structure which is made by the high strength film material (PVC or Teflon) supported by the strengthening members (steel frame, steel column and cable). Membrane structures are divided into two categories: tensioned membrane structures and inflatable membrane structures. The tensioned membrane structure is formed by tension of support columns, steel frames or cables (Fig. 3.27 a). The aerated membrane structure is filled with air at a certain pressure to form a certain shape (Fig. 3.27 b). Membrane structure is adapt to deformation, it also has light weight, good seismic performance, convenient production, construction speed, low cost.

（5）张弦结构

张弦结构是由刚性构件（或结构）通过撑杆与柔性索连接而成的一种预应力复合钢结构（也可称为弦支结构，见图 3.28）。张弦结构能充分发挥刚性和柔性材料的力学特性；同时通过施加预应力不仅可以提高结构的刚度，还能使结构处于自平衡状态，减小支座的水平推力，形成较大跨度的承载体系。

张弦结构可以分为平面张弦结构和空间张弦结构。平面张弦结构属于平面受力体系，如单向张弦梁或张弦桁架。上海浦东国际机场（见图 3.29 a）、南京国际博览中心（见图 3.29 b）均采用了单向张弦梁结构形式。空间张弦结构是将图 3.28 中的刚性上弦替换为空间结构（网架、网壳、空间桁架等），基于张拉整体概念合理布置下部撑杆和柔性索形成空间受力体系。空间张弦结构在大跨度建筑中有广泛的应用，常见的结构形式有张弦空间桁架、弦支网架、弦支穹顶等，如图 3.29 c 所示的中国国家奥体中心羽毛球馆即采用了弦支穹顶结构。

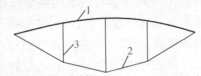

1—刚性构件（或结构）Rigid member (structure)；
2—索 Cable；
3—撑杆 Strut

图 3.28　张弦结构的基本形式
Fig. 3.28　Composition of cable-supported structure

(a) 上海浦东国际机场
Shanghai Pudong International Airport

(b) 南京国际博览中心
Nanjing International Expo Center

(c) 国家奥林匹克羽毛球馆
Olympic Badminton Gym

图 3.29　张弦结构的应用
Fig. 3.29　Applications of cable supported structure

(5) String structure

String structure is a prestressed composite steel system which is made up of rigid upper members (or structure), cable and middle strut (Fig.3.28, also known as cable-supported structure). String structure can take full advantage of mechanical properties of both rigid and flexible materials. At the same time, by applying prestress, the rigidity of the structure can be improved. We can also make the structure to be in the self-equilibrium state, and reduce the horizontal thrust of the support, and realize the bearing system with a large span.

The string structure can be categorized as plane string structure and space string structure. The plane string structure belongs to the plane stress system, such as the single-direction beam string structure or the truss string structure. Shanghai Pudong International Airport and Nanjing International Expo Center (Fig.3.29 a,b) both adopt the single-direction beam string structure. Spatial string structure is a space force system, which replaces rigid upper chord by space structure (grid, latticed shell, space truss, etc), according to the concept of reasonable arrangement of tensegrity strut and cable. Space string structure is widely used in large span buildings. The common structure forms are spatial truss string, string truss, suspended dome, etc. The National Olympic Badminton Gym adopts the suspended dome structure (Fig.3.29 c).

（6）薄壳结构

薄壳结构是空间曲面形状的结构（见图3.30），可以建造出优美的建筑屋顶形式。薄壳结构的优点是受力均匀，能充分利用材料强度，又能将承重与围护功能结合在一起。薄壳结构按曲面的形式分为圆柱筒壳、双曲抛物面壳、球面薄壳、折板等。

（7）其他新结构

轻质高强材料的应用是实现大跨度结构的前提条件。根据增强复合材料的不同，纤维增强复合材料(FRP)分为玻璃纤维增强复合材料(GFRP)，碳纤维增强复合材料(CFRP)及芳纶纤维增强复合材料(AFRP)等。纤维增强复合材料具有轻质、高强和耐腐蚀性能好等特性，是建造大跨度结构的理想材料。图3.31为一种FRP编织网结构的示意图。三亚体育场总建筑面积8.8万余平方米，体育场屋盖为轮辐式索桁架结构，平面投影短轴224 m、长轴261.8 m，中心开口短轴134 m，长轴171.8 m。为提

图 3.30　薄壳结构
Fig. 3.30　Thin-shell structures

（6）Thin-shell structure

The thin-shell structure is a curved thin-wall structure（Fig.3.30）that creates a graceful roof form for the building, and it has the advantages of even force distribution, making full use of material and integrating the functions of load-bearing and building envelope. The thin-shell structure can be classified into cylindrical shell, hyperbolic paraboloid shell, spherical shell and folded plate etc.

图 3.31　FRP 编织网结构示意图
Fig. 3.31　Sketch of FRP weaved structure

（7）New structure types

The application of light-weight and high-strength materials is a prerequisite for the realization of long-span structures. According to the different reinforcing materials, fiber reinforced polymer(FRP) is divided into glass fiber reinforced composites (GFRP), carbon fiber reinforced composites (CFRP) and aramid fiber reinforced composites (AFRP). With the characteristics of light weight, high strength and good corrosion resistance, it is an ideal material for the construction of long-span structures. Fig. 3.31 is a schematic diagram of an FRP braided mesh structure. Sanya Stadium has a total construction area of more than 88 000 square meters. The roof of the stadium is a hub-and-spoke cable truss structure. The short axis of the

升结构抗风能力,屋盖结构布置内环交叉索;为降低内环交叉索对锁夹形成的不平衡力,该项目创新应用了强模比远高于普通钢索的CFRP平行板索,以新材料、新构件解决了结构问题(见图3.32)。

　　玻璃结构也是一种新型的结构形式,提高建筑物的通透性,能以立体的方式展现无障碍的视觉效果,突破常规形态限制,让设计师发挥艺术想象力,最大程度张显建筑物的个性风采,提供特定场合的建筑文化和人文环境(见图3.33)。但是用于制作结构的玻璃通常非普通玻璃。

图3.32　三亚体育中心体育场纤维结构实例
Fig.3.32　Fiber structure of Sanya Stadium

图3.33　张家界玻璃桥面
Fig.3.33　The glass bridge in Zhangjiajie

plane projection(Fig.3.32) is 224 meters, the long axis is 261.8 meters, the short axis of the center opening is 134 meters, and the long axis is 171.8 meters. In order to improve the wind resistance capacity of the structure, the roof structure is arranged with inner ring cross cables. In order to reduce the unbalance force caused by the intersecting cable of the inner ring to the lock clip, the project innovates the application of CFRP parallel plate cable whose strong modulus ratio is much higher than that of ordinary steel cable, and solves the structural problem with new materials and new components.

　　Glass structure is also a new type of structural form which improves the permeability of the building, it can show barrier-free visual effect in a three-dimensional way, break through the conventional form restrictions, let designers give play to their artistic imagination, show the personality of the building to the greatest extent, provide a specific occasion of architectural culture and cultural environment(Fig.3.33). But the glass used to make structures is usually not ordinary glass.

3.5 多层和高层建筑

多层和高层建筑多用于居民住宅、商场、办公楼和旅馆等民用建筑中。随着经济建设的不断发展，越来越多的多高层建筑被建造出来，为城市增添了现代化色彩。

多层和高层建筑属于不同的建筑类型，其区分的界限各个国家有不同的规定。我国的区分方法依据《民用建筑设计通则》(GB 50352—2005)，如表3.1所示。

3.5.1 多层建筑

多层建筑常用于住宅、教学楼、旅馆等民用建筑中(见图3.34)，其结构形式主要有框架结构和混合结构。框架结构一般是钢筋混凝土框架结构，是由钢筋混凝土梁和柱通过刚性节点连接而成的承重结构体系(见图3.35)。

表 3.1 我国多层和高层建筑划分规定
Tab. 3.1 Classification rules of the multi-storey and tall buildings in China

建筑类型 Building types		划分规定 Classification rules
住宅 residential buildings	多层 multi-storey buildings	多于2层且少于十层 more than 2 storeys and less than 10 storeys
	高层 tall buildings	十层及以上 more than 10 storeys
住宅以外的民用建筑 civil engineering except residential buildings	多层 multi-storey buildings	高于10 m且低于24 m more than 10 m and less than 24 m
	高层 tall buildings	24 m及以上(不包括建筑高度大于24 m的单层公共建筑) more than 24 m (except the single storey buildings which are more than 24 m)

注:建筑高度超过100 m的为超高层建筑。
Note：The buildings with more than 100 meters height are called super tall buildings.

3.5 Multi-storey buildings and tall buildings

Multi-storey and high-rise buildings are mostly used in residential buildings, shopping malls, office buildings, hotels and other civil buildings. With the continuous development of economic construction, more and more multi-high-rise buildings have been built, which adds a modern color to urban construction.

Multi-storey and high-rise buildings belong to different building types, and the boundaries of their differentiation have different regulations in different countries. The classification method in China is based on the *General Principles of Civil Building Design* (GB 50352—2005), as shown in Tab.3.1.

3.5.1 Multi-storey buildings

Multi-storey buildings are often used in residential buildings, teaching buildings, hotels and other civil buildings (Fig. 3.34). Their structural forms mainly include frame structure and mixed structure. Frame structure is generally reinforced concrete frame structure, which is a load-bearing structure system formed by the connection of reinforced concrete beams and columns through rigid joints (Fig. 3.35).

框架结构具有相对空间大、布置灵活等优点。混合结构是指砖砌墙体和钢筋混凝土楼板形成的房屋结构,利于不规则的、有多功能要求的建筑平面设计。

3.5.2　高层建筑

高层建筑在国外已经有 100 多年的发展历史。1885 年,美国根据现代钢框架结构原理建造了共 11 层的芝加哥家庭保险公司大厦(见图 3.36 a),成为近代高层建筑的开端。

图 3.34　多层建筑
Fig. 3.34　Multi-storey buildings

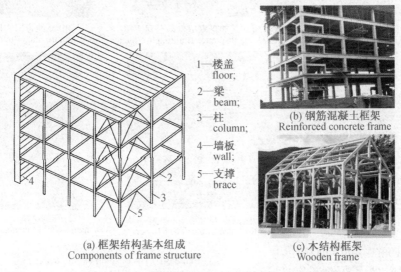

1—楼盖
floor;
2—梁
beam;
3—柱
column;
4—墙板
wall;
5—支撑
brace

(b) 钢筋混凝土框架
Reinforced concrete frame

(c) 木结构框架
Wooden frame

(a) 框架结构基本组成
Components of frame structure

图 3.35　框架结构
Fig. 3.35　Frame structures

The frame structure has the advantages of large space and flexible arrangement. Mixed structure refers to the building structure formed by brick wall and reinforced concrete floor slab, which is conducive to irregular and multifunctional architectural plane design.

3.5.2　Tall buildings

The development of the tall building has more than one hundred years in foreign countries. In 1885, Home Insurance Building in Chicago was built with 11 storeys according to the principle of the modern steel frame structure in American　(Fig.3.36 a), and it is the beginning of modern tall buildings.

1931 年,纽约建造了著名的帝国大厦(见图 3.36 b),地上建筑高 381 m,共有102 层,成为曼哈顿的地标建筑。

20世纪50年代以后,轻质高强材料有所应用、新的抗侧力结构有了一定的发展、电子计算机和相应的结构分析软件得到普遍应用,可快速完成复杂的抗风、抗震结构分析。同时,这促进了新的施工方法的不断涌现,使得高层建筑结构得到迅速的发展。1972年,纽约建造了著名的世贸中心大楼,该建筑由两座塔式摩天大楼组成,计110层,高402 m(见图3.36 c),2001年9月11日遭恐怖袭击,大楼坍塌。

到目前为止,世界上最高的建筑为位于阿联酋的哈利法塔(见图3.2)。位于我国上海的环球金融中心是位列世界第三的高楼,共10层,总高492 m。我国高层建筑的总数位于世界第一。

高层建筑的结构形式主要有框架剪力墙结构、剪力墙结构和筒体结构等。

(a) 芝加哥家庭保险公司大厦
Home Insurance Building in Chicago

(b) 纽约帝国大厦
Empire State Building

(c) 纽约世界贸易中心大楼
World Trade Center Twin Tower

(d) 环球金融中心
Shanghai World Financial Center

图 3.36　世界高层建筑
Fig. 3.36　Tall buildings in the world

上海中心大厦

In 1931, the famous Empire State Building (Fig. 3.34 b), which is 381 meters above the ground and has 102 storeys, was built in New York. It is a landmark building in Manhattan.

Since the 1950s, light and high strength materials have been applied, new lateral force resistant structures have been developed to a certain degree, and electronic computers and corresponding structural analysis software have been applied, which can quickly complete the analysis of complex wind-resistant and aseismatic structures. At the same time, they promoted the emergence of new construction methods, making the high-rise building structure have a rapid development. In 1972, New York built the famous World Trade Center, which consists of two towers with 110 floors and a height of 402m (Fig.3.36 c). The World Trade Center was completely collapsed by the terrorist attacks on September 11, 2001.

So far, the world's tallest building is the Burj Khalifa in the United Arab Emirates (Fig. 3.2). The World Financial Center in Shanghai is the third tallest building in China, with a total height of 492m and 10 floors. The total number of tall buildings in China ranks first in the world.

The structure forms of tall building mainly include frame-shear wall structure, shear wall structure and tube structure, etc.

（1）剪力墙结构

剪力墙结构采用钢筋混凝土墙板承受水平荷载和竖向荷载作用,主要功能是承受水平剪力作用。钢筋混凝土墙体的抗剪能力很强,可有效地抵抗水平荷载,故称剪力墙。剪力墙结构的整体性强,抗侧刚度大,水平荷载作用下的侧向位移小,因此多用于高层建筑结构中,如朝鲜的柳京饭店(见图3.37),但是剪力墙结构对建筑平面和空间布置要求较高,难以满足大空间建筑功能要求。

（2）框架-剪力墙结构

框架-剪力墙结构是由框架和剪力墙两种结构体系组合成的一种结构形式,即在框架结构体系的基础上,在合适的地方布置一定数量的抗侧力结构墙——剪力墙(见图3.38),以弥补框架结构抗侧力不足的缺陷,同时又可以保留框架结构空间布置灵活的优点,充分发挥两种结构形式的优势。

（3）筒体结构

制作简单几何形体的钢筋混凝土封闭筒,可用于超高层建筑结构中,它具有十分优越的力学特性,能够

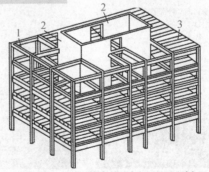

图 3.37　柳京饭店
Fig. 3.37　Ryugyong Hotel

（1）Shear wall structure

The shear wall structure adopts reinforced concrete wallboard to bear the horizontal load and vertical load, and its main function is to bear the horizontal shear force. Reinforced concrete wall has strong shear ability which can effectively resist horizontal load, so it is called shear wall. The shear wall structure has strong integrity, large la-teral stiffness and small la-teral displacement under horizontal load, so it is mostly used in high-rise building structures, such as Ryugyong Hotel in The Democratioc People's Republic of Korea （Fig.3.37）. However, the shear wall structure has high requirements on building plane and spatial layout, and it is difficult to meet the functional require-ments of large-space buildings.

1—框架 frame; 2—剪力墙 shear wall; 3—板 floor

图 3.38　框架-剪力墙结构
Fig. 3.38　Frame-shear wall structure

（2）Frame-shear wall structure

Frame-shear wall structure is composed of frame and shear wall two structure systems, a kind of structure form that is on the basis of frame structure systems, which is in the right place to decorate a certain amount of resistance to lateral force wall—shear wall structure (Fig.3.38), and to make up for the insufficient lateral force of frame struc-ture defects, at the same time it can keep frame structure space decorate flexible bit and give full play to the advantages of the two structural forms.

（3）Tube structure

The reinforced concrete closed cylinder used for making simple geometry can be used

承受更大的竖向和水平荷载作用,以及更高的整体性和更大的抗侧刚度。筒体结构可以是剪力墙薄壁筒或密柱框筒。

筒体结构可分为框筒结构、筒中筒结构、框架-核心筒、多重筒体和束筒结构(见图3.39)。

筒体结构的空间刚度极大、抗扭性能也很好,是超高层建筑结构的主要形式之一(见图3.40)。

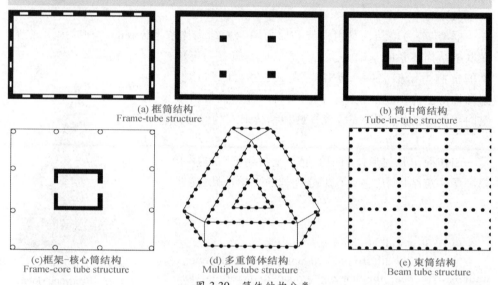

(a) 框筒结构
Frame-tube structure

(b) 筒中筒结构
Tube-in-tube structure

(c)框架-核心筒结构
Frame-core tube structure

(d) 多重筒体结构
Multiple tube structure

(e) 束筒结构
Beam tube structure

图 3.39　筒体结构分类
Fig. 3.39　Classification of tube structure

(a) 西尔斯大厦（束筒结构）
Sears Tower(beam tube structure)

(b) 合肥海顿国际广场（框筒结构）
Hefei Haydn International Plaza (frame-tube structure)

(c) 广州国际会展中心（筒中筒结构）
Guangzhou International Convention Centre (tube-in-tube structure)

图 3.40　筒体结构的应用
Fig. 3.40　Application of tube structure

super tall building structures. It has very superior mechanical properties which can withstand greater vertical and horizontal load action, higher integrity and greater lateral stiffness. The cylinder structure can be shear wall thin-walled cylinder or dense column frame cylinder.

Tube structure can be classified into frame-tube structure, tube-in-tube structure, frame-core tube, multiple tube structure and beam tube structure (Fig.3.39).

The tubular structure is one of the main forms of super high-rise building structures with great spatial stiffness and good torsional resistance (Fig.3.40).

（4）巨型结构

巨型结构是由大型结构单元组成的主结构和其他结构单元组成的次结构共同工作的一种结构形式。巨型结构按主要结构的承力体系可分为巨型桁架结构、巨型框架结构、巨型悬挂结构和巨型分离结构等。在力学性能上，巨型结构具有传力明确、施工速度快、节约材料等优点，在超高层建筑中有很多应用（见图3.41）。

此外，随着可持续发展要求的提高，需要在传统的房屋建筑结构体系中融入环保和节能技术，因而形成了一些新的"绿色"结构体系，可参见第10章绿色建筑中的相关内容。

(a) 汉考克中心
John Hancock Center

(b) 匹兹堡钢铁大厦
U.S. Steel Tower (Pittsburgh)

(c) 香港汇丰银行大厦
HSBC Main Building

图 3.41 巨型结构的应用
Fig 3.41 Application of mega structure

(4) Mega structure

Mega-structure is a structure formed by a main structure consisting of large structural units and a secondary structure consisting of other structural units. According to the load-bearing system of the main structure, the mega-structure can be divided into mega-truss structure, mega-frame structure, mega-suspension structure and mega-separation structure, etc. In terms of mechanical properties, mega-structures have the advantages of clear force transmission, fast construction speed and relatively saving materials, etc., which have been widely used in super high-rise buildings (Fig. 3.41).

In addition, with the improvement of sustainable development requirements, traditional building structure system needs to integrate environmental protection and energy saving technology, which help to form a number of new green structure systems that can be found in the relevant content of green building in chapter 10.

3.6 特种结构

特种结构是指具有特种用途的工程结构,包括高耸结构、海洋工程结构、管道结构和容器结构等,下面对部分结构进行介绍。

3.6.1 烟囱

烟囱是工业与民用建筑中常见的一种构筑物,是一种排烟用高耸结构,用以改善燃料的燃烧条件,通过静电除尘等环保设施,减轻烟气对大气的污染。

烟囱的外形一般是圆锥体,截面为底大上小构造。烟囱按建造的材料可分为砖砌烟囱、钢筋混凝土烟囱等(见图3.42),一般为自立式烟囱。

3.6.2 水塔

水塔是用于储水和配水的高耸结构物,是给排水工程中常用的构筑物。水塔由水箱、塔身和基础三部分组成。

| (a) 砖烟囱 | (b) 钢筋混凝土烟囱 | (c) 钢烟囱 | (d) 拉线式烟囱 |
| Brick chimney | Reinforced concrete chimney | Steel stack | Guyed chimney |

图 3.42 烟囱种类
Fig. 3.42 Chimney types

3.6 Special structures

Special structures are constructed for specific purposes, such as high-rising structures, ocean engineering structures, pipeline structures and vessel structures. Several special structures are introduced in this section.

3.6.1 Chimney

Chimney is a structure commonly used in industrial and civil buildings. It is a tall structure used for smoke exhaust and improving the combustion conditions of fuel, through electrostatic dust removal and other environmental protection facilities. Chimney can reduce the smoke pollution to the atmosphere.

The shape of the chimney is generally a cone, with a large base and a small section. Chimneys can be divided into brick chimneys, reinforced concrete chimneys, etc. (Fig.3.42) according to the construction materials. Generally, they are free-standing chimneys.

3.6.2 Water tower

The water tower is a high-rising structure which is used for the storage and distribution of water, and it is widely used in water supply engineering. The water tower is usually composed of water tank, tower body and foundation.

水塔按制作材料分为钢筋混凝土水塔、钢结构水塔、砌体结构水塔。水箱的形式有圆柱壳式、倒锥壳式、球形、箱形等(见图 3.43),其中前两种形式在我国应用最多。为增强稳定性,塔身通常设有支筒和支架,多数为钢筋混凝土钢架或钢构架。

3.6.3 筒仓

筒仓是贮存颗粒状和粉状松散材料(如稻谷、面粉、水泥、碎煤、矿粉等)的立式容器。筒仓的平面形状有正方形、矩形、多边形和圆形等,圆形筒仓因仓壁受力合理,在实际工程中应用最多。筒仓根据所用材料可分为钢筋混凝土筒仓、钢筒仓、砖砌筒仓和塑料筒仓等(见图 3.44),其中钢筋混凝土筒仓由于造价低、耐久性好、施工方便及抗冲击性能良好等优点在我国应用较多。

图 3.43 水箱的形式
Fig. 3.43 Types of water tank

图 3.44 筒仓类型
Fig. 3.44 Types of silo classified by materials

The water tower is classified into reinforced concrete water tower, steel water tower and brick-concrete water tower by manufacturing materials. The shapes of water tanks are various, such as cylinder, turbination, sphere, box (Fig.3.43), and the cylindrical water tank and turbinate water tank are widely used in China. The structural forms of the tower body are generally tube or bracket for stability, while the bracket usually adopts reinforced concrete frame or steel frame.

3.6.3 Silo

Silo is an upright container which is used to store granular and loose powder material(such as paddy, flour, cement, crushed coal, and ore powder). The flat-shapes of silo are usually square, rectangle, polygon and circle, and the circular silo is mostly used for its reasonable force distribution. According to the material used, the silo can be classified as reinforced concrete silo, steel silo, brick silo and plastic silo(Fig.3.44) etc. Reinforced concrete silos are most widely used in China for its advantages of low cost, good durability, convenient construction and good performance on impact resistance.

按照筒仓的贮料高度与直径或宽度的比例关系,可将筒仓划分为浅仓和深仓两类。浅仓主要用于短期贮存,深仓主要用于长期贮存。

3.6.4 电视塔

电视塔是用于广播电视发射传播的构筑物。由于发射电视信号的位置越高,其传播的范围就越大,因此电视转播塔越建越高。为了充分发挥转播塔功能,周边可设计观光旅游等设施,或设置观景台、餐厅。电视塔通常是一个地方的标志性建筑。

目前,世界上最高的电视塔是2012年建成的东京晴空塔(高634 m),我国最高的电视塔是广州塔(高600 m),建成于2009年,是世界第三高电视塔。世界第二高的电视塔是美国的KVLY电视塔,该塔高628 m,桅杆结构,也是第二高自立式电视塔(见图3.45)。

图 3.45　电视塔

Fig. 3.45　Television towers

Silo can be classified into bunker and deep bin according to the relationship between the height of storage in the silo and its diameter or width. The bunker is mainly used for short-term storage, while the deep bin is mainly used for long-term storage.

3.6.4 Television tower(TV tower)

TV tower is used for broadcasting and television transmission structures. The higher the position of the transmitted television signal, the greater the range of its transmission, so that the television tower is built higher and higher. In order to give full play to the function of the broadcast tower, it's a good idea to design sightseeing and tourism facilities nearby or up a viewing platform and restaurant. TV tower is actually a landmark building of a place.

The Tokyo sky tower is the highest TV tower in the world so far which was built in 2012, and its height is 634 m. The highest TV tower in China is the Guangzhou tower which was constructed in 2009. The height of Guangzhou tower is 600 m which ranks third in the world. The second tallest building is the KVLY tower (Fig. 3.45)in America which is 628 m in height and mast structured, it ranks second among the self-standing towers.

3.6.5 风力发电塔架

为了解决能源问题带来的困扰，同时由于环境保护的需要，绿色能源如太阳能、风能、潮汐能等越来越受到关注。我国风能资源丰富，有巨大的发展潜力。近些年，全国各地兴建了很多风力发电塔，风力发电技术得到迅速发展。

风力发电系统主要由塔架、发电机、齿轮增速器、桨叶、联轴器等组成，与土木工程相关的主要是塔架和塔架基础部分，其成本约占总成本的 20%。按照结构形式主要可分为格构式塔架和锥筒型塔架(见图 3.46)。格构式塔架的用钢量小、造价低，但占地面积大、施工周期长。锥筒型塔架的优点是构造简单、制作和施工方便等，但也存在整体稳定性差、钢材利用率低等缺点。

图 3.46 风力发电塔结构形式
Fig. 3.46 Structure type of wind turbine tower

3.6.5 Wind turbine tower

In order to solve the energy problems, as well as the need for environmental protection, green energy such as solar, wind, tidal power, etc. is getting more and more attention. Wind resource is rich and has great development potential in China. In recent years, a lot of wind power towers have been built, and wind power technology has been rapidly developed in our country.

Wind power generation system mainly consists of a tower, a generator, a gear speed increaser, blades and the coupling. Tower and tower foundation are most relevant to civil engineering and the cost of building accounts for about 20% of the total cost. Towers can be divided into lattice towers and cone towers (Fig.3.46). Lattice towers need a small amount of steel, low cost, but occupy large area, and take long construction period. The cone tower has the advantages of simple structure, convenient production and construction, but it has poor stability, and the overall steel utilization rate is low.

3.6.6 立体停车场

立体停车场是用来存放车辆的场所,包括机械或机械设备系统(见图 3.47),主要由钢构架、回转台、输送车或升降电梯、监控操作台及辅助设备(消防、配电、防盗机构)几大部分组成。最早的立体车场建于 1918 年,是美国芝加哥市的一家宾馆的停车场。目前,立体停车场在我国很多城市得到应用,相关技术已经比较成熟。

3.7 建筑结构设计

建筑结构设计是对建筑空间骨架的设计,是确保建筑物和构筑物安全的技术活动,因此必须按照结构的组成规律、力学分析方法和结构设计的基本原则和要求进行,目的是使得所使用的建筑结构物具有安全可靠性。我国的《建筑结构可靠度统一设计标准》(GB 50068—2018)对结构设计工作给出如下规定。

图 3.47 立体停车场
Fig. 3.47 Stereo garages

3.6.6 Stereo garage

Stereo garage is a machine or mechanical system used for access and storage of vehicles (Fig.3.47), and it is composed of steel frame, turret, transfer car or elevator, control console and accessory equipments (such as fire protection equipment, electrical distribution plant and security facilities). The earliest stereo garage is a hotel parking garage in Chicago which was built in 1918. Currently, the stereo garages are extensively applied in many cities in China in order to ease the traffic pressure, and the relevant technique is increasingly mature.

3.7 Design for building structures

Architectural structural design is the design of the architectural space skeleton, which is a technical activity to ensure the safety of buildings and structures. So it must be in accordance with the law of the composition of structure, mechanical analysis methods and the basic principles and requirements of structural design, to make the building structures used safety and reliability. China's *Unified Design Standard for Reliability of Building Structures* (GB 50068—2018) provides the following provisions for structural design work:

① 在正常施工和正常使用时,能承受可能出现的各种作用。

② 在正常使用时,具有良好的工作性能。

③ 在正常维护条件下,具有足够的耐久性能。

④ 在设计规定的偶然事件发生时及发生后,仍能保持必需的整体稳定性。

建筑工程的结构设计步骤:方案设计→力学模型→结构分析→内力组合→截面设计→施工图绘制。

3.7.1 方案设计

结构方案设计是在建筑设计的基础上进行的结构选型、结构布置及结构构建参数计算等。

① 结构选型主要是根据建筑物的功能要求,确定结构体系和选择结构形式,通过比较来确定方案的合理性及其相应的经济技术指标,最终选择合适的结构方案。

② 结构布置是指结构构件在平面上的摆放位置, 根据结构的使用功能和所选择的结构体系来确定。

③ 主要结构构件的尺寸包括构件截面的形状和尺寸、构件的长度等,根据结构平面布置方案和所选择的结构类型及材料来确定,一般均有经验参考,通过承载能力和刚度条件加以控制。

① The designed structure can withstand all kinds of actions in the condition of normal construction and regular service.

② The designed structure should have good performance under regular service conditions.

③ The designed structure should have adequate durability under regular maintenance.

④ The designed structure should be able to maintain the required stability during and after the accidental events mentioned in the Codes.

The design procedure of building structures can be generally divided into the following stages: schematic design→mechanical model→structural analysis→combination of internal forces→cross section design→construction drawing.

3.7.1 Schematic design

Structural scheme design is based on the architectural design, structural selection, structural layout and the calculation of structural construction parameters.

① Structural selection is mainly based on the functional requirements of the building to determine the structural system and choose the structural form, through comparison to determine the rationality of the program and the corresponding economic and technical indicators, and finally choose the appropriate structural program.

② Structural layout refers to the placement of structural components on the plane, which is determined by the use of the structure function and the selected structural system.

③ The size of the main structural component, including the shape and size of the component section, the length of the component, its determination method is based on the structure layout scheme and the selected structure type and material. Generally there is experience for reference, the size can be controlled through the bearing capacity and stiffness conditions.

3.7.2 结构分析

结构分析是结构设计的重要内容,主要包括以下几个重要方面:

① 结构模型的建立。通过对实际结构的力学模型简化来建立结构分析模型。简化的过程中要给出一些假设条件,反映结构的实际受力情况,且要满足结构力学分析模型的要求,同时应尽可能的简单,以方便分析和计算。用于进行结构分析的力学模型类型较少,但是工程结构的实际情况是十分复杂的,因此结构模型的简化是一件十分重要的工作。

② 荷载的统计和施加作用。根据建筑物的功能和性质,以及其在整个服役期限内可能遭受的各类荷载作用情况,按照荷载规范的规定,计算荷载数值的大小,并按照简化后的力学模型作用到相应的位置。

③ 计算分析。完成计算简图后,按照确定的计算理论和方法,对结构进行内力分析,一般情况下采用静力学弹性分析,对于复杂的情况采用非线性动力分析等。目前大多用大型结构分析软件进行分析,也有对应于不同专业的专用软件。

3.7.3 构件设计

根据结构模型力学分析的计算结果,得到构件截面内力的大小,按照不同结构的设计规范进行结构构件设计,包括杆件的截面设计、节点设计、配筋设计等。

3.7.2 Structure analysis

Structure analysis is an important part of structure design which mainly includes the following aspects:

① Establishment of structural model. By simplifying the mechanical model of the actual structure, the structural analysis model is established. In the process of simplification, some assumptions should be given to reflect the actual stress situation of the structure, and to meet the requirements of the structural mechanics analysis model, and as simple as possible to facilitate the analysis and calculation. There are few types of mechanical models used for structural analysis, but the actual situation of engineering structures is very varied, so the simplification of structural models is very important work.

② The statistics and effect of load. According to the function and nature of the building and all kinds of loads that may be subjected to during the whole service life, the value of the building is calculated in accordance with the provisions of the load code, and the corresponding position is acted on in accordance with the simplified mechanical model.

③ Computational analysis. After the calculation diagram is completed, according to the determined calculation theory and method, the internal force analysis of the structure is carried out. In general, it is static elastic analysis, and nonlinear dynamic analysis is used for complex situations. At present, most of the large structure analysis software is used, and there are some specific software which is corresponding to different professions.

3.7.3 Member design

According to the calculation results of the mechanical analysis of the structural model, the size of the internal force of the section of the component is obtained. According to the design specifications of different structures, the structural component design is carried out, including the section design of the bar, the joint design, the reinforcement design, etc.

3.7.4　施工图绘制

设计的最后一个阶段是绘制施工图,要求图纸表达正确、规范、简明和美观。

3.8　未来展望

在今后很长一段时间内,钢筋混凝土结构及钢结构仍将是不可替代的主要建筑结构形式,型钢混凝土和钢管混凝土等组合结构的应用也会越来越广泛。结构形式的创新很大程度上取决于新型结构材料的研究成果,随着土木工程科学技术的进步,新材料、新结构形式将不断涌现。未来土木工程的发展方向如下:

① 完善工程设计理论和方法,准确地建立结构分析模型并提高计算的精确度,对复杂荷载作用效应给出科学的分析计算结果;采用大型计算软件并开发智能化设计,使结构设计更具科学合理性。

② 创新新型结构体系和建造技术满足现代建筑的多功能要求,超高层和大跨度建筑空间的建筑将会不断被推出,采用参数化设计的新概念建筑也将更加丰富多彩。

③ 建筑结构实现机械工程、信息工程技术的综合应用,实现智能化结构设计和工程应用,创新设计新概念和新方法。

3.7.4　Construction drawing

The final stage of structural design is to make the construction drawing which is required to be correct, canonical, concise and beautiful.

3.8　Future prospects

For a long time in the future, reinforced concrete structure and steel structure will still be the main building structure forms that cannot be replaced, and the application of steel reinforced concrete and concrete filled steel tube and other composite structures will be more and more widely. The innovation of structural forms largely depends on the research results of new structural materials. With the progress of civil engineering science and technology, new materials and new structural forms will continue to emerge. The future direction of civil engineering development is as follows:

① Improve the theory and method of engineering design, accurately establish the structural analysis model and improve the accuracy of calculation, give scientific analysis and calculation results of complex load effect, adopt large computing software and develop intelligent design, make the structure design more scientific and rational.

② Innovate new structural systems and construction technologies to meet the high functional requirements of modern buildings. The construction of super-tall buildings and long-span building spaces will be continuously introduced, and new concept buildings using parametric design will become more colorful.

③ Building structure combines the comprehensive application of mechanical engineering and information engineering technology to realize intelligent structural design and engineering application, and innovate new concepts and new methods of design.

注:本章除图3.28外,其余图片、视频资料均来源于网络。
Note: In this chapter, the pictures and video materials are from webs except Fig.3.28.

知识拓展
Learning More

相关链接　Related Links
A. 土木在线
B. 中华钢结构论坛
C. 仿真科技论坛

小贴士　Tips
(1) 除本章3.4节中列出的结构形式外,大跨度建筑的结构形式还包括折板结构、蒙皮结构等;特种结构还包括海洋钻井平台、压力容器、核电站等,有兴趣的同学可查阅相关的书籍和网站进行了解。

Besides the structure forms introduced in section 3.4, there are also many other structure forms, such as the long-span structure also includes folded plate structure and stressed skin structure, and the special structure also includes ocean drilling platform, pressure vessel, nuclear power station. If you have interest in these subjects, you can refer to the relevant books and websites.

(2) 对工程结构进行分析设计时,需要用到相关的软件, 如PKPM,SAP2000,3D3S,ANSYS,ABAQUS等,有兴趣的同学可查阅相关的书籍和网站了解与学习。

When analyzing or designing a building structure, we may use the relevant software, such as PKPM, SAP2000, 3D3S, ANSYS, ABAQUS. If you want to study the above software, you can refer to the relevant books and websites.

(3) 目前,与结构工程专业相关的注册工程师种类主要有注册结构工程师、注册监理工程师和注册建造师,相关内容和要求可参考国家住房和城乡建设部网站http://www.mohurd.gov.cn/。

At present, registered engineers of structural engineering mainly include registered structural engineer, certified supervision engineer and certified architect. If you want to get more information, you can refer to the web of ministry of housing and urban-rural of China: http://www.mohurd.gov.cn/.

(4) 结构工程设计过程中,需要参照相应的国家规范或规程,如《建筑结构荷载规范》《混凝土结构设计规范》《钢结构设计规范》等,有兴趣的同学可查阅相关的资料进行学习。

When constructing a building, we must abide by the national Codes, such as *Load Code for the Design of Building Structures*, *Code for Design of Concrete Structures*, *Code for Design of Steel Structures*. If you have interest in the context of the Codes, you can refer to relevant books or websites.

思考题　Review Questions
(1) 尝试以一个实际建筑为例,寻找其中的基本构件。
Try to find the basic structural members in an actual building.

(2) 尝试按照建筑结构所用材料对建筑结构进行分类，描述每种结构类型的优缺点及其应用范围。

Try to classify the building structure by construction materials, and point out the characteristics of each structural type and the range of its application.

(3) 你对哪种结构形式最感兴趣？你周围的建筑物都有哪些结构形式？

Which kind of structure are you most interested in? What kinds of structures do the buildings around you have?

参考文献
References

［1］丁大钧,蒋永生.土木工程概论[M].2版.北京:中国建筑工业出版社,2010.

［2］叶志明.土木工程概论[M].3版.北京:高等教育出版社,2009.

［3］江见鲸,叶志明.土木工程概论[M].北京:高等教育出版社,2001.

［4］郑晓燕,胡白香.新编土木工程概论[M].2版.北京:中国建材工业出版社,2012.

［5］吕志涛.新世纪我国土木工程活动与预应力技术的展望[J].东南大学(自然科学版),2002, 32(3):457–459.

［6］董莪,黄林青.土木工程概论[M].北京:中国水利水电出版社,2011.

［7］曹双寅,吴京.工程结构设计原理[M].4版.南京:东南大学出版社,2012.

［8］陈志华.张弦结构体系[M].北京:科学出版社,2013.

［9］张毅刚.张弦结构的十年(一):张弦结构的概念及平面张弦结构的发展[J].工业建筑, 2009,39(10): 105–113.

［10］张毅刚.张弦结构的十年(二):双向张弦结构和空间张弦结构的应用和发展[J].工业建筑,2009,39(11): 93–99.

［11］莫特罗.张拉整体:未来的结构体系[M].薛素铎,刘迎春,译.北京:中国工业出版社,2007年.

［12］惠卓,秦卫红,吕志涛.巨型建筑结构体系的研究与展望[J].东南大学(自然科学版), 2000(4): 1–8.

［13］王肇民,邓洪洲,董军.高层巨型框架悬挂结构体系抗震性能研究[J].建筑结构学报, 1999,20(1): 23–30.

［14］马跃强.风力发电塔系统整体建模与模态分析研究[J].石家庄铁道大学学报(自然科学版),2010,23(4): 21–25。

［15］齐玉军.FRP编织网结构受力性能及设计方法研究[D].北京:清华大学, 2011.

［16］Palanichamy M S. Basic Civil Engineering[M]. 3rd ed. New York: Mc GrawHill, 2005.

［17］Narayanan R S, Beeby A W. Introduction to Design for Civil Engineers[M]. New York Rout-ledge,2001.

［18］Scott J S. Dictionary of Civil Engineering[M]. 4th ed. Penguin Books Publisher,1991.

第4章　交通与水利工程

Chapter 4　Traffic Engineering and Hydraulic Engineering

先导案例

Guide Case

　　战国时期(公元前256年)秦国蜀郡太守李冰率众修建的都江堰水利工程(见图4.1),位于四川成都平原西部都江堰市西侧的岷江上,距成都56 km。该大型水利工程至今依旧在灌溉田畴,是造福人民的伟大工程。其以年代久、无坝引水为特征,成为世界水利文化的鼻祖。这项工程主要由鱼嘴分水堤、飞沙堰溢洪道、宝瓶口进水口三大部分和百丈堤、人字堤等附属工程构成,科学地解决了江水自动分流(鱼嘴分水堤四六分水)、自动排沙(鱼嘴分水堤二八分沙)、控制进水流量(宝瓶口与飞沙堰)等问题,消除了水患。

　　都江堰有这么几个特点:

　　① 都江堰引来的水,惠及下游川西平原40多个县,10 000多平方公里,使得1 000多万亩田地旱涝保收,从此四川出现沃野千里。"天府之国"由此得名。

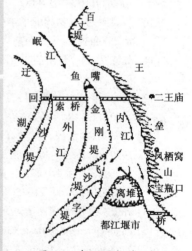

图 4.1　都江堰水利工程
Fig. 4.1　Dujiangyan Irrigation Work

　　Li Bing as prefect of Shu prefecture of Qing (256 BC) state built the Dujiangyan Irrigation Work (Fig.4.1), on the Minjiang River in the west of Dujiangyan city in Sichun, about 56 km away from Chengdu. It is a great irrigation project, and still irrigates fields. It is the originator of world irrigation culture, marked by history and diversion. The ancient project consists of three major parts, namely the Yuzui water-dividing Dike, the Feishayan Spillway and the Baopingkou Water Inlet as well as some sub-projects, like BaiZhang Dike, Renzi Dike and so on. It solved the river automatic diversion, automatic flushing and control of influent flow, which eliminated the flood.There are many features of Dujiangyan as bellow.

　　First, the water from Dujiangyan Irrigation Work benefit more than forty counties, more than ten million acres of land. The vast fertile land appears in Sichuan. Hence the name "Land of Abundance".

②都江堰虽是 2000 多年前修筑的,但至今仍然发挥功效,这在世界历史上绝无仅有。

③都江堰运用道家思想,因势利导,朴实无华。虽然修建工程耗费了十几年的时间,且以后历朝历代每年都要组织人对内江进行掏挖疏浚,但是相比于它发挥的功效,其成本收益还是比较大。

本章第一节阐述水利工程的内涵,介绍常见的水工建筑物。陆上交通分为公路和铁路,本章第二节与第三节分别介绍道路工程和铁路工程的结构组成。机场工程是为航空运输服务的一系列建筑体系的组合,本章第四节从机场的功能区划和结构组成来介绍机场工程。

4.1　水利工程

4.1.1　水资源特点

作为一种独特的自然资源,水资源有优势也存在利用缺点。由于水利工程的本质目的是发挥水资源优势并克服水资源利用时的缺点,所以在学习水利工程之前需要了解水资源的特点(见图 4.2)。

4.1.2　水利工程的内涵

(1) 水利工程的概念

水利工程是通过对自然界的水资源进行控制和调配,以实现除水害、兴水利、护水源的目的而兴建的工程(见图 4.3)。

Second, it has been playing a role in flood irrigation for over two thousand years, which is unprecedented in the history of the world.

Third, based on the Taoism, Dujiangyan improve the occasion. Although it took more than ten years to build, the cost is relatively small, compared with the effect it played.

The first part introduces the content of hydraulic engineering. The land transportation includes two modes of traffics, the road and the railway. The second and third part introduce the content of road engineering and railway engineering respectively. The airport engineering is a composite construction system in the service of aviation transportation, which will be shown in the fourth part.

4.1　Hydraulic engineering

4.1.1　The characteristics of water resources

As a kind of unique natural resources, water resources have advantages and disadvantages. The essential purpose of hydraulic engineering is enhancing the advantages of water resources and overcoming the disadvantages of water resources utilization. Therefore, we need to comprehend the characteristics of water resources before learning hydraulic engineering (Fig.4.2).

4.1.2　Connotation of the hydraulic engineering

(1) The conception of hydraulic engineering

Projects, which are used to control and redeploy water resources for mitigating water disasters, promoting water conservation and protection of water resources, protecting the environment of water resources, are called hydraulic engineering(Fig.4.3).

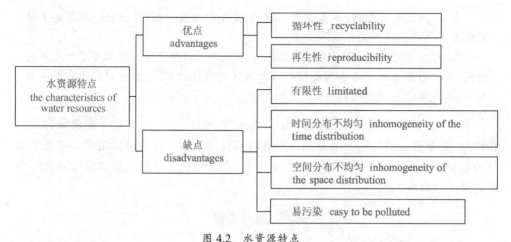

图 4.2 水资源特点
Fig. 4.2 The characteristics of water resources

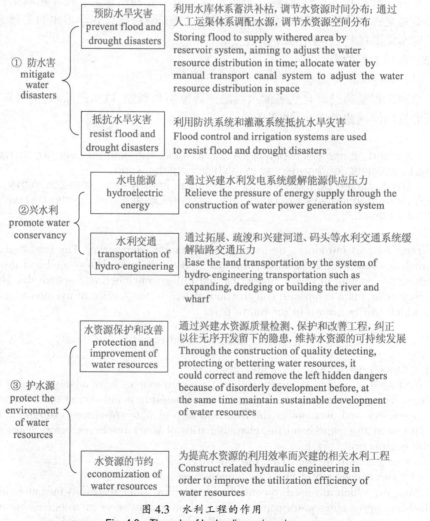

图 4.3 水利工程的作用
Fig. 4.3 The role of hydraulic engineering

(2) 水利工程的层次

水利工程按其规模自下而上可分为水工建筑物、水利枢纽、体系化综合水利工程 3 个层次：① 水工建筑物是水利工程中采用的各种单体建筑物的统称。② 水利枢纽是在水系的某一区域内由若干个不同功能的水工建筑物协同工作所组成的有机综合体。③ 体系化综合化水利工程是若干个水利枢纽和水工建筑物及天然水源有序结合组成的综合性、跨区域工程，如三峡工程、南水北调工程。

4.1.3　水工建筑物

(1) 水工建筑物分类

水工建筑物按功能可以分成 6 类(见图 4.4)。

① 挡水建筑物，即用来拦截水流、抬高水位及调节蓄水量的建筑物。

② 泄水建筑物，即用于宣泄水库、渠道的多余洪水，排放泥沙和冰凌的建筑物。

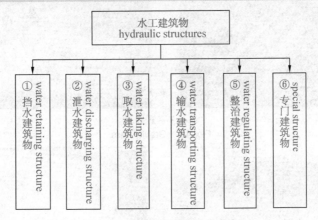

图 4.4　水工建筑物按功能分类
Fig. 4.4　The functional classification of hydraulic structures

(2) The level of hydraulic engineering

According to its scale, hydraulic engineering is classfied into three levels. ① Hydraulic structure is the single building in hydraulic engineering. ② Hydro-junction is complex which is made up of several hydraulic structures with different functions.③ Integration and systematism hydraulic engineering is the comprehensive and inter-regional engineering which is made up of several hydraulic structures and hydro-junctions, such as Three Gorges Dam and South-to-North water transfer project.

4.1.3　Hydraulic structures

(1) The classification of hydraulic structures

According to its function,hydraulic structures can be divided into six categories (Fig.4.4).

① Water retaining structure is used to intercept water, raise the water level and regulate the water storage.

② Water discharging structure is used to discharge extra flood in reservoirs and channels, and let out sediment and ice.

③ 取水建筑物,即用以从水库或河流引取各种用水的水工建筑物。

④ 输水建筑物,即用以将水输送到用水地的水工建筑物。

⑤ 整治建筑物,即用以改善河流的水流条件、稳定河槽及为防护河流、水库、湖泊中的波浪和水流对岸坡冲刷的建筑物。

⑥ 专门建筑物,即为灌溉、发电、过坝等需要兴建的建筑物。

(2)常见的水工建筑物

① 坝。大坝按坝体材料可分土石坝和混凝土坝。混凝土坝又可细分为重力坝(见图 4.5 所示)、拱坝(见图 4.6)和支墩坝(见图 4.7)。

图 4.5　重力坝
Fig. 4.5　Gravity dam

图 4.6　拱坝
Fig. 4.6　Arch dam

③ Water taking structure is a hydraulic structure that is used to take all kinds of water from reservoirs or rivers.

④ Water transporting structure is a hydraulic structure that is used to transport water to where water is needed.

⑤ Water regulating structure is used to improve river conditions, stable channels as well as protect the slope from erosion of rivers, reservoirs, waves and currents in the lakes.

⑥ Special structure is used to irrigate, generate power, or pass the dam and so on.

(2) Common hydraulic structures

① The category of a dam includes earth-rock dam and concrete dam by material. The category of concrete dam includes gravity dam (Fig.4.5), arch dam (Fig.4.6) and buttress dam(Fig.4.7).

大坝是重要的挡水建筑物,有时也兼有泄水功能,即通过坝顶溢流(允许坝顶溢流的大坝称为溢流坝)和坝身泄水孔泄水。

② 河岸式溢洪道。溢流坝设有溢流通道,但当坝体不适合溢流,或不能满足大量泄洪要求时,需在坝体外的河谷两岸适当位置单独设置溢洪道,称为河岸式溢洪道。河岸式溢洪道形式多样,图 4.8 所示为井式溢洪道。

图 4.7   支墩坝
Fig. 4.7   Buttress dam

图 4.8   井式溢洪道
Fig. 4.8   Shaft spillway

The dam is one of the most important water retaining structures. The dam also sometimes combines drainage function by crest overflow and its draining hole. The dam that allows crest overflow is called overflow dam.

② Overflow dam has its overflow channels. While the dam type is not suitable for overflow or cannot meet the requirements of large amount of flood discharging, spillway is set up separately at suitable location on both sides of the valley out of dam. That is called chute spillway. Chute spillway has diverse forms. Fig.4.8 shows a shaft spillway.

③ 堤。沿江河、沟渠、湖、海岸、水库等边缘修筑的挡水建筑物称为堤,图4.9所示为河堤。此外,在引水枢纽中还有一种导流堤,其作用是将河流导入取水建筑物。

④ 渡槽、倒虹吸管、水工涵洞和水渠。

渡槽是用以输送渠道水流跨越河渠、溪谷、洼地和道路的架空水槽,见图4.10。渡槽两端与渠道连接,是一种重要的输水建筑物。

倒虹吸管是用以输送渠道水流穿过河渠、溪谷、洼地、道路的压力管道(见图4.11)。倒虹吸管两端与渠道连接,也是一种重要的输水建筑物。

图 4.9   河堤
Fig. 4.9   Dike

图 4.10   渡槽
Fig. 4.10   Aqueduct

③ Water retaining structures at the edge of rivers, ditches, lakes, coasts and reservoirs are called dikes(Fig.4.9). In addition, there is a kind of diversion dike in water diversion hydro-junction, which can guide the water flow into water taking structures.

④ Aqueduct, inverted siphon, hydraulic culvert, ditch.

图 4.11   倒虹吸管
Fig. 4.11   Inverted siphon

Aqueduct(Fig.4.10) is an overhead flume used to transport water of ditches across rivers, valleys, depressions and roads. Both sides of aqueduct are connected with ditches, and aqueduct is an important water transporting structure.

Inverted siphon is a penstock used to transport water of ditches across rivers, valleys, depressions and roads (Fig.4.11). Both sides of inverted siphon are connected with ditches, and it is an important water transporting structure.

水工涵洞是公路或铁路与沟渠相交的地方,使水从路下通过的水工建筑(见图4.12)。水工涵洞是一种常见的输水建筑物。

水渠是最常见的输水建筑物,是具有自由水面的人工水道(见图4.13)。

⑤ 其他。其他常见的水工建筑物还有水闸、水工隧洞等。水闸是一种利用闸门挡水和泄水的低水头水工建筑物。水工隧洞是穿山开挖建成的封闭式输水道。

图 4.12　水工涵洞
Fig. 4.12　Hydraulic culvert

图 4.13　水渠
Fig. 4.13　Ditch

Hydraulic culvert is a hydraulic structure at the intersection of roads (or railways) and ditches(Fig.4.12). It is a common transporting structure.

Ditch is the most common transporting structure, and it is the man-made ditches with free water surface(Fig.4.13).

⑤ Other common hydraulic structures include sluice, hydraulic tunnel, etc. Sluice is a low head hydraulic structure, retaining water and draining water by gate. Hydraulic tunnel is a closed water pipeline building through the mountains.

4.1.4 水利枢纽分类

水利枢纽按功能可分为蓄水枢纽、引水枢纽和泵站枢纽,如图 4.14 所示。

① 蓄水枢纽。蓄水枢纽的主要水工建筑物的组成和功能如图 4.15 所示。除此之外,根据开发目标的不同,蓄水枢纽中还可能包含专门建筑物,以满足通航和水力发电的需求。

为合理、有效地利用水资源,蓄水枢纽中也可能包含一些整治建筑物。

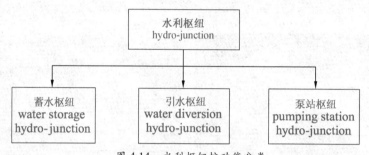

图 4.14 水利枢纽按功能分类
Fig. 4.14 The functional classification of hydro-junction

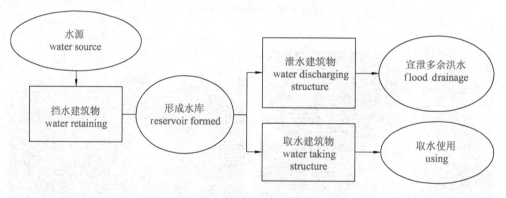

图 4.15 蓄水枢纽的组成及功能
Fig. 4.15 Parts and functions of water storage hydro-junction

4.1.4 The classification of hydro-junction

According to its function, the hydro-junction can be classified into water storage hydro-junction, water diversion hydro-junction and pumping station hydro-junction, as shown in Fig.4.14.

① Water storage hydro-junction. Parts of water storage hydro-junction and its functions are shown in Fig.4.15. Besides, special structures which meet the demands of navigation and water power generation according to the difference of development targets are included.

In order to use water resources reasonably and effectively, regulating structures are included in water storage hydro-junction.

　　② 引水枢纽。引水枢纽包括有坝引水枢纽和无坝引水枢纽,处于水渠系统和自然水源交汇的渠首河段。无坝引水枢纽的主要水工建筑物组成和功能如图 4.16 所示。除此之外,根据开发目标的不同,引水枢纽中还可能包含专门建筑物,以满足通航和水力发电的需求。为合理、有效地利用水资源,引水枢纽中也可能包含一些整治建筑物。

　　③ 泵站枢纽。为将低处水抽送到高处的水工建筑物综合体称为泵站枢纽,多以泵站及水闸为主体。

4.1.5　水利工程的设计标准

　　任何工程建筑都必须兼顾安全性和经济性要求,水利工程也不例外。安全性和经济性是一对矛盾体,为了正确处理好两者的关系,分等定级系统必须引入水利工程的设计中。

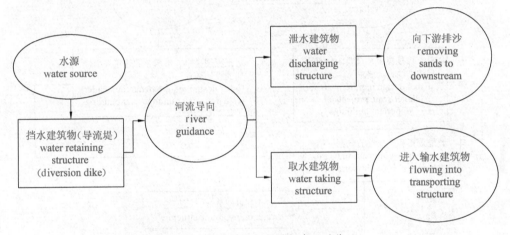

图 4.16　无坝引水枢纽的组成及功能
Fig. 4.16　Parts of no-dam water diversion hydro-junction

　　② Water diversion hydro-junction. Water diversion hydro-junction can be divided into no-dam and dam water diversion hydro-junction. It lies in the intersection of river and canal which is called canal head. Fig.4.16 introduces the main parts and functions of a no-dam water diversion hydro-junction. Besides, the existence of special structures can meet the demands of navigation and water power generation depending on the different development targets. In order to use water resources reasonably and effectively, regulating structures are included in water diversion hydro-junction.

　　③ Pumping station hydro-junction. Pumping station hydro-junction is a complex structure used to pump water to higher places.

4.1.5　Design standard of hydraulic engineering

　　Safety and economy should be considered at the same time in any engineering construction, and there is no exception in hydraulic engineering. Safety and economy are a pair of contradictory entity. In order to correctly handle the relationship between safety and economy, ranking and grading system must be initiated into the design of hydraulic engineering.

　　水利工程设计时,需将水利枢纽工程分等,并在此基础上对水工建筑物分级(见图4.17所示)。对不同等级的水工建筑物,按照不同的设计要求进行设计,以达到既安全又经济的目的。

4.2　道路工程

　　道路是一种带状的三维空间人工构筑物(见图4.18、图4.19),它包括路基、路面、桥梁、涵洞、隧道等结构实体。道路的设计一般从几何和结构两大方面进行考虑。道路的结构设计一般要求用最小的投资,使其在自然力及车辆荷载的共同作用下,在使用期限内保持良好状态,满足使用要求。

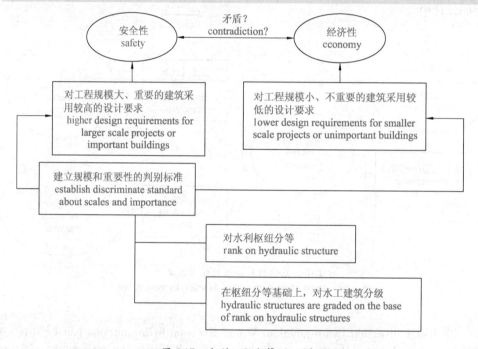

图4.17　水利工程分等、级示意图
Fig. 4.17　Schematic diagram of ranking and grading of hydraulic engineering

It is necessary to rank hydro-junction project, and on this basis, we can rank on hydraulic structure, as shown in Fig.4.17, in order to achieve the goal of safety and economy, different ranking hydraulic structure is designed according to different design requirements.

4.2　Road engineering

The road is a banded three-dimensional space of artificial structure, which includes subgrade, pavement, bridge, culvert tunnels and other structural entities (Fig. 4.18 and Fig.4.19). Usually, the design of the road proceeds in two points, namely, geometry and structure. Under the combined action of natural force and vehicle loading during service life, the road structure is hoped to be in good state with minimized investment to satisfy the operating requirements.

4.2.1　道路路基

　　路基是行车部分的基础,它由土、石按照一定尺寸、结构要求建筑成带状的土工结构物。道路路基的横断面一般有 3 种形式:路堤、路堑和半填半挖,如图 4.20 所示。路基的几何尺寸由高度、宽度和边坡组成。其中,路基高度由路线纵断面设计确定,路基宽度根据设计交通量和公道路等级而定,路基边坡根据路基整体稳定性的要求而定。

图 4.18　高速公路
Fig. 4.18　Expressway

图 4.19　市政道路
Fig. 4.19　Municipal road

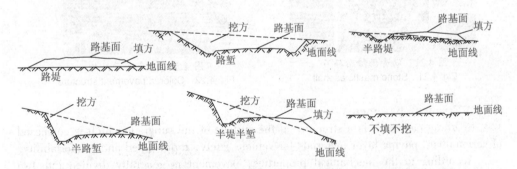

1—路堤 embankment; 2—路基面 subgrade surface; 3—填方 fill; 4—地面线 ground line;
5—挖方 excavation; 6—路堑 cutting; 7—半路堤 half of the embankment; 8—半路堑 half of the cutting;
9—半堤半堑 half cutting and half filling; 10—不填不挖 unfilling and unexcavation

图 4.20　路基的断面形式
Fig. 4.20　Cross section of subgrade

4.2.1　Roadway subgrade

　　Subgrade is a banded geotechnical structure which consists of soil and stone according to certain size and structural requirements. Subgrade is the foundation of driving. There are three kinds of forms about the cross-section of subgrade: embankment, cutting and cut-and-fill, as shown in Fig.4.20. The geometric parameters of subgrade include height, width and slope. The height is determined by the route longitudinal profile design, the width is determined by the design of traffic volume and road grade, and the slope is determined by the requirements of embankment stability.

4.2.2 道路路面

道路路面是用各种坚硬材料分层铺筑而成的路基顶面的结构物，以供车辆安全、迅速和舒适地行驶。

路面按其力学性能主要分为柔性路面和刚性路面两大类。此外还有一类是半刚性路面，由无机结合料(水泥、石灰)、水硬性材料(稳定土、砂、砾石)和工业废料(粉煤灰、矿渣等)组成，此类材料后期强度增长较快，最终强度比柔性路面强度高，但比刚性路面低。

近年来发展起来的沥青玛蹄脂碎石(stone mastic asphalt,SMA)路面和彩色路面也引起越来越多的关注(见图4.21，图4.22)。

SMA是一种新型沥青混合料，由沥青、纤维稳定剂、矿粉和少量细集料组成的沥青玛蹄脂填充间断级配粗集料而成。这种结构能全面提高沥青混合料和沥青路面的使用性能，从而减少维修养护费用，延长使用寿命。彩色路面具有美观，与景观搭配，且车流引导效果好等优点。

图 4.21 沥青玛蹄脂碎石
Fig. 4.21 Stone mastic asphalt

图 4.22 彩色路面结构
Fig. 4.22 Colored pavement structure

4.2.2 Roadway pavement

Roadway pavement is a structure at the surface of the subgrade which is composed of various hard paving layer materials for vehicle safety, rapid speed and comfortability.

According to the mechanical properties, pavement is generally divided into two categories, namely, flexible pavement and rigid pavement. There is another type named as semi-rigid pavement, which mainly includes inorganic binder (cement, lime), hydraulic material (stabilized soil, sand, gravel) and industrial waste (such as fly ash, slag). The strength of this pavement grows faster in the late period, and its ultimate strength is stronger than that of flexible pavement, but lower than that of rigid pavement。

SMA (stone mastic asphalt, called SMA for short) pavement and colored pavement, which have developed in recent years (Fig.4.21 and Fig.4.22), have attracted more and more attention.

SMA is a new kind of asphalt mastic filled with novel asphalt mixture gradation of coarse aggregate, which is composed of asphalt, fiber stabilizer, powder and a small amount of fine aggregate. This structure improves the performance of asphalt mixture and asphalt pavement, and reduces the maintenance cost with the extension of the service life. Colored pavement has the advantages of beauty, landscape collocation and efficient guidance of traffic flow.

4.2.3　道路排水结构

为了确保路基稳定,免受地面水和地下水的侵害,道路还应修建专门的排水设施。地面水的排除系统按其排水方向不同,分为纵向排水和横向排水。纵向排水有边沟(见图 4.23)、截水沟和排水沟等结构物;横向排水有桥梁、涵洞(见图 4.24)、路拱、过水路面、透水路堤和渡水槽等结构物。

随着高等级公路建设的迅速发展,路面结构的内部排水系统显得日益重要。路面内部水的排出是一个完善的系统工程,一方面要加强路面结构本身的防水措施、提高路面材料的抗水害能力,另一方面要加强内部排水设施设计。二者是相辅相成的,只预防不排出或只排出不预防,均解决不了路面内部水对路面结构的损害。

图 4.23　道路边沟
Fig. 4.23　Roadway side ditch

图 4.24　涵洞
Fig. 4.24　Culvert

4.2.3　Roadway drainage structure

In order to ensure the stability of the subgrade and protect it against the surface water and groundwater, drainage facilities should be built specifically. According to the different drainage directions, surface water exclusion system is divided into longitudinal drainage and transverse drainage. Longitudinal drainage includes the side ditch (Fig. 4.23), the intercepting ditch, the drainage ditch, etc., while the transverse drainage consists of bridges, culverts(Fig.4.24), roadway crowns, overflow pavement, permeable embankment, flume, etc.

With the rapid development of the construction of high level highways, internal drainage system in the pavement structure becomes more and more important. Internal water discharge is a perfect system construction. We should enhance the waterproof measures of pavement structure and strengthen the moisture resistance of pavement material. On the other hand, we should strengthen the interior drainage design. The two aspects are complementary, any one cannot solve the damage to pavement structure without combining the two.

4.2.4　道路特殊结构物和附属结构

道路的特殊结构物有隧道、高架桥、悬出路台、防石廊、挡土墙和防护工程等,如图 4.25 所示。

一般在道路上,除了上述各种基本结构外,为了保证行车安全、便捷还需要设置交通管理设施(见图 4.26)、服务设施和环境美化设施等附属结构。

(a) 公路隧道
Vehicular tunnel

(b) 高架桥
Viaduct

图 4.25　公路特殊结构物
Fig. 4.25　Special structures of highway

(a) 标识立柱
Identifications of the colum

(b) 路面标线
Pavement marking

图 4.26　交通管理设施结构
Fig. 4.26　Construction of traffic management facilities

4.2.4　Special structure of Roadway and accessories

Special structures of roadway include tunnel, viaduct, hanging out embankment, anti-stone veranda, retaining wall and protection engineering etc, as shown in Fig.4.25.

In order to ensure the traffic safety and convenience, other accessories like traffic management facilities (Fig.4.26), service facilities, landscaping facilities are required, besides the basic structures mentioned above.

交通安全设施是为了保证行车安全和发挥道路作用,在各级道路的急弯、陡坡等路段设置的安全设施,如护栏、护网、护柱,如图 4.27 所示。

4.2.5　高速公路

高速公路的建设情况反映着一个国家和地区的交通发达程度,乃至经济发展的整体水平,如图 4.28 所示。我国拥有世界上路程最长的高速公路,总里程超过 150 000 km(2020 年)。

目前国际上对高速公路还没有一个公认的定义。我国对高速公路的基本定义如下:

① 一般能适应 120 公里/小时或者更高的速度,路面有 4 个以上车道的宽度;

② 中间设置分隔带,采用沥青混凝土或水泥混凝土高级路面,设有齐全的标

(a) 护栏 Guardrail　　　(b) 护网 Protective screen　　　(c) 护柱 Guard post

图 4.27　交通安全设施结构
Fig. 4.27　Construction of traffic safety facilities

(a) 高速公路 (中国)　　　(b) 不限时高速公路 (德国)　　　(c) 高速公路立交 (美国)
Expressway(China)　　　Unlimited expressway(Germany)　　　Expressway interchange (America)

图 4.28　世界各地的高速公路
Fig. 4.28　The expressway around the world

Traffic safety facilities (Fig.4.27), such as guardrail, protective screen and guard post, play important roles in the driving safety of roadway. They are usually set up at the sharp turn and steep slope.

4.2.5　Expressway

The development of expressway reflects the traffic development of a region or country, and even the overall level of economic development (Fig.4.28). The amount of expressway in China is the largest all over the world, which exceeded 150 000 km in 2020.

At present, expressway has not been defined officially. In China, expressway is defined as follows:

① the surface of the expressway uses asphalt and cement concrete which will be adapted to the speed of 120 km/h or more, and the road has more than four lanes with

志、标线、信号及照明装置;

③ 禁止行人和非机动车在路上行走,与其他线路采用立体交叉、行人跨线桥或地道的形式交汇。

4.3 铁路工程

铁路是供火车等交通工具行驶的轨道。铁路运输是一种陆上运输方式,以机车牵引列车在两条平行的铁轨上行走。广义的铁路运输工具包括磁悬浮列车、城市轻轨和地下铁路等,或称轨道交通。铁路运输的最大优点是运量大、安全可靠、速度快、成本低、对环境污染小,基本不受气候影响,能耗远远低于航空和公路运输,是现代运输体系中的主干力量。

铁路按照牵引动力分为电力牵引、内燃牵引及蒸汽牵引 3 种;按照轨距分为标准轨距铁路(轨距为 1 435 mm)、宽轨铁路(1 600 mm)和窄轨铁路(762 mm 或 1 000 mm)3 种;按任务和运量各国铁路一般分为若干等级,有些国家的铁路分为干线、支线和山区线,中国国家铁路划分为Ⅰ级、Ⅱ级、Ⅲ级及地方铁路。

轨距、牵引动力种类和铁路等级不但体现铁路的性能,而且也决定铁路上各种建筑物的标准、总的工程投资和运营支出。

the separation zone;

② indicating devices are set up on the road such as mark line, signal and lighting devices;

③ pedestrians and non-motor vehicles are forbidden on the road, people and cars use grade separation pedestrian flyover and tunnel to cross it at intersections.

4.3 Railway engineering

The railway is a track for the smooth running of trains and other transportation means. Railway transportation is one of modes of land transportation, dragging the trains to travel on two parallel tracks by the engine. Generally, railway transportation, also called rail transit, includes maglev, urban light rail and underground railway etc. The railway transportation has the advantages of abundant transportation volume, high security, high speed, low cost, less environmental pollution, little weather effect, with lower energy consumption than aviation and highway transportation. It is the main force in the transportation system nowadays.

Railway can be divided into three types according to dragging power, namely, electric power drag, combustion drag and steam drag. It can also be divided into three types according to the rail distance, namely, standard-rail-distance railway(the rail distance is 1 435 mm), wide-rail-distance railway (the rail distance is 1 600 mm) and narrow-rail-distance railway (the rail distance is 762 mm or 1 000 mm). Railway can be classified into various classes according to the mission and transportation volume, such as main line, branch line and mountain line in some countries, as well as Ⅰ-class-railway Ⅱ-class-railway Ⅲ-class-railway and regional railway.

The rail distance, the kind of dragging power and the rank of railway not only affect the performance of the railway, but also determine the standard of buildings beside the railway, the total investment and service cost.

4.3.1　基本结构组成

铁路是由线路、路基、线上结构三部分构成的。此外,属于铁路工程的还有桥梁、涵洞、隧道、车站设施、机务设备、电力供应等结构设施。

我国铁路的等级和主要技术指标见表4.1。

4.3.2　路基及断面形式

铁路路基是一种土石结构,承受并传递轨道重力及列车动态作用,是轨道的基础,也是保证列车运行的重要建筑物。路基处于各种地形地貌、地质、水文和气候环境中,有时还遭受各种灾害,如洪水、泥石流、崩塌、地震、氯盐等有害化学作用及冻融作用等。

表 4.1　我国铁路等级和主要技术指标
Tab. 4.1　The railway grades and major technical indexes

等级 Grade	I	II	III
路网中的作用 function in the route net	骨干 backbone	骨干、联络、辅助 backbone, contact or assistance	地区性 regional
远期年客货运量/万吨 forward in the traffic/10^4 t	≥1 500	750 ~ 1 500	≤750
最高行车速度/(km/h) the highest speed/(km/h)	120	100	80
最大坡度/‰ the biggest gradient/‰	6(12)	12(15)	15(20)
最小(转弯)半径/m the shortest radius /m	1 000(400 ~ 350)	800(350 ~ 300)	600(300 ~ 250)

注:括号外数值为一般地段,括号内数值为困难地段。

Note：The value of common areas is outside the brackets, and the parentheses values are for difficult areas.

4.3.1　Basic components of the structure

The railway consists of route, roadbed and structures on the line. Besides, structures and facilities such as bridges, culverts, tunnels, station facilities, maintenance equipment, power supplying system also belong to the railway construction.

The grades and the main technical indexes of railway are shown in the Tab.4.1.

4.3.2　Subgrade and the forms of fracture surface

R ailway subgrade is a kind of soil-rock structure, enduring and passing orbital gravity and dynamic effect from trains, which is the basis of the orbit, as well as the important structure to ensure the normal operation of trains. Subgrade is subject to various kinds of topography, geology, hydrology and climate environment, sometimes also suffering from all kinds of disasters, such as floods, landslides, collapse, earthquake, harmful chemical action such as chlorine salt and freeze-thaw process.

（1）铁路路基的断面形式

铁路路基断面形式有路堤、半路堤、路堑、半路堑、半填半挖等，与公路路基断面形式基本相同，如图4.29所示。

（2）路基的稳定性

路基的稳定性是指路基抵抗列车动荷载及各种自然力影响所出现的道砟陷槽、翻浆冒泥和路基剪切滑动与拱起等作用的能力。道路(铁路)设计中必须对路基的稳定性进行验算。

4.3.3　线上结构

（1）轨枕

轨枕又称枕木，是铁路配件的一种。轨枕既要支承钢轨，又要限定钢轨的位置，还要把钢轨传递来的巨大压力传递给道床，因此它必须具备一定的柔韧性和弹性，列车经过时，轨枕可以适当变形以缓冲压力。

路堤 Embankment　　路堑 Cutting　　半路堑 Half cutting

图 4.29　铁路路基的断面形式

Fig. 4.29　The forms of fracture surface in railway

（1）The forms of fracture surface in railway

The forms of fracture surface in railway include embankment, half embankment, cutting, half cutting, half filling and half digging, etc., like the forms of fracture surface in highway, as shown in the Fig.4.29.

（2）The stability of roadbed

The stability of roadbed refers to its resistance ability of the collapse of road, mud pumping and shear sliding and arching of roadbed affected by dynamic load from trains and all kinds of natural force. The stability of roadbed is supposed to be calculated in the design.

4.3.3　Structures on the line

（1）Rail sleeper

Rail sleeper is a kind of railway accessories. Rail sleeper not only supports and limits the rail, but also transfers the great pressure from the steel rail to the bed of the rail, so it must have certain flexibility and elasticity to provide some appropriate deformation to buffer the pressure while the train is passing.

轨枕分为木枕、钢筋混凝土轨枕(见图 4.30)和预应力轨枕(见图 4.31)。

(2) 钢轨

钢轨的作用在于引导机车车辆的车轮前进,承受车轮压力,并将其传递到轨枕上,如图 4.32 所示。在电气化铁道或自动闭塞铁路区段,钢轨还可兼做轨道电路之用。

钢轨的断面形状采用具有最佳抗弯性能的工字形断面,有轨头、轨腰、轨底三部分组成。

我国铁路上使用的钢轨有 75,60,50,43,38 kg/m 等几种(见图 4.33)。钢轨的标准长度为 12.5 m 和 25.0 m 两种,特重型、重型轨采用 25.0 m 的标准长度钢轨。近年来大力发展的时速不小于 250 km/h 客运专线的钢轨标准长度为 100 m。

图 4.30 钢筋混凝土轨枕
Fig. 4.30 Reinforced concrete rail sleeper

图 4.31 预应力轨枕
Fig. 4.31 Prestressing force rail sleeper

图 4.32 钢轨
Fig. 4.32 Steel rail

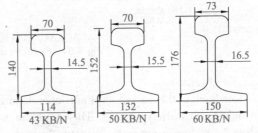

图 4.33 43,50,60 kg 钢轨截面图(单位:mm)
Fig. 4.33 Sectional view of steel rail of 43,50 and 60 kg

Sleepers are divided into wooden sleepers, reinforced concrete sleeper (shown in Fig. 4.30) and prestressing force rail sleeper(Fig. 4.31).

(2) Steel rail

Steel rail can guide the wheels, bear the pressure from the wheel and transfer the pressure to the rail sleeper (Fig.4.32). Besides, it can be used as the track circuit in electrified railway or automatic block railway section.

I-beam is adopted in the shape of the steel rail section, which consists of rail head, rail waist and rail base. I-beam has the optimal bending capacity.

The types of steel rail applied in China are 75,60,50,43,38 kg/m(Fig.4.33), etc. The standard length of steel rail is 12.5 m and 25.0 m. Extra-heavy-mode and heavy-mode adopt the 25.0 m standard steel rail. In recent years, the 100 m standard steel rail is applied in the passenger transport line whose speed is over 250 km/h.

（3）道床

道床通常指的是铁路轨枕下面，路基面上铺设的石碴（道碴）垫层，其主要作用是支撑轨枕，把轨枕上部的巨大压力均匀地传递给路基面，并固定轨枕的位置，阻止轨枕纵向或横向移动，大大减少路基的变形，同时缓和机车车轮对钢轨的冲击，还具有排水功能。

道床分为普通有碴道床、沥青道床和混凝土整体道床。有碴道床通常由具有一定粒径、级配和强度的硬质碎石堆集而成。沥青道床是为了改善普通石碴道床的散体特性而加入乳化沥青或沥青砂浆的结构形式。整体道床多为现浇钢筋混凝土结构，常用于高铁、不易变形的隧道内或桥梁上（见图 4.34）。

高速铁路采用的无碴轨道是整体道床的一种，它是以混凝土或沥青砂浆取代散粒道碴道床而组成的轨道结构形式（见图 4.35、图 4.36），具有轨道稳定性高，刚度均匀性好，结构耐久性强和维修工作量显著减少等特点。

(a) 有碴道床
Ballast bed

(b) 混凝土整体道床
Concrete ensemble bed

图 4.34　铁路道床
Fig. 4.34　Roadbed of railway

（3）Roadbed

Roadbed is the ballast cushion layer laying on the surface of the road foundation and below the railway sleeper. The main functions of roadbed are supporting the rail sleeper, transferring the great pressure from the rail sleeper to the surface of foundation of road, fixing position of the rail sleeper to prevent sleeper from the longitudinal or lateral movement, greatly reducing the transformation of roadbed, as well as relieving the shock to the steel rail caused by the vehicles at the same time. Furthermore, it has the function of drainage.

Roadbed is divided into ordinary ballast bed, asphalt bed and concrete ensemble bed. Ballast bed usually consists of hard rocks which have certain particle size, grading and strength. Asphalt bed is the ballast bed mixed with emulsified asphalt or asphalt mortar in order to improve the character of direct prose style of ordinary ballast bed. Ensemble bed often uses the structure of cast-in-place reinforced concrete, and it is usually used in high-speed rails and tunnel or bridges which are not easy to generate deformation(Fig.4.34).

Ballastless roadbed, used in most high-speed railways, is a kind of concrete ensemble bed, consisting of concrete or asphalt mortar instead of particulate ballast bed (Fig.4.35 and Fig.4.36). It has the advantages of high stability, perfect uniformity of rigidity, high durability of structure and great reduction of maintenance.

目前高速铁路上应用比较成熟的无砟轨道结构主要有挡肩板式、无挡肩板式、双块式轨枕埋入式、双块式轨枕压入式、框架板式等。我国通过消化、吸收与自主设计研发结合,也形成了具有自主知识产权的结构形式,并先后在秦沈客运专线、赣龙铁路和遂渝综合试验段等工点上进行了铺设,设计、制造和充填层材料等关键技术已接近世界先进水平。

(4) 道岔

道岔是一种使机车车辆从一股道转入另一股道的线路连接设备,通常在车站、编组站附近大量铺设(见图 4.37)。有了道岔,可以充分发挥线路的通过能力,即使是单线铁路,铺设道岔,修筑一段大于列车长度的叉线,也可以对开列车。

图 4.35 无砟轨道
Fig. 4.35 Ballastless roadbed

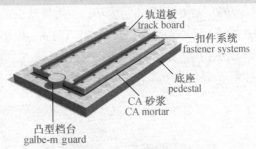

图 4.36 无砟轨道道床结构图
Fig. 4.36 Structure chart of the ballastless roadbed

(a) 普通铁路
Common railway

(b) 高速铁路
High-speed railway

图 4.37 铁路道岔
Fig. 4.37 Turnout of railway

At present, ballastless bed structure in high-speed railways consists of mainly retaining plate, non-retaining plate, double block sleepers embedded, double block sleepers indentation and frame plate. By digestion, absorption and design independently, China formed a structure with independent intellectual property rights, which used in Qinhuangdao-Shenyang passenger special line, Ganlong railway and Suiyu expressway. Part of the key technologies like design, manufacture and filling material is close to the world advanced level.

(4) Turnout

Turnout is a kind of wiring devices transferring the vehicle from one track to another, which is usually paved nearby the station and the marshalling station(Fig.4.37). It can give full play to the capacity of passing trains with turnout, even if it is a single-track railway. When a turnout, whose distance is more than that of the train, is paved, it can have the capacity of passing trains.

4.3.4 高速铁路

铁路现代化的一个重要标志是大幅度地提高列车运行速度。一般来讲,铁路的速度分级为:① 常速:时速 100~120 km;② 中速:时速 120~160 km;③ 准高速和快速:时速 160~200 km;④ 高速:时速 200~400 km;⑤ 特高速:时速 400 km 以上。

高速铁路是发达国家于 20 世纪 60—70 年代逐步发展起来的一种城市与城市之间的运输工具。

高速铁路具有载客量高、输送能力大、速度快、安全性好、正点率高、舒适方便、能源消耗低、环境影响小、经济效益好等优点。中国是世界上高速铁路发展最快、系统技术最全、集成能力最强、运营里程最长、运营速度最高、在建规模最大的国家(见图 4.38)。截至 2020 年 12 月, 中国铁路营业里程已达 14.63 万km 以上, 其中时速 200~350 km 的高速铁路将达 3.8 万 km。中国高速铁路网至少包括 5 种类型的线路:① "四纵四横"客运专线;② 城际客运系统;③ 经提速改造后的既有线、完善路网布局;④ 西部开发性新线;⑤ 海峡西岸铁路。

图 4.38　中国的高速铁路
Fig. 4.38　High speed railway in China

4.3.4 High-speed railway

An important symbol of the modernization of railway is the great improvement of train speed. Generally, the speed of railway can be divided into: ① normal speed: 100~120 km/h; ② medium speed: 120~160 km/h; ③ rapid speed: 160~200 km/h; ④ high speed: 200~400 km/h; ⑤ special high speed: more than 400 km/h.

High-speed railway is a transportation tool developed in the developed countries during 1960 s to 1970 s.

High-speed railway has the advantages of high volume, high capacity of transportation, high speed, good safety, great convenience, lower energy consumption, small environmental effect and good economic benefit. In the field of high-speed railway, China has the most abundant technology, the longest running mileage, the largest scope of construction in the world (Fig.4.38). At the end of 2020, the running mileage of high speed railway in China has been more than 146 300 kilometers, among which 38 000 kilometers can ensure the trains run at 200 to 350 km/h. China's high-speed railway network consists of 5 types. ① "Four vertical and four horizontal" high-speed rail passenger lines. ② Intercity railway passenger transportation system. ③ The speed-up existing railway. ④ The new railway for western development. ⑤ The railway for the west coast.

4.3.5　磁悬浮铁路

磁悬浮铁路是一种新型的交通运输系统，利用电磁系统产生的推力将车辆托起，使整个列车悬浮在导轨上，利用电磁力进行导向，利用直线电机将电能直接转换成动力推动列车前进(见图 4.39)。

与传统铁路相比，磁悬浮铁路由于消除了轮轨之间的接触，因而无摩擦力，线路垂直荷载小，适合高速运行。目前，磁悬浮铁路的最高试验速度为 552 km/h。磁悬浮列车及其工作原理如图 4.40 所示。

磁悬浮铁路系统由线路高架、导轨和轨道系统等结构体系和磁悬浮列车组成。高架线路结构主要有钢或钢筋混凝土梁，车站一般为钢结构形式，如图 4.41、图 4.42 所示。

图 4.39　磁悬浮列车和轨道(上海)
Fig. 4.39　Maglev train and track(Shanghai)

图 4.40　磁悬浮列车工作原理
Fig. 4.40　Working principle of maglev train

图 4.41　高架线路(钢筋混凝土)
Fig. 4.41　Elevated lines (reinforced concrete)

图 4.42　车站结构
Fig. 4.42　The station structure

4.3.5　Magnetic levitation railway

Magnetic levitation railway is a new type of transportation using electromagnetic repulsion system to suspend the entire train on rails, pushing the train with electromagnetic force, and driving the train by electrical energy converted from motor(Fig.4.39).

Compared with the traditional railway, magnetic levitation railway is frictionless because of the elimination of contact between the wheel and the rail. The vertical load acting on the line is small, which is suitable for high-speed operation. The current maximum test speed is 552 km/h. The working principle of maglev train is shown in Fig.4.40.

Magnetic levitation railway system is composed of its structure system, such as the line elevated track, rail and maglev train.

The elevated line is in the form of steel or reinforced concrete, and the station is generally in the form of steel structure(Fig.4.41 and Fig.4.42).

导轨和轨道系统主要由导轨梁、车轮支撑轨、推进线圈、悬浮和导引线圈等结构组成,如图 4.43 和图 4.44 所示。

4.3.6 城市轻轨与地下铁道

地下铁道简称地铁,狭义上专指以地下运行为主的城市铁路系统或捷运系统(见图 4.45)。广义上,由于许多此类的系统为了配合修筑的环境,可能也会有地面化的路段存在,因此通常涵盖了各种地下与地面上的轨道交通运输系统。地铁和轻轨不单纯是走地下和走地上的区别的,一般区分这两个概念要根据铁路的运输能力、车辆大小来判断。

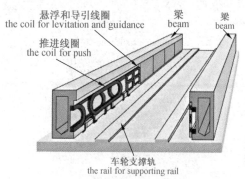

图 4.43 轨道系统结构组成
Fig. 4.43 Track system structure

图 4.44 磁悬浮导轨
Fig. 4.44 Magnetic tract

(a) 上海地铁 (3 号线)
The Shanghai subway (line 3)

(b) 重庆轻轨 (3 号线)
Chongqing light rail transit (line 3)

图 4.45 城市轨道交通系统
Fig. 4.45 City rail transit system

The rail system is mainly supported by rail beams, wheel-rails, propulsion coils, levitation and guidance coil structures, etc.(Fig.4.43 and Fig.4.44)

4.3.6 City light rail transit and subway

The underground railway (also named as subway) specifically refers to the operation of the main city underground railway system or rapid transit system (Fig.4.45). Generally, it also includes the up-ground line for compromising with the construction environment. Therefore, it covers rail transportation system of underground and on the ground. The subway and the light rail are not only classified by the locations, but also the railway transport capacity and the size of the vehicle.

地铁和轻轨作为城市轨道交通,具有噪音和干扰少、节约能源和土地、污染少、运量大、速度快、准时、促进城市发展等优点。然而,它们也有投资大、建设周期长,对水灾、火灾和恐怖主义等抵御能力弱等不足。由于经济发展和城市交通的需要,目前我国的城市轨道交通建设规模庞大,"十三五"期间,国内城市轨道运营新增里程达4 000 km。

地下铁道一般沿城市主要街道布置,在市区和郊区修建,因而施工方案的选取应充分考虑地铁对城市交通、建筑物拆迁及沿线管线的影响,即从技术、经济等方面综合考虑。地下铁道的修建方法很多,概括起来主要有两大类,即明挖法和暗挖法。

(1) 明挖法

明挖法也叫基坑法,是先将隧道部位的岩(土)体全部挖除,然后修建地下结构,再进行回填的施工方法,适用于浅埋轨道交通(见图 4.46)。

(a) 明挖法基坑开挖
The excavation with open-cut method

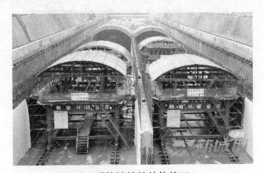

(b) 明挖法地铁结构施工
The metro construction with open-cut method

图 4.46　明挖法地下结构
Fig. 4.46　The underground structure of open-cut method

Metro and light rail, whose advantages are less noisy and interrupting, high speed and large capacity, energy-saving, pollution-reducing, and promoting the development of the city. At the same time, it needs large investment and has a long construction period, and it is weak when faced flood, fire and terrorism. Due to the demand of economic development and the city traffic, the subway construction now is proceeding in several cities. During the "Thirteenth Five-Year Plan", the new mileage of urban rail transit line reached 4 000 km.

Normally, the metro goes along the main street in the city, extending from the urban to the suburban areas, so its effect on city traffic, building demolition and utilities along the line should be fully considered in construction plan, which means making a comprehensive consideration from the point of technology, economy, and other aspects. There are several methods applied into metro construction, which can be generally divided into open-cut method and undermining method.

(1) Open-cut method

The open-cut method is also called foundation pit excavation method. At first, dig out the soil then construct the underground structure, and backfill at last(Fig.4.46).

明挖法具有施工简单、快捷、经济、安全的优点,缺点是对周围环境的影响较大。城市地铁工程发展初期都把它作为首选的开挖技术,目前常作为地铁车站的施工方法。

在城市繁忙地带修建地铁车站时,往往占用道路,影响交通。当地铁车站设在主干道上,而交通不能中断,且需要确保一定交通流量要求时,可选用盖挖法。盖挖法是指由地面向下开挖至一定深度后,将顶部封闭,其余的下部工程在封闭的顶盖下进行施工(见图 4.47)。根据工程实际情况盖挖法又可分为盖挖顺作法、盖挖逆作法和盖挖半逆作法。

(2) 暗挖法

当城市轨道交通工程埋深超过一定限度后,明挖法不再适用,而要改用暗挖法,即不挖开地面,采用在地下挖洞的方式施工。矿山法和盾构法等均属暗挖法。

① 矿山法是指主要用钻眼爆破方法开挖断面来修筑隧道及地下工程的施工方法,也是目前暗挖法中最常用的一种方法,因借鉴矿山开拓巷道的方法,故名矿

(a) 盖挖法施工图
Cut and cover method

(b) 盖挖法 (支撑柱和覆盖板)
Cut and cover method (support column and the cover plate)

图 4.47 盖挖法地下结构
Fig. 4.47 The underground structure in covered excavation

It is suitable for shallow subway, which has the advantages of simple construction, being fast, cost-saving, safe and the disadvantages of great impact on surrounding environment. In the early stage of construction development, it is deemed as a prior construction method. At present, it is usually applied to construction of metro station.

The construction of subway station in busy areas often causes the traffic problems. When the subway station is located in the main street, and the traffic cannot be interrupted, the cut and cover excavation method can be more suitable. Cut and cover excavation method is a simple construction method for shallow tunnels where a trench is excavated and roofed over with an overhead support system, and the rest structure is to be built below the roof (Fig.4.47). Three basic forms of cut-and-cover tunnel are available, namely, cover-excavation method, covered top-down excavation and cover excavation semi-inverse method.

(2) Undermining method

When the depth of metro construction exceeds a certain limit, the open-cut is no longer applicable. It is more suitable to use tunneling method. Both mining method and shield method belong to this one.

① Mining method is one of the most commonly used methods in underground engineering and tunnel construction. It uses drilling and blasting method to extract geological materials and construct tunnel subsequently. It derives from the method used in mine roadway in mine engineering.

山法。

但地铁施工多在浅部松软土层中进行,因此在传统矿山法和新奥法(见图 4.48 a)的基础上发展出了"浅埋暗挖法"(见图 4.48 b)。

采用矿山法进行暗挖施工时,隧道结构一般分为初期支护和二次衬砌。开挖之后立即进行的支护形式称为初期支护(见图 4.49 a),初期支护一般有喷射混凝土、喷射混凝土加锚杆、喷射混凝土锚杆与钢架联合支护等形式。二次衬砌(见图 4.49 b)是指在隧道已经进行初期支护的条件下,用混凝土等材料修建的内层衬砌,以达到加固支护,优化路线防排水系统,美化外观,方便设置通信、照明、监测等设施的目的。

(a) 光面爆破 (新奥法)
The smooth blasting (NATM)

(b) 台阶开挖 (浅埋暗挖法)
The step excavation (shallow tunneling method)

图 4.48　暗挖法
Fig. 4.48　Undermining method

(a) 初期支护施工
The intital support construction

(b) 二次衬砌施工
The secondary lining construction

图 4.49　衬砌结构
Fig. 4.49　Lining structure

While the subway construction is normally carried out in shallow soft soil, the "shallow tunneling method" (Fig.4.48 b) is developed based on the traditional mining method and new austrian tunneling method(Fig.4.48 a).

The tunnel structure is generally divided into initial support and secondary lining in mining tunnel. Initial support which consists of shotcrete, rock bolt, steel combined supporting (Fig.4.49 a)will be carried out immediately after soil extraction. The internal structure will be constructed by concrete after the initial support so as to support reinforcement, optimize the drainage system, smooth appearance, and being convenient to set communication, lighting, and monitoring facilities(Fig.4.49 b).

② 盾构法是暗挖法施工中的一种全机械化施工方法,它是将盾构机械在地下推进,通过盾构外壳和管片支承四周围岩以防止发生隧道坍塌,同时在开挖面前方用切削装置进行土体开挖,通过出土机械运出洞外,靠千斤顶在后部加压顶进,并拼装预制混凝土管片,形成隧道结构的一种机械化施工方法。盾构法修建如图4.50所示。

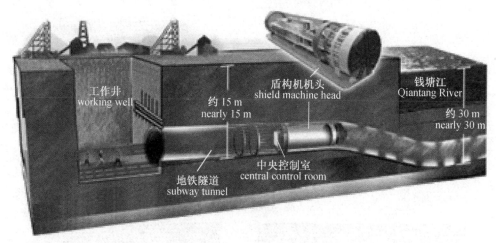

图 4.50　盾构法修建地铁隧道(杭州地铁 1 号线过钱塘江)
Fig. 4.50　Subway construction shield（Hangzhou Metro Line 1 over the Qiantang River）

② A tunneling shield is a protective structure used in the excavation of tunnels through soil that is too soft or fluid to remain stable during the time it takes to line the tunnel with a support structure of concrete. In effect, the shield serves as a temporary support structure for the tunnel while being excavated. In the front of the shield a rotating cutting wheel is located. Behind the cutting wheel there is a chamber. Depending on the type of the TBM, the excavated soil is either mixed with slurry (so-called slurry TBM) or left as-is （earth pressure balance or EPB shield). The choice for a certain type of TBM depends on the soil conditions （Fig.4.50). Systems for removal of the soil (or the soil mixed with slurry) are also present.

Behind the chamber there is a set of hydraulic jacks supported by the finished part of the tunnel which is used to push the TBM forward. Once a certain distance is excavated （roughly 1.5~2 meters), a new tunnel ring is built by the erector. The erector is a rotating system which picks up precast concrete segments and places them in the desired position.

Behind the shield, inside the finished part of the tunnel, several support mechanisms which are parts of the TBM can be found: dirt removal, slurry pipelines if applicable, control rooms, rails for transport of the precast segments, etc.

　　盾构法施工具有施工速度快、洞体质量稳定、对周围建筑物影响较小等特点,适合在软土地基段施工。常见的盾构机有泥水加压盾构、土压平衡盾构、异型盾构等(见图 4.51)。泥水加压盾构结构如图 4.52 所示。

图 4.51　盾构机
Fig. 4.51　The shield machine

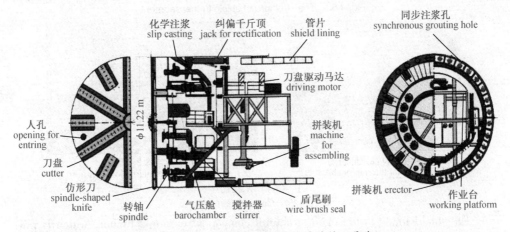

图 4.52　泥水加压盾构结构图(上海地铁 8 号线)
Fig. 4.52　Slurry shield structure diagram (Shanghai Metro Line 8)

　　Shield tunneling has the characteristics of quick construction, stable excavation face, minimal impact on the surrounding buildings. Therefore, it is suitable for soft layer. The common types of TMB are slurry shield, earth pressure balance shield and other special-shaped shield (Fig.4.51). The slurry shield structure diagram is shown in Fig.4.52.

盾构管片是盾构隧道的主要结构构件(见图 4.53),是隧道的最外层屏障和永久衬砌结构,承担着抵抗土层压力、地下水压力以及一些特殊荷载的作用。管片质量直接关系到隧道的整体质量和安全,影响隧道的防水性能及耐久性能。

4.4 机场工程

民航运输和经济发展是相辅相成的,当代的国际中心城市,如纽约、上海、巴黎、东京,无一不是重要的国际航空枢纽中心。近年来,我国的机场建设逐渐提速,截至"十三五"末,全国机场的数量达到 241 个,机场密度达到 0.25 个/万平方公里(见图 4.54)。

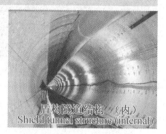

图 4.53 盾构隧道管片结构
Fig. 4.53 The structure of shield tunnel segment

(a) 北京首都国际机场
Beijing Capital International Airport

(b) 上海浦东国际机场
Shanghai Pudong International Airport

图 4.54 民航机场
Fig. 4.54 Civil aviation airport

The shield lining is the main structural component of shield tunnel, which is protective screen outer most layer of the tunnel and permanent structure to resist the soil pressure, groundwater pressure and other certain load (Fig.4.53). The overall tunnel quality, safety, waterproofing performance and durability of tunnel depend on the lining quality.

4.4 Airport project

The construction of civil aviation transportation and the development of economy are complementary to each other. International central cities such as New York, Shanghai, Paris, Tokyo are all major international aviation hubs. In recent years, airport construction in China has accelerated gradually, and at the end of "Thirteenth Five-Year Plan", the number of airport reached about 241. Airport density reached to 0.25 per ten thousand square kilometers(Fig.4.54).

4.4.1　机场的分类和组成

（1）机场的分类

机场,亦称飞机场、空港,较正式的名称是航空站,是专供飞机起降活动的飞行场。机场大小各不相同,除了跑道之外,通常还设有塔台、停机坪、航空客运站、维修厂等设施,并提供机场管制服务、空中交通管制等其他服务。机场一般分为军用和民用两大类(这里主要介绍民用机场),我国把大型民用机场称为空港,把小型机场称为航站。

按机场规模和旅客流量可将机场分为 3 种类型:国际机场、干线机场和支线机场。

（2）民航机场的组成

机场主要由飞行区、地面运输区和候机楼区 3 个部分组成。飞行区是飞机活动的区域;地面运输区是车辆和旅客活动的区域;候机楼区是旅客登记的区域,也是飞行区和地面运输区的接合部位。

① 飞行区分空中部分和地面部分。空中部分指机场的空域,包括进场和离场的航路;地面部分包括跑道(见图 4.55)、滑行道、停机坪和登机门,以及一些为维修和空中交通管制服务的设施和场地, 如机库、塔台、救援中心等。

图 4.55　跑道
Fig. 4.55　Runway

4.4.1　The classification and composition of airport

（1）The classification of airport

Airport, also known as airdrome or aerodrome, or officially the air station, is special for aircraft taking off and landing. Besides the runway, airport also includes tower, apron, aviation passenger station, maintenance factory and other facilities. It provides airport services, air traffic control and other services. The airport is generally divided into two types, military and civilian. The civil airport is mainly introduced in this chapter. The large civil airport is called aerodrome, and the small airport is called terminal in China.

According to the size and the passenger flow, the airport can be divided into international airport, trunk airport and regional airport.

（2）The composition of airport

The airport is mainly composed of flight area, ground transportation area and waiting area. The flight area is for aircraft. The ground transportation area is for vehicle and passengers, and the waiting area is the connection area of flight area and ground transportation area with the function of check-in.

① Flight area is divided into aerial part and ground part. The aerial part refers to the airport airspace, including the approach and departure route; the ground part includes runway (Fig.4.55), taxiway, apron, boarding gate, and some facilities and venues for maintenance and the air traffic control service, such as the hangar, tower, and rescue center.

② 候机楼区包括候机楼建筑本身(见图 4.56)及候机楼外的登机坪和旅客出入通道。它是地面交通和空中交通的结合部,是机场对旅客服务的中心地区。

③ 地面运输区是城市进出空港的通道(见图 4.57),大型城市为了保证机场交通的通畅,修建了从市区到机场的专用高速公路,甚至还开通地铁和轻轨交通,方便旅客出行。

4.4.2 跑道结构

跑道结构是机场飞行区的主体,直接提供飞机起飞滑跑和着陆滑跑的路径。跑道材质可以是沥青或混凝土,简易的机场跑道还可以是平整的草、土或碎石地面。跑道要有一定的长度、宽度、坡度、平坦度,以及结构强度和摩擦力等。

4.4.3 机坪与机场净空区

机场的机坪主要有等待坪和掉头坪。前者供飞机等待起飞或让路而临时停放之用,通常设在跑道端附近的平行滑行道旁边;后者则供飞机掉头使用,当飞行区不设平行滑行道时,应在跑道端头部设掉头坪。

图 4.56 候机楼
Fig. 4.56 Waiting area

图 4.57 进出机场的磁悬浮铁路
Fig. 4.57 Maglev train for in and out of the airport

② Waiting area (Fig.4.56) includes the terminal as well as the boarding ramp and the passenger channels. It is a combination of ground traffic and air traffic, as well as the central area of the airport passenger service.

③ Ground transportation area is the channel of the city for import and export(Fig. 4.57). In order to ensure the smooth traffic of the airport, most of large cities have built special highway from the urban area to the airport, and even built the subway and light rail traffic for travel convenience.

4.4.2 Track structure

Track structure is the main airport zone, directly providing service for aircraft take-off and landing. The track line can be made of asphalt or concrete, and simple airport runway can also be flat grass, soil or gravel ground. The runway must have a certain length, width, slope, flatness, structural strength and friction, etc.

4.4.3 The apron and the airfield clearance zone

The airport apron includes a waiting area and a U-turn area. The former is for the plane waiting for take-off or temporary parking, and it is usually located in the parallel taxiway next to the end of runway. The latter is for U-turn of the plane. When the flight zone is not provided with a parallel taxiway, the U-turn area is needed at the end of runway.

机场净空区是指飞机起飞着陆涉及的范围,保证在起飞和降落的低高度飞行时没有地面障碍物妨碍导航和飞行(见图 4.58)。

4.4.4 航站区

航站区主要由候机楼、站坪及停车场组成。航站楼的结构设计涉及位置、形式、建筑面积等要素。

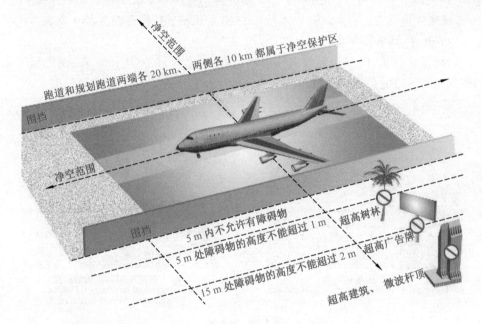

注:① 跑道和规划跑道两端各 20 km、两侧各 10 km 都属于净空保护区:
　　20 kilometers on both ends of runway and 10 kilometers on both sides of runway are
　　airfield clearance zone;
② 5 m 内不允许有障碍物: no obstacles within 5 meters;
③ 5 m 处障碍物的高度不超过 1m: The height of obstacle is less than 1m within 5 m;
④ 15 m 处障碍物高度不超过 1m: The height of obstacle is less than 1 m within 15 m;
⑤ 超高树木: super-high trees;
⑥ 超高广告牌: sky sign;
⑦ 超高建筑: super-high construction

图 4.58　机场净空区示意图
Fig. 4.58　Sketch map of airport clearance area

　The airfield clearance zone is the place where aircraft takes off and lands on, ensuring that no obstacle will hinder the ground navigation and flight when the plane is taking off and landing at low altitude as shown in Fig.4.58.

4.4.4　Terminal area

　The terminal area is mainly composed of the air-terminal, the station site and the parking lot. The structural design of the terminal involves location, form, construction area, etc.

(1) 候机楼

候机楼是航站区的主要建筑物(见图 4.59)。候机楼的设计,不仅要考虑其功能,还要考虑环境、艺术氛围及民族(或地方)风格等。候机楼一侧连着机坪,另一侧与地面交通系统相联系。其基本功能是安排好旅客、行李的流程,为其改变运输方式提供各种设施和服务,使航空运输安全有序。

旅客候机楼的基本结构设施应包括:① 车道;② 公共大厅;③ 安全检查设施;④ 政府机构;⑤ 候机大厅;⑥ 行李处理设施(行李分检系统和行李提取系统);⑦ 机械化代步设施(人行步道,自动扶梯);⑧ 登机梯;⑨ 旅客信息服务设施等。

(2) 站坪、机场停车场与货运区

站坪或称客机坪,是设在候机楼前的机坪,供客机停放、上下旅客、完成起飞前的准备和到达后的各项作业使用。

(a) 昆明机场候机楼
The terminal building of Airport in Kunming

(b) 上海浦东国际机场候机楼
The terminal building of Pudong
International Airport in Shanghai

图 4.59　候机楼
Fig. 4.59　Terminal building

(1) Air-terminal

The air-terminal is the main building of the terminal area, as shown in Fig.4.59. Its feature, the environment, the artistic and national (or local) style need to be considered when being designed. The air-terminal one side connects with the apron and the other side connects with the ground transportation system. Its fundamental function is to arrange passengers, baggage, and to provide facilities and services for their transportation to make sure the safety and order of the air transportation.

The basic structural facilities of the passenger terminal include: driveway, public hall, safety inspection facilities, federal government agencies, waiting hall, baggage handling facilities (baggage sorting system and baggage reclaim system), mechanized transport facilities (pedestrian space, escalator), airstairs, passenger information service facilities, etc.

(2) Standing platform, airport parking lot and freight area

Standing platform, also named as the passenger platform is located in front of the terminal. It's a platform for aircraft parking, passengers' alighting and boarding, the departure preparation and miscellaneous actions after arriving.

机场停车场设在机场候机楼附近。停车场建筑面积主要根据高峰小时车流量、停车比例等确定。

机场货运区供货运办理手续、飞机装卸货、临时存储等使用，主要包括业务楼、货运库、装卸场及停车场。货运区应离开旅客航站区及其他建筑物适当距离，以便将来发展。

4.5　未来展望

① 交通工程具有系统性、综合性、交叉性、社会性、超前性和动态性。未来交通工程将从五个方面展开：工程（Engineering）、法规（Enforcement）、教育（Education）、能源（Energy）和环境（Environment）。涉及技术层面的主要从智能交通、城市交通一体化枢组和地下交通网建设三个方面展开。

② 水利工程具有如下特点：一是有很强的系统性和综合性；二是对环境影响较大；三是工作条件恶劣；四是效益具有随机性；五是技术复杂。

③ 水资源配置格局优化是国家碳中和碳达峰战略的重要举措之一。国家将论证启动"朔天运河"工程，其投资规模将是三峡工程的 28 倍。无疑，这项工程具有示范性，其综合效益将十分显著。

Airport parking lot is near the the airport terminal. The construction area is mainly based on peak hour traffic flow, traffic parking ratio, etc.

Airport freight area, which provides space for cargo handling procedures, loading or unloading cargo, temporary storage service, etc., includes the freight business building, loading and unloading of field and parking lot. The freight area should keep an appropriate distance from the passenger terminal area and other buildings for its future development.

4.5　Future Prospect

① Traffic engineering is systematic, comprehensive, intercrossingity, sociality, advanced and dynamic. Future traffic engineering will be crried out from five aspects: engineering, enforcement, education, energy and environment. The technical aspects include intelligent transportation, integrated urban transportation hub and underground transportation network construction.

② Water conservancy projects possess the following characteristics: First, strong systematic and comprehensive; second, great impact on the environment; third, bad working conditions; fourth, the randomness of benefit; fifth, complex technology.

③ Optimization of water resources allocation pattern is one of the important measures of national carbon neutralization and carbon peak strategy. The state will demonstrate and start the Shuotian Canal Project, whose investment scale will be 28 times that of the Three Gorges Project. Undoubtedly, this project has demonstration character, its comprehensive benefit will be remarkable.

水利工程的发展趋势:一是防治水灾的工程措施与非工程措施进一步结合,非工程措施占据越来越重要的地位;二是水资源的开发利用进一步向综合性、多目标发展;三是水利工程的作用,不仅要满足日益增长的人民生活和工农业生产发展需要,而且要更多地为保护和改善环境服务;四是大区域、大范围的水资源调配工程,如跨流域引水工程,将进一步发展;五是由于新的勘探技术、新的分析计算和监测试验手段以及新材料、新工艺的发展,复杂地基和高水头水工建筑物将随之得到发展,当地材料将得到更广泛的应用,水工建筑物的造价将会进一步降低;六是水资源和水利工程的统一管理、统一调度将逐步加强。

The development trend of water conservancy projects is mainly as follows: first, engineering measures and non-engineering measures for flood prevention and control are further combined, and non-engineering measures are more and more important; second, the development and utilization of water resources is further integrated and multi-objective; third, the role of water conservancy projects is not only to meet the growing needs of people's life and the development of industrial and agricultural production but also to protect and improve the environment; fourth, widely regional and large ranged water resources allocation projects, such as cross-basin diversion projects, will be further developed; fifth, complex foundation and high head hydraulic structures will be developed, local materials will be more widely used, and the cost of hydraulic structures will be further reduced due to the development of new exploration technology, new analytical calculation and monitoring test means, new materials and new processes; and sixth, the unified management and operation of water resources and water conservancy projects will be gradually strengthened.

注:本章图片均来源于网络。
Note: In this chapter, all pictures are from webs.

知识拓展
Learning More

相关链接　Related Links
(1) 中国公路网
(2) 中华铁道网
(3) 中国机场网
(4) 中华人民共和国水利部官网

小贴士　Tips
(1) 李冰与都江堰 Li bing and Dujiangyan

李冰(约公元前 302 年—公元前 235 年),号陆海,战国时代著名的水利工程专家。公元前 256 年—公元前 251 年被秦昭王任为蜀郡(今成都一带)太守。其间,李冰治水,创建了奇功,其建都江堰的指导思想,就是道家"道法自然"、"天人合一"的思想。

Li Bing (from 302 BC to 235 BC), named Lu Hai, is an irrigation project expert. From 256 BC to 251 BC, Li Bing is the prefect of Shu prefecture appointed by the king of Qin. Li Bing prevented the flood by using the Taoism.

李冰 Li Bing

都江堰 Dujiangyan

（2）詹天佑和人字形铁路

詹天佑（1861—1919 年），字眷诚，号达朝，中国近代铁路工程专家。12 岁留学美国，1878 年考入耶鲁大学土木工程系，专习铁路工程。1905—1909 年主持修建我国自建的第一条铁路——京张铁路，创造"竖井施工法"和"人"字形线路，震惊中外，有"中国铁路之父""中国近代工程之父"之称。

Jeme Tien Yow and Chevron railway

Jeme Tien Yow(1861–1919), whose courtesy name is Juancheng, and styled Dachao, was a distinguished Chinese railroad engineer. Educated in Yale, he was the chief engineer responsible for construction of the Imperial Peking-Kalgan Railway （Beijing to Zhangjiakou）, which is the first railway constructed in China without foreign assistance. For his contributions to railroad engineering in China, Zhan Tianyou is still known as the "Father of China's Railroad" and "Father of engineering in modern China".

詹天佑
Zhan Tianyou

京张铁路中的"人"字形线路段
Jingzhang railway in the herringbone line segments

（3）国际机场、干线机场和支线机场

国际机场包括干线机场和支线机场，指供国际航线用，并设有海关、边防检查、卫生防疫、动植物检疫和商品检验等联动机构的机场。干线机场是指省会、自治区首府及重要旅游、开发城市的机场。支线机场是指省、自治区内地面交通不便的地方所建的机场，其规模通常较小。

International airport, main airport,regional airport

International airport can accommodate international flights, typically equipped with customs, immigration facilities, quarantine of animals and plants, commodity inspection. Main airport is an airport located in the provincial capital, the capital of the autonomous region and the important tourist cities. Regional airport is an airport located in the area where the transport is not well developed within the province and autonomous regions. Normally its size is small.

思考题　Review Questions

(1) 三峡工程中采用了哪些常见的水工建筑物？

What common hydraulic structures are used in Three Gorges Dam?

(2) 在山区修建高速公路,必须建造哪些结构设施？

What structural facilities should be built when constructing an expressway in mountain area?

(3) 高速铁路与普通铁路有哪些区别,其结构有哪些特殊要求？

What are the differences between high-speed railway and ordinary railway, and what are the special requirements of its structure?

(4) 简述我国的城市轨道交通发展现状和对轨道结构的要求。

Please make a brief account of urban rail transit development in China and its requirements on track structures.

(5) 你认为未来民用机场的结构设计应该向什么方面发展。

What do you think of the development direction of the structural design in civil airport?

参考文献
References

[1] 陈学军. 土木工程概论[M].3 版.北京:机械工业出版社,2016.

[2] Swamy H C M. Elements of Civil Engineering[M]. Laxmi Publications,2008.

[3] 叶志明.土木工程概论[M].3 版.北京:高等教育出版社,2009.

[4] 全国一级建造师执业资格考试用书编写委员会.民航机场工程管理与实务[M].北京:中国建筑工业出版社,2019.

[5] 中国交通年鉴社.中国交通年鉴[M].北京:中国交通年鉴出版社,2012.

[6] 佟立本.高速铁路概论[M].5 版.北京:中国铁道出版社,2017.

[7] 倪福全.农业水利工程概论[M].北京:中国水利水电出版社,2011.

[8] 朱宪生,冀春楼.水利概论[M].郑州:黄河水利出版社,2004.

第5章 市政工程

Chapter 5 Municipal Engineering

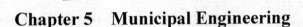

先导案例 青岛——中国最不怕淹的城市

Guide Case Tsingtao——City with the Best Waterlogging Prevention in China

2011 年的盛夏,中国接连被雨水重创。北京、武汉、长沙、成都、扬州……当许多城市被暴雨击瘫时, 青岛却表现出了不一样的风采:在经历持续降雨 21 小时、最强降雨 40 分钟、降雨量超过 100 mm 后,道路上的雨水在 10 分钟内被排得干干净净,不负"中国最不怕淹的城市"的称号。

青岛的地下排水系统为何如此神奇? 这得益于 100 多年前德国人的设计。目前青岛的部分排水管道是 1900 年德国强租胶州湾时所建。其特点如下:① 首次采用了雨污分流的排水系统;② 按区域设有排水泵站,由泵站将污水集中加压排入远海; ③ 采用的管道截面呈上宽下窄的鹅蛋形,下面较窄的部分被贴上了瓷片,如图 5.1 所示。这样的设计在

The summer in 2011 witnessed a successive hit on China by rain. When many cities, such as Beijing, Wuhan, Chengdu, Yangzhou, were shattered Tsingtao showed a different scene. Although Tsingtao experienced continuous rainfall for 21 hours, when the strongest rainfall lasted for 40 minutes, with more than 100 mm precipitation, it took about 10 minutes to clean the rain on the road. This proves that Tsingtao has the best waterlogging prevention in China.

图5.1 德国人设计的青岛排水管道
Fig. 5.1 Tsing Tao drainage pipeline designed by German

The reason for the good performance of the underground drainage system is that part of the system is built by Germans in 1900 when Germany rented Jiaozhou Bay. There are some characteristics: ① the system separates sewage and rain for the first time; ② set drainage pumping stations according to regions and by these stations sewage is intensively pressed into the sea; ③ as shown in Fig. 5.1, the pipe cross-section is egg-shaped and the narrow part is labeled with tiles.

水流较少时可保持流速和动力,淤泥也能被一起冲走;水流较多时,能够实现快速流过、迅速排水。④雨水管道中设置了一种 h 形的"雨水斗"机关(类似家中卫生间下水管的反水阀),雨水进入雨水斗后,脏物被沉淀到左边的"斗"中,而质量较轻的雨水则顺着右边的管道排走。如此一来,杂物既容易清理,也不会造成整个排水管道的堵塞。

如此匠心独用、高瞻远瞩的排水系统设计使得青岛市在 100 多年内没有出现过内涝,也避免了暴雨对城市造成威胁。

市政工程是指市政基础设施建设工程。市政基础设施是指在城市区、镇(乡)规划建设范围内设置的、基于政府责任和义务为居民提供有偿或无偿公共产品和服务的各种建筑物、构筑物、设备等的总称。市政工程是指城市建设中的各种公共交通设施和给水、排水、燃气、城市防洪、环境卫生及照明等基础设施建设,它们是城市生存和发展必不可少的物质基础,是提高人民生活水平和对外开放的基本条件。

5.1 城市道路和桥梁

公共交通是城市的大动脉,也是城市实现其效用的一个重要途径,它的发展直接影响着城市的经济发展和社会发展。城市公共交通设施主要是指道路和桥梁。

This kind of design allows that the sludge can be washed away at low flow time and water can be ejected rapidly if the flow is large. ④ An h-shape "water bucket" organ (like the counter-valve set in our bathrooms) is set in the rainwater pipes, which allows the dirty precipitated to the bucket on the left and rain water flows away at the right side. It ensures the pipes will not be jammed by the dirty which can be cleaned easily.

The special drainage system design makes Tsingtao away from waterlogging for over 100 years. It also eliminates the threat of storm to the city.

Municipal engineering refers to the municipal infrastructure project, which is the general term of buildings, structures and equipment that are set up within the scope of planning and construction in the urban area or town, in order to provide residents with paid or unpaid public services based on the government responsibility and obligation. Municipal engineering is infrastructure construction which includes all kinds of public transport facilities, water supply and drainage, gas, flood control, environmental health, lighting and so on. It is not only the indispensable basis of city survival and development, but also the basic condition of improving people's living standards and society developments.

5.1 Urban roads and bridges

Public transportation is not only the artery of a city, but also plays an important role in achieving its utility, which directly affects the city's economic and social development. Urban public transport facilities mainly refer to road and bridge.

5.1.1　城市道路

城市道路是建在城市范围内,供车辆及行人通行并具备一定技术条件和设施的道路,如图5.2所示。

城市道路按其在城市道路系统中的地位、交通功能及对沿线建筑物的服务功能分为快速路、主干路、次干路和支路,见表5.1。

① 快速路是指为较高车速的远距离交通而设置的重要城市道路。

② 主干路是指在城市道路网中起骨架作用的道路。

③ 次干路是指城市中数量较多的一般交通道路,同时具有服务功能。

④ 支路是指城市道路网中干路以外联系次干路或者供区域内部使用的道路,用以解决局部地区交通和群众的使用要求。

图5.2　城市道路
Fig. 5.2　Urban road

5.1.1　Urban road

Urban road is built within the scope of the city for vehicles and pedestrians, which has certain technical conditions and facilities, as shown in Fig.5.2.

According to its status in urban road system, transportation function and service function, the urban road can be divided into expressway, arterial road, secondary trunk road and branch road, as shown in Tab.5.1.

表5.1　城市道路分类
Tab.5.1　Classification of urban road

① Expressway is an important city road for high speed transportation over a long distance.

② The arterial road plays a major role in urban road network.

③ The secondary trunk road is the common road in urban city and plays the service function.

④ In order to meet the requirements of local transportation and the crowed, branch road is constructed to connect with secondary trunk road or used within the specific area of urban road network.

5.1.2 城市桥梁

城市桥梁指在城市范围内,修建在河道上的桥梁、道路与道路立交、道路跨越铁路的立交桥及人行天桥,包括永久性桥和半永久性桥。城市桥梁是城市道路的重要组成部分,如图 5.3 所示。

跨河桥的主要作用是跨越城市中的河流,将河流两端的道路连接起来。在设计跨河桥时需考虑道路的线型和河道排泄洪水的能力,对桥梁高度和下部的净空进行限制。

立交桥可以使平交路口的车流在不同高程上跨越,从空间上分开,各行其道,互不干扰,大大提高了车速和路口的通行能力。它可以充分利用城市空间,是大中型城市的一种重要交通方式。立交桥设计时需综合考虑路、桥之间的协调配合和桥下净空的要求。

人行天桥一般建造在车流量大、行人稠密的地段,或者交叉口、广场及铁路的上方。人行天桥只允许行人通过,用于避免车流和人流平面相交时的冲突,保障人们安全穿越,提高车速,减少交通事故。

跨河桥 River bridge

立交桥 Overpass

人行天桥 Pedestrian bridge

图5.3 城市桥梁类型
Fig. 5.3 Classifications of city bridge

5.1.2 Municipal bridge

Municipal bridges refer to bridges, overpasses, pedestrian bridge above rivers, highways and railways, including permanent and semi-permanent ones, which are the essential part of the city's transport system (Fig.5.3).

River bridge is mainly used to connect two adjacent elements between two rivers. The design of pavements and flood discharge ability should be concerned. In addition, there are limitations about the bridge height and space under the bridge.

Overpass is a crossing of two highways or railroads at different levels where clearance to traffic on the lower level is obtained by elevating the higher level. Overpass benefits citizens and relieve the traffic congestion. The combination of highway and bridge as well as the limitations of under clearance need to be considered.

Pedestrian bridge is generally set up in crowed areas, or intersections, squares and the railways. Pedestrian bridge is designed only for pedestrians to ensure citizens' safety, increase car speed, and decrease the number of traffic accidents.

5.2　城市给排水工程

　　人类的生产、生活离不开水。可用水的及时供应与废水的及时排出是城镇得以正常运转的重要保障，而实现这一目的的一个重要条件就是设置给排水管道，其工程质量不仅影响城镇功能的充分发挥，而且对人居健康、道路交通、水环境保护和城市安全都有直接的影响。城市内的水循环如图 5.4 所示。

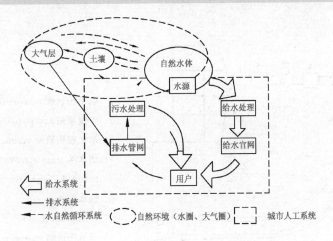

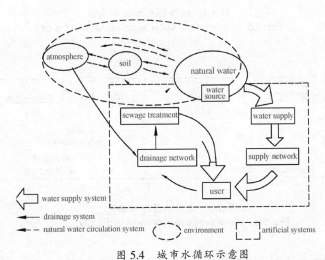

图 5.4　城市水循环示意图
Fig. 5.4　Schematic of water cycle in the city

5.2　Urban water supply and drainage engineering

　　Our daily life and industrial production are inseparable from water. The installation of drainage pipelines is one of measurements to guarantee water supply and wastewater discharge timely. The quality of project has direct relation with the health of the residence, transportation, environment and city safety. Urban water cycle is shown in Fig. 5.4.

5.2.1　城市给水工程

给水工程的基本任务是安全可靠、经济合理地供应城乡人民生活、工业生产、保安防火、交通运输、建筑工程、公共设施、军事建设等各项用水,满足用户对水量、水质和水压的要求。给水工程主要由给水水源、取水构筑物、原水管道、给水处理厂和给水管网组成,同时具有收集和输送原水、改善水质的作用,城市给水系统如图 5.5 所示,水质净化流程如图 5.6 所示。

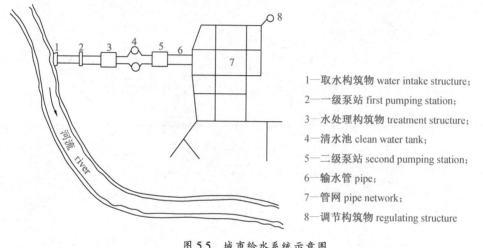

1—取水构筑物 water intake structure;

2—一级泵站 first pumping station;

3—水处理构筑物 treatment structure;

4—清水池 clean water tank;

5—二级泵站 second pumping station;

6—输水管 pipe;

7—管网 pipe network;

8—调节构筑物 regulating structure

图 5.5　城市给水系统示意图
Fig. 5.5　Schematic of water supply system in the city

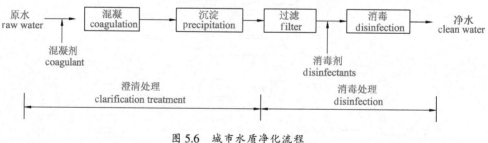

图 5.6　城市水质净化流程
Fig. 5.6　Water purification process in the city

5.2.1　Water supply engineering

The main function of water supply engineering is to ensure water supply for living, industrial production, fire security, transportation, public facilities, military construction, etc. The supply must be safe, reliable, economical. Meantime, water supply engineering should meet the local requirements about water quality and relative properties. Generally, a whole water supply engineering consists of water source, water intake structures, water pipelines, treatment plants and water distribution network, with the function of collecting water, transporting water, improving water quality. The sketch map is shown in Fig.5.5 and Fig.5.6.

5.2.2　城市排水工程

排水工程为排除人类生活污水和生产中的各种废水及多余地面水的工程,由排水管系(或沟道)、废水处理厂和最终处理设施组成,通常还包括抽升设施(如排水泵站)。

①排水管系是收集和输送废水(污水)的管网,有合流管系和分流管系两种。合流管系只有一个排水系统,雨水和污水用同一管道排出。分流管系有两个排水系统,雨水系统收集雨水和冷却水等污染程度很低、不经过处理可直接排入水体的工业废水,其管道称雨水管道;污水系统收集生活污水及需要处理后才能排入水体的工业废水,其管道称污水管道。

②废水处理厂包括沉淀池、沉沙池、曝气池、生物滤池、澄清池等设施及泵站、化验室、污泥脱水机房、修理工厂等建筑。废水处理的一般目标是去除悬浮物和改善耗氧性,有时还进行消毒和进一步处理。

③最终处理设施视不同的排水对象设有水泵或其他提水机械,将经过处理厂处理满足规定排放要求的废水排入水体或土壤。

常用的排水管渠有钢筋混凝土管、陶土管、金属管、浆砌砖石管渠等。另外,为了排除污水,除管渠本身外,还需在管渠系统上设置某些附属构筑物,包括雨水口、连接暗井、检查井、跌水井、水封井、倒虹管、冲洗井、防潮门、出水口等。排水管渠在建成通水后,为保证其正常工作,必须经常进行养护和管理。排水管渠内常见污物淤塞管道、过重的外荷载、地基不均匀沉陷或污水的侵蚀作用等现象,使管渠损坏、开裂或被腐蚀。

5.2.2　Urban drainage engineering

The drainage works are designed to eliminate wastewater and graywater, and puddles excess on the ground. Their main components are drainage system (or channels), wastewater treatment plant and disposal facilities, pumps (such as drainage pumps).

① The drainage system is used to collect and transport wastewater or sewage, which can be divided into combined systems and separate systems. Combined system is composed of a single drainage pipe to pour over rainwater and sewage together. While the separate system can be divided into two drainage systems: one is rainwater collections, which can be used to collect rainwater, cooling water, and wastewater with fewer chemical particles. The other is sewage collections, which can be used to collect wastewater, graywater after treatment.

② The wastewater treatment plant includes sedimentation pool, settling basin, aeration tank, biological filter, clarification pool and other facilities, such as pumping station, sludge dewatering room, laboratory. The general objective of wastewater treatment is to remove harmful particles and improve oxidation in water, and sometimes need disinfected and post-processing.

③ The final disposal facility. Drainage pumps and relative facilities have been designed and installed according to plans and local requirements. The processed wastewater is discharged into water or land when it meets specifications and requirements.

Sewer pipes are usually made of reinforced concrete, clay, metal, bricks, etc. Moreover, some auxiliaries have been installed for the purpose of eliminating the sewage, including rain inlets, connecting staples, inspection wells, drop wells, water wells, inverted siphon, flushing well, moisture-proof valves, and water outlets. In order to guarantee the performance, maintenance and management are needed after construction. The common failures of drainage systems are dirt blockage, external loads, uneven subsidence of foundation or wastewater erode which make pipeline damaged、cracked or eroded.

5.3 城市燃(煤)气管道和热力管道安装工程

5.3.1 城市燃(煤)气工程

城市燃气输配系统按照用气量和所需压力将燃气输送和分配到城市各类用户。城市燃气输配系统一般由门站、燃气管网、储气设备、调压设施、管理设施、监控设施组成,如图 5.7、图 5.8 所示。城市燃气输配系统的各种设施,应能满足各类用户的小时最大用气量,并能适应其波动情况。输配系统应该保证不间断地、可靠地给用户供气,且运行管理安全,维修检测方便。

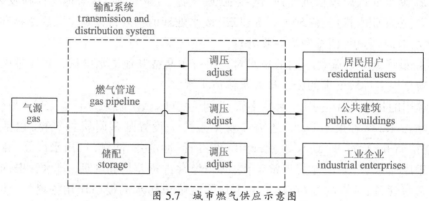

图 5.7　城市燃气供应示意图
Fig. 5.7　Schematic of gas supply system in the city

图 5.8　城市燃气管道
Fig. 5.8　Gas pipelines in the city

5.3 Urban gas pipelines and heat pipelines installation engineering

5.3.1　Urban gas supply engineering

Urban gas distribution system is designed for gas transportation and distribution in accordance with the required volume and pressure. Generally, the urban gas distribution system consists of valves, gas pipelines, storage tanks, pressure regulating facilities, management facilities and monitoring facilities, as shown in Fig.5.7 and Fig.5.8. All kinds of facilities of the urban gas distribution system should be able to ensure users' maximum hourly consumption, and can be adapted to changes of consumption. Gas distribution should be carried on continuously and efficiently.

此外,还应考虑在检修或发生故障时,可关断某些部分管段而不致影响全系统的工作。在一个输配系统中,宜采用标准化和系列化的站室、构筑物和设备,采用的系统方案应具有最大的经济效益,并能分阶段地建造和投入运行。

与其他管道相比,燃气管道有特别严格的要求,因为管道漏气可能导致火灾、爆炸、中毒等事故。燃气管道的压力越高,管道接头脱开、管道本身出现裂缝的可能性也越大。管道内燃气压力不同时,对管材、安装质量、检验标准及运行管理等要求亦不相同。我国城镇燃气管道按燃气设计压力(MPa)分为七级(见表 5.2)。

燃气输配系统各种压力级别的燃气管道之间应通过调压装置连接。当有可能超过最大允许工作压力时,应设置防止管道超压的安全保护设备。

表 5.2　城镇燃气管道设计压力分级
Tab.5.2　Ratings of designed pressure of gas pipeline in the city

名　称 Name		压　强 Pressure p/MPa
高压燃气管道 high-pressure gas pipeline	A	$2.5 < p \leqslant 4.0$
	B	$1.6 < p \leqslant 2.5$
次高压燃气管道 secondary-pressure gas pipeline	A	$0.8 < p \leqslant 1.6$
	B	$0.4 < p \leqslant 0.8$
中压燃气管道 medium pressure gas pipeline	A	$0.2 < p \leqslant 0.4$
	B	$0.01 < p \leqslant 0.2$
低压燃气管道 low-pressure gas pipeline		$p \leqslant 0.01$

The management and maintenance of gas distribution system should be considered. In addition, turning off some part of the pipes cannot affect the whole performance of the system when maintenance, repair works are needed. In a distribution system, serialized station room and equipment should be standardized. The design of system should eliminate the costs, and can be constructed and put into operation in stages.

Compared with other pipelines, gas pipelines are particularly strict in case of fire, explosion, poisoning, etc. caused by gas leakage. The pressure of gas pipelines has a direct relationship with the performance of connections and pipes. The work of installation, construction, inspection and management should keep in pace with the standards and requirements according to types and pressure. In China, gas pipeline are divided into seven levels according to the pressure(Tab.5.2).

Gas distribution system is needed to balance different pressure levels among pipes. Sometimes, safety protection device should be installed to prevent overpressure in the pipelines.

5.3.2 城市热力管道工程

热力管道是输送蒸汽或过热水等热能介质的管道,如图 5.9 所示。热力管道的特点是输送的介质温度高、压力大、流速快,在运行时会给管道带来较大的膨胀力和冲击力。因此在管道安装中应解决好管道材质、管道伸缩补偿、管道支吊架、管道坡度及疏排水、放气装置等问题,以确保管道的安全运行。

热力管道的敷设分为地上敷设和地下敷设,通常选用钢管,并尽量采用焊接连接;安装时,水平管道要具有一定的坡度。蒸汽管道的坡向最好与介质流向相同,这样管内蒸汽同凝结水流动方向相同,避免噪声。热水管道的坡向最好与介质流向相反,这样管内热水与气流方向相同,减少了热力流动的阻力,也有利于排气,防止噪声。

图 5.9 城市热力管道
Fig. 5.9 Heat pipe in the city

5.3.2 Urban heat pipeline installation engineering

Heat pipe is designed to transport steam or hot water, as shown in Fig.5.9. The characteristics of heat pipe are high temperature, high pressure and high flow rate of the transported medium which accelerate expansion force and impact force of pipes. Therefore, materials, expansion compensation, supports, the slope and drainage, bleeding device of pipelines should be taken into account for the purpose of safe operation.

The heat pipes are usually located on the ground or underground, and the pipes are made of steel. The method of connection is welding, and a certain slope is needed when horizontal pipelines are installed. The slope of the steam pipe should match the medium flow dimension, so that it can eliminate noise and make steam and water flow in the same direction. The slope of the hot water pipe should be on the opposite of the medium flowing, so that flow direction of the hot water and air are the same which could reduce thermal resistance, discharge exhaust and eliminate noise.

热力管道的每段管道最高点和最低点分别安装排气和泄水装置。方形补偿器水平安装时,与管道坡度和坡向一致;垂直安装时,最高点应安装排气阀,最低点应安装排水阀,便于排水与放气。热力管道的排水、放气装置如图 5.10 所示。

热力管道的温度变形要充分利用转角管道进行自然补偿,如图 5.11 所示。自然补偿不能满足要求时, 要加设补偿器补偿, 常用的补偿器有 L 形补偿器、Z 形补偿器、方形式补偿器、球形补偿器等,使用时要按照敷设条件采用维修工作量小且价格较低的补偿器。

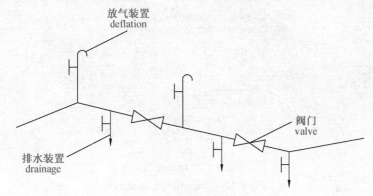

图 5.10　热力管道的排水、放气装置
Fig. 5.10　Drainage and deflation in heat pipe

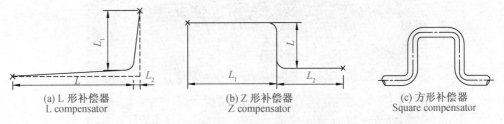

图 5.11　热力管道补偿器
Fig. 5.11　Heat pipe compensator

The exhaust outlet device should be installed at the highest point and the lowest point of pipes respectively. Horizontal installation of square compensator should be consistent with pipe slope; exhaust valves should be installed at the highest point, and drain valves should be installed at the bottom in the vertical installation. The distribution of pipes is shown in Fig.5.10.

The sharp directions and corners of pipes can eliminate deformation caused by temperature, as shown in Fig.5.11. Additional compensators are needed when natural compensation cannot meet requirements. Square compensator, wave compensator, sleeve compensator, and spherical compensator are commonly used according to the conditions and limitations.

5.4 综合管廊

综合管廊是建于城市地下用于容纳两类及以上城市工程管线的构筑物及附属设施,如图 5.12 所示。

入廊管线包括给排水舱、燃气舱、热力舱、电力舱、电信舱等,如图 5.13 所示。

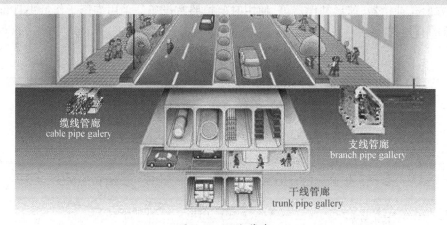

图 5.12 综合管廊

Fig. 5.12 Integrated pipe gallery

图 5.13 入廊管线

Fig. 5.13 Corridor pipeline

5.4 Integrated pipe gallery

The comprehensive pipe gallery is a structure and ancillary facilities built under the city to accommodate two or more types of urban engineering pipelines, as shown in Fig. 5.12.

Corridor pipelines including water supply and drainage cabins, gas cabins, thermal cabins, power cabins, telecommunications cabins, etc., are showed in Fig. 5.13.

根据布置的管线种类可将综合管廊分为以下三类：

(1) 干线管廊

用于容纳城市主干工程管线,采用独立分舱方式建设的综合管廊(一般 2 个舱以上)。

(2) 支线管廊

用于容纳城市配给工程管线,采用单舱或双舱建设的综合管廊。

(3) 缆线管廊

采用浅埋沟道方式建设,设有可开启盖板,但其空间不能满足人员正常通行要求,用于容纳电力电缆和通信电缆。

5.4.1　综合管廊结构分类

根据《城市综合管廊工程技术规范》(GB 50838—2015)要求,综合管廊的结构设计使用年限为 100 年,结构安全等级为一级。

综合管廊结构类型分现浇混凝土综合管廊结构和预制拼装综合管廊结构两种。现浇混凝土综合管廊结构为采用现场整体浇筑混凝土的综合管廊。预制拼装综合管廊结构为工厂内分节段浇筑成型,现场采用拼装工艺施工成为整体的综合管廊。

5.4.2　综合管廊特点

① 综合管廊缓解了直埋管线存在的各种问题:a. 检修及敷设管线需不断破挖

According to the types of pipelines arranged, the integrated pipeline gallery can be divided into the following three categories:

(1) Trunk pipe gallery

A comprehensive pipeline gallery constructed with independent sub-cabin methods for accommodating urban backbone engineering pipelines. (Generally more than 2 cabins)

(2) Branch pipeline gallery

It is used for accommodating urban distribution engineering pipelines and adopts a single-cabin or double-cabin integrated pipe gallery.

(3) Cable pipe gallery

It is constructed in a shallow buried trench with an openable cover plate but its space cannot meet the requirements for normal passage of personnel, and is used to accommodate power cables and communication cables.

5.4.1　Structural classification of integrated pipe gallery

According to the Technical Specification for *Urban Comprehensive Pipe Gallery Engineering* (GB 50838—2015), the structural design service life of the comprehensive pipe gallery is 100 years, and the structural safety level is level 1.

The structure types of the comprehensive pipe gallery are divided into two types: cast-in-place concrete comprehensive pipe gallery and prefabricated composite pipe gallery structure. The structure of the cast-in-place concrete comprehensive pipe gallery is a comprehensive pipe gallery that adopts on-site integral casting of concrete. The prefabricated and assembled integrated pipe gallery structure is cast in sections in the factory, and the on-site assembly process is used to construct a whole integrated pipe gallery.

5.4.2　Features of comprehensive pipe gallery

① The integrated pipe gallery alleviates the various problems of the directly buried pipeline, such as: a. Overhaul and laying of pipelines require continuous excavation of

路面;b. 各种管线分属不同部门管理,信息不畅,重复建设;c. 直埋管线与土壤接触,易造成管线腐蚀、损坏;d. 电力线缆占地大,影响城市规划及市容,且高压线易造成电磁辐射污染。

② 综合性强。综合管廊人廊范围涵盖了给水、雨水、污水、再生水、天然气、热力、电力、通信等城市工程管线。

③ 自动化程度高。管廊运营采用信息化管理,安装有感应器和探测器,运行状况及时反映到主控室,各种管线每一段的运行情况一目了然。人员或动物一进入管廊即被发现并标明其所在位置。

④ 综合管廊建设采用整体规划,可实现规模化、网络化,并使城市功能更加完善,地下空间有序管理。

⑤ 与管廊同步建设的地下空间设施如交通等将强化统筹,城市抗风险能力大大提高。

5.4.3 综合管廊规划

当遇到下列情况之一时,宜采用综合管廊形式规划建设:

① 交通运输繁忙或地下管线较多的城市主干道及配合轨道交通、地下道路、城市地下综合体等建设工程地段。

the road surface; b. various pipelines are managed by different departments, with poor information and repeated construction; c. directly buried pipelines are in contact with the soil, which may cause corrosion and damage to the pipelines; d. power cables occupies a large area, which affects urban planning and city appearance, and high-voltage lines are likely to cause electromagnetic radiation pollution.

② Strong comprehensiveness. The comprehensive pipeline corridor covers urban engineering pipelines such as water supply, rainwater, sewage, reclaimed water, natural gas, heat, electricity, and communications.

③ High degree of automation. The operation of the pipe gallery adopts information management, and sensors and detectors are installed. The operation status is reflected in the main control room in time, and the operation status of each section of various pipelines is clear at a glance. Persons or animals are found as soon as they enter the corridor and their location is marked.

④ The construction of the integrated pipe gallery adopts overall planning, which can realize scale and network, and make the city function more perfect, and the underground space is managed in an orderly manner.

⑤ The underground space facilities, such as transportation, constructed simultaneously with the pipe gallery will strengthen the overall planning, and the city's anti-risk ability will be greatly improved.

5.4.3 Planning of comprehensive pipe gallery

When encountering one of the following situations, it is advisable to adopt the form of integrated pipe gallery for planning and construction:

① City arterial roads with heavy traffic or many underground pipelines, and construction sites that cooperate with rail transit, underground roads, and urban underground complexes.

② 城市核心区、中央商务区、地下空间高强度成片集中开发区、重要广场、主要道路的交叉口、道路与铁路或河流的交叉处、过江隧道等。

③ 道路宽度难以满足直埋敷设多种管线的路段。

④ 重要的公共空间。

⑤ 不宜开挖路面的路段。

综合管廊的标准断面形式应根据容纳的管线种类及规模、建设方式、预留空间等确定,应满足管线安装、检修、维护作业所需要的空间要求。

① 天然气管道应在独立舱室内敷设。

② 热力管道采用蒸汽介质时应在独立舱室内敷设。

③ 热力管道不应与电力电缆同舱敷设。

④ 110 kV 及以上电力电缆不应与通信电缆同侧布置。

⑤ 给水管道与热力管道同侧布置时,给水管道宜布置在热力管道下方。

⑥ 进入综合管廊的排水管道应采取分流制,雨水纳入综合管廊可利用结构本体或采用管道方式,污水应采用管道排水方式,宜设置在综合管廊底部。

⑦ 综合管廊每个舱室应设置人员出入口、逃生口、吊装口、进风口、排风口、管线分支口等。

② City core areas, central business districts, high-strength concentrated development zones of underground space, important squares, intersections of major roads, intersections of roads and railways or rivers, tunnels across the river, etc.

③ The width of the road is difficult to meet the requirements of the sections where multiple pipelines are directly buried.

④ Important public space.

⑤ Sections of pavement are not suitable for excavation.

The standard section form of the integrated pipe gallery shall be determined according to the type and scale of the pipelines to be accommodated, construction methods, reserved space, etc., and shall meet the space requirements required for pipeline installation, overhaul, and maintenance operations.

① The natural gas pipeline should be laid in an independent cabin.

② When the heating power pipeline adopts steam medium, it should be laid in an independent cabin.

③ The heating pipeline should not be laid in the same warehouse as the power cable.

④ The power cables of 110 kV and above should not be arranged on the same side as the communication cables.

⑤ When the water supply pipeline is arranged on the same side as the heating pipeline, the water supply pipeline should be arranged under the heating pipeline.

⑥ Drainage pipes entering the integrated pipe gallery shall adopt a diversion system, and rainwater shall be incorporated into the integrated pipe gallery to use the structural body or the pipeline method. The sewage shall be drained by pipes and should be set at the bottom of the integrated pipe gallery.

⑦ Each compartment of the integrated pipeline gallery shall be provided with personnel entrances and exits, escape openings, hoisting openings, air inlets, exhaust outlets, and pipeline branch openings.

⑧ 综合管廊管线分支口应满足预留数量、管线进出、安装敷设作业的要求。

5.5 城市防洪和防汛工程

洪灾是十分复杂的灾害系统,它的诱发因素极为广泛,水系泛滥、风暴、地震、火山爆发、海啸等都可以引发洪灾。在各种自然灾难中,洪水造成的死亡人口占因自然灾难死亡人口总数的75%,经济损失占因自然灾难造成全部损失的40%。为了尽量减少洪水造成的危害,保护城市的工业生产和人民生活财产安全,有必要根据城市的总体规划和流域的防洪规划,认真做好城市或工厂的防洪规划,以提高城市的防洪能力。城市防洪设施主要包括堤防、内行洪排水设施、水库及其他设施,详细分类见表5.3所示。

① 堤防工程。通过增加河流两岸大堤的高度和稳定性,提高河道安全泄洪量,避免洪水对城区造成危害,如图5.14 a 所示。

表5.3 城市防洪设施分类
Tab.5.3 Types of urban flood control facilities

⑧ The branch ports of the integrated pipeline gallery shall meet the requirements of reserved quantity, pipeline entry and exit, and installation and laying operations.

5.5 Urban flood control projects

The flood disaster is a very complex system, because of its extremely broad predisposing factors, such as flooding, storm, earthquake, volcanic eruptions and tsunami. Among all natural disasters, death of the population due to floods accounted for 75%, and economic losses accounted for 40% of the total loss. In order to minimize the harm caused by the floods, and to protect the industrial production in cities, people's lives and pro-perties, it is necessary to do a good job in flood control planning for cities or factories according to the overall planning of the city and program of flood control. Urban flood control facilities mainly include the dikes, within the flood passage drainage, reservoirs, and other facilities, and their classification is shown in Tab.5.3.

① Dike project. The purpose of dike project is to increase the safe discharge volume and avoid the flood of urban hazards by increasing the height of the river levee and its stability, as shown in Fig.5.14 a.

② 整治河道和护岸。对弯曲河道进行裁弯取直,对淤积河道进行疏浚,加深河床以加大河道过水能力,降低水位,缩短河流里程(见图 5.14 b)。

在河岸因水流冲刷容易造成河岸坍塌、影响河岸稳定和建筑物安全的地段采用护岸措施。

(a) 城市防洪堤
Levee

(b) 河道整治
River training

(c) 城市防洪闸
Floodgate

(d) 生物工程防洪
Bio-engineering flood control

(e) 管渠排涝
Tube drainage

(f) 洪水预报系统
Flood forecasting system

图 5.14　城市防洪设施
Fig. 5.14　Urban flood control facilities

② River regulation and bank protection. Straightening the meandering sections, and dredging the siltation river can deepen the riverbed, so as to increase the water capacity of the river, lower the water level and shorten the river mileage(Fig.5.14 b).

The revetment measures should be taken to the locations where the river bank collapse and the river bank stability and building safety affected by the riparian water erosion.

③ 防洪闸。河口城市和临江河城市,汛期外水水位高,往往形成江(河、湖、海)水倒灌,影响河流泄洪而造成洪涝灾害。在下流出口处设防洪闸,是防止洪水、海水倒灌的一个重要措施,如图5.14 c所示。

④ 分(蓄)洪区和水库。在流经城市的河流上游修建水库拦洪蓄水,或将洪水引入低洼地,或用分洪道分洪,均可减轻下游城市的洪水压力。

⑤ 生物工程措施。结合小流域治理,在流域上植树种草,增强流域下渗蓄水能力,从而减少进入河道中的径流和泥沙,起到蓄水防洪作用,如图5.14 d所示。

⑥ 山洪和泥石流的拦蓄、排导工程。在山坡上修建谷坊、塘堰、梯田,可以拦截泥沙、减少山洪危害,同时避免诱导泥石流发生。修建排洪沟、泥石流排导沟,可将山洪和泥石流引导至保护区范围以外。

⑦ 排涝措施。城市内涝可通过修建管渠排涝,一般采用自流排泄、高水排泄的方法,以上方法不能解决时修建泵站抽排,如图5.14 e所示。

⑧ 其他非工程措施。如洪水预报、洪水警报、蓄滞洪区管理、洪水保险、河道清障、灾后救济等,如图5.14 f所示。

完善配套的城市防洪排涝设施是城市经济持续快速发展的重要保障。防洪工作需要根据自然规律协调安排,要合理地投入,并取得可能获得的最大收益,要进行科学规划,确定城市的防洪标准,修建城市防洪工程体系;保证城市在发生规划标准的

③ Floodgates. In the estuary and riverside city, the high water level of the flood results in the intrusion of the river (lake, sea) water, which affects river flood discharge and causes the floods. Setting floodgate in the downstream can prevent floods which is an important measure, as shown in Fig.5.14 c.

④ Flood diversion and reservoirs are built to control flood, or draw the flood into low-lying land and using flood diversion in the upper reaches of the river can reduce the flood pressure.

⑤ Bio-engineering measures. Combining small watershed management with planting trees and grass in the basin can increase the infiltration basin water storage capacity, thereby reducing runoff and sediment which flow into the river, as shown in Fig.5.14 d.

⑥ Storing and exhaust engineering for flash floods and mudslides . Construction of check dams, ponds, terraces on the hillside can intercept sediment and reduce flash flood hazards, as well as avoid the occurrence of induced mudslides. The construction of flood discharge trench and the mudslides exhaust ditch can lead flash floods and landslides out of the protected areas.

⑦ Drainage measures. General construction of pipe drainage can solve urban water logging. Using gravity excretion, high water excretion can achieve the goals. If the problems cannot be solved through methods above, we will build a pumping station as shown in Fig.5.14 e.

⑧ Other non-engineering measures, such as flood forecasting, flood warning, flood storage management, flood insurance, river wrecker, post-disaster relief, are shown in Fig.5.14 f.

At the same time, the complete set of city flood control and drainage facilities is the important guarantee for the city sustained and rapid economic development. Flood prevention work should harmonize with the laws of nature. Reasonable investment should

常遇和较大洪水时国家经济和社会活动不受影响,遇到超标准的大洪水和特大洪水时,有预定的分蓄行洪区和防洪措施,国家经济和社会不发生动荡,不影响国家长远计划的完成。

目前我国许多城市的防洪排涝能力不足,为克服这一缺陷,并充分利用雨水资源,住建部、财政部、水利部等部门于 2015 年开展了"海绵城市"的试点建设。海绵城市是指通过在城市地下建设海绵体使城市能够像海绵一样,在适应环境变化和应对自然灾害等方面具有良好的"弹性",下雨时吸水、蓄水、渗水、净水,需要时将蓄存的水"释放"并加以利用,如图 5.15 所示。海绵城市建设应遵循生态优先等原则,将自然途径与人工措施相结合,在确保城市排水防涝安全的前提下,最大限度地实现雨水在城市区域的积存、渗透和净化,促进雨水资源的利用和生态环境保护。首批试点城市有河北迁安市、吉林白城市、江苏镇江市、浙江嘉兴市等 16 座城市。

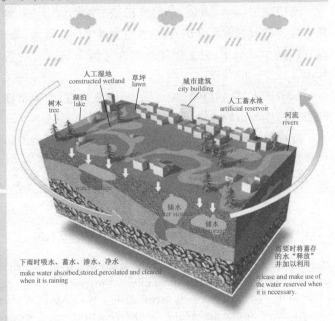

图5.15　海绵城市水循环原理
Fig. 5.15　Water cycle of Sponge city

be made in the premise, so as to obtain the maximum benefit. The flood control standards and the urban flood control systems should be made through scientific planning. Make sure that the economic and social activities will not be affected when large flood happened.

Building the flood area and flood control measures in case of the large floods and devastating floods can make the economy, social life as well as the completion of the national long-term plan not be affected.

Currently, flood control and drainage capacity in many Chinese cities is insufficient. To overcome the deficiency and make full use of rainwater resources, Ministry of Housing and Urban-Rural Development, Ministry of Finance and Ministry of Water Resources launched the "sponge city" pilot construction in 2015. The sponge city means that cavernosum is built underground which ensures the city "flexible" to adapt to environmental changes and respond to natural disasters. As is shown in Fig.5.15, when it is raining, the cavernosum can make water absorbed, stored, percolated and cleaned. The structure can also release the water stored when it is necessary. Sponge city construction should follow the principle of ecological priority. This structure can also combine the natural approach with artificial measures. On the premise of ensuring the security of the city drains, we should do our best to realize the rainwater stored, percolated and cleaned in urban area underground to promote the use of rainwater resources and protect the ecological environment. The first batch of pilot cities contain Qian'an in Hebei Province, Baicheng in Jilin Province, Zhenjiang in Jiangsu Province, Jiaxing in Zhejiang Province and other 12 cities.

5.6 海绵城市

作为城市发展理念和建设方式转型的重要标志,我国海绵城市建设"时间表"已经明确且"只能往前,不可能往后"。全国已有 130 多个城市制订了海绵城市建设方案。确定的目标核心是通过海绵城市建设,使 70%的降雨就地消纳和利用。围绕这一目标确定的时间表是到 2030 年,80%的城市建成区达到这个要求。如果一个城市建成区有 100 平方公里的话,至少有 80 平方公里在 2030 年要达到这个要求。

5.6.1 海绵城市的定义

海绵城市(eco-sponge city)是指城市能够像海绵一样,在适应环境变化和应对自然灾害等方面具有良好的"弹性",下雨时吸水、蓄水、渗水、净水,需要时将蓄存的水"释放"并加以利用,最大限度地维持或恢复城市开发前的自然水文循环,如图 5.15 所示。

基本原则:规划引领,生态优先,安全为重,因地制宜,统筹建设。

5.6.2 海绵城市的建设

海绵城市建设的本质是解决城镇化与资源环境的协调和谐问题,目标是让城市弹性适应环境变化与自然灾害,要求将快排式的传统雨洪管理模式转变为"渗、滞、蓄、净、用、排"的新模式,要求城市开发前后的水文特征基本不变。

5.6 Sponge city

As an important symbol of the transformation of urban development concepts and construction methods, the "timetable" of my country's sponge city construction has been clear and "can only move forward, not backward." More than 130 cities across the country have formulated plans for the construction of sponge cities. The core of the determined goal is to make 70% of the rainfall absorbed and utilized on the spot through the construction of sponge cities. The timetable set around this goal is that by 2030, 80% of urban built-up areas will meet this requirement. If a city has a built-up area of 100 square kilometers, at least 80 square kilometers will meet this requirement in 2030.

5.6.1 Definition of Sponge City

An eco-sponge city refers to a city that, like a sponge, has good "elasticity" in adapting to environmental changes and responding to natural disasters. It absorbs, stores, seeps, and purifies water when it rains, and stores water when needed. The stored water is "released" and used to maximize the maintenance or restoration of the natural hydrological cycle before urban development. As shown in Fig.5.15.

Basic principles: planning guidance, ecological priority, safety first, measure adjustment to local conditions, overall planning.

5.6.2 Construction of Sponge City

The essence of sponge city construction is to solve the coordination and harmony between urbanization and resources and environment. The goal is to make the city adapt to environmental changes and natural disasters flexibly. The new model of "utilization and drainage" requires that the hydrological characteristics before and after urban development remain basically unchanged.

① 渗：减少路面、屋面、地面等硬化地表面积，雨水就地下渗。

由于城市下垫面过硬，到处都是水泥，改变了原有自然生态本底和水文特征，因此，要加强自然的渗透，把渗透放在第一位。其好处在于，可以避免地表径流，减少从水泥地面、路面汇集到管网里的水量，同时，涵养水源，补充地下水，还能通过土壤净化水质，改善城市微气候。而渗透雨水的方法多样，主要是改变各种路面、地面铺装材料，改造屋顶绿化，调整绿地竖向，从源头将雨水留下来然后渗下去。采用的技术为透水铺装，如图 5.16 所示。

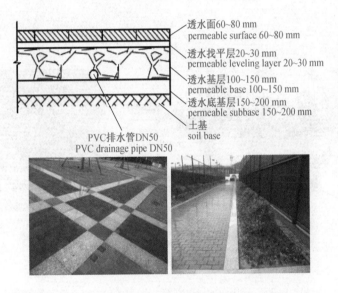

透水面60~80 mm
permeable surface 60~80 mm

透水找平层20~30 mm
permeable leveling layer 20~30 mm

透水基层100~150 mm
permeable base 100~150 mm

透水底基层150~200 mm
permeable subbase 150~200 mm

土基
soil base

PVC排水管DN50
PVC drainage pipe DN50

图 5.16　透水铺装原理图

Fig. 5.16　Schematic diagram of permeable paving

① See page: Reduce the hardened surface area of roads, roofs, grounds, etc., and rainwater seeps underground.

Because the underlying surface of the city is too hard and there is cement everywhere, it has changed the original natural ecological background and hydrological characteristics. Therefore, it is necessary to strengthen natural penetration and put penetration in the first place. The advantage is that it can avoid surface runoff, reduce the collection of groundwater from the cement floor and road surface into the pipe network, at the same time, conserve groundwater, supplement the lack of groundwater, and can also purify the water quality through the soil and improve the urban microclimate. There are many ways to infiltrate rainwater. The main ones are to change various pavement and ground paving materials, transform roof greening, adjust the vertical direction of green space, and keep rainwater from the source and then "seep" it down. The technology used is permeable paving, as shown in Fig. 5.16.

②滞：延缓峰现时间,降低排水强度,缓解雨洪压力。

其主要作用是延缓短时间内形成的雨水径流量。例如,通过微地形调节,让雨水慢慢地汇集到一个地方,用时间换空间。通过"滞",可以延缓形成径流的高峰。具体形式总结为四种:雨水花园、生态滞留池、渗透塘和人工湿地。

●雨水花园 雨水花园是指在园林绿地中种有树木或灌木的低洼区域,由树皮或地被植物作为覆盖。它可以将雨水滞留下渗透补充地下水并降低暴雨地表径流的洪峰,还可通过吸附、降解、离子交换和挥发等过程减少污染。其中浅坑部分能够蓄积一定的雨水,延缓雨水汇集的时间,土壤能够增强雨水下渗,缓解地表积水现象。蓄积的雨水能够供给植物利用,减少绿地的灌溉水量,如图 5.17 所示。

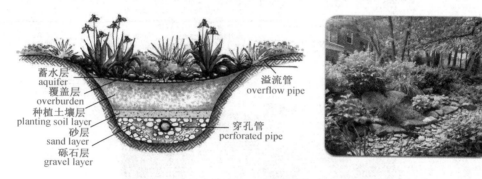

图 5.17 雨水花园

Fig. 5.17 Rainwater Garden

② Stagnation: delay the peak time, reduce the drainage intensity, and alleviate the risk of rain and flood.

Its main function is to delay the runoff of rainwater formed in a short period of time. For example, through micro-topography adjustment, rainwater can be collected slowly to one place, and time can be used to change space. Through "stagnation", the formation of the peak of runoff can be delayed. The specific forms are summarized into three types: rain garden, ecological retention pond, infiltration pond, and constructed wetland.

● Rain garden Rain garden refers to a low-lying area where trees or shrubs are planted in the garden green space, covered by bark or ground cover plants. It replenishes groundwater and reduces the peaks of surface runoff from rainstorms by retaining rainwater. It can also reduce pollution through processes such as absorption, degradation, ion exchange, and volatilization. The shallow pits can accumulate a certain amount of rainwater, delay the collection time of rainwater, and the soil can increase rainwater infiltration and alleviate the phenomenon of surface water accumulation. The accumulated rainwater can be used by plants and reduce the amount of irrigation water for green spaces, as shown in Fig.5.17.

● 生态滞留区 概念上来讲，生态滞留区就是浅水洼地或景观区利用工程土壤和植被来存储和治理径流的一种形式，治理区域有草地过滤、砂层和水洼面积、有机层或覆盖层、种植土壤和植被。生态滞留区对于土壤的要求和工程技术上的要求不同于雨水花园，形式根据场地位置不同也较为多样，如生态滞留带、滞留树池等。

植草沟具有输水功能和一定的截污净化功能，适用于径流量小及人口密度较低的居住区、工业区或商业区、公园、停车场及公共道路两边，可以代替路边的排水沟或者雨水管渠系统。植草沟沟顶宽 0.5~2 m，深度 0.05~0.25 m，边坡（垂直:水平）1:3~1:4，纵向坡 0.3%~5%。植草沟可设置在雨水花园、下凹式绿地前作为预防处理，如图 5.18 所示。

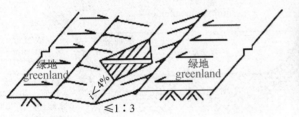

图 5.18 生态滞留带
Fig. 5.18 Ecological retention zone

● Ecological retention area Conceptually, ecological retention area is a form of shallow water depression or landscape area using engineering soil and vegetation to store and control runoff. The governance area includes grass filtration, sand layer and puddle area, organic layer or mulch, planting soil and vegetation . The requirements for soil and engineering technology in ecological retention areas are different from rain gardens, and the forms are also more diverse depending on the location of the site, such as ecological retention areas and retention tree ponds.

Zhi cao ditch has the function of water conveyance, and has a certain function of interception and purification. It is suitable for residential areas, industrial or commercial areas, parks, parking lots and public roads with small runoff and low population density. It can replace roadside drainage ditches or rainwater pipe systems. The width of the top of Zhi cao ditch is 0.5~2 m, the depth is 0.05~0.25 m, the side slope (vertical:horizontal) is 1:3~1:4, and the longitudinal slope is 0.3%~5%. It can be set up in front of rain gardens and recessed green spaces as preventive treatment, as shown in Figure 5.18.

● 渗透塘 渗透塘又称渗水洼塘,即利用天然或人工修筑的池塘或洼地进行雨水渗透,补充地下水。渗透塘能有效地削减径流峰值,但渗透塘护坡需要种植耐湿植物。若渗透塘较深(超过 60 cm),护坡周边就要种植低矮灌木,形成低矮绿篱,消除安全隐患。同时,整个渗透塘系统必须形成微循环才能防止水体腐败, 如图 5.19 所示。

● 人工湿地 人工湿地是一个综合的生态系统,它应用生态系统中物种共生、物质循环再生原理,以及结构与功能协调原则,将雨水花园、生态滞留池收集的雨水进行集中的净化。其具有缓冲容量大、处理效果好、工艺简单、投资省、运行费用低等特点,极其适合在海绵城市建设中应用,如图 5.20 所示。

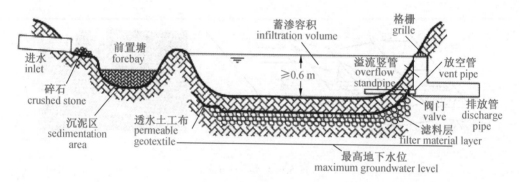

图 5.19 渗透塘

Fig. 5.19 Permeation pond

● Infiltration pond Infiltration ponds are water-seepage depressions, that is, natural or artificial ponds or depressions are used to infiltrate rainwater to supplement groundwater. Infiltration ponds can effectively reduce the peak runoff. However, the slope protection of the infiltration pond needs to plant moisture-tolerant plants. If the Infiltration pond is deeper (more than 60 cm), low shrubs should be planted around the slope protection to form a low green hedge to eliminate potential safety hazards. At the same time, the entire Infiltration pond system needs to form a microcirculation to prevent water body corruption, as shown in Fig. 5.19.

● Constructed wetland Constructed wetland is a comprehensive ecosystem. It applies the principles of species symbiosis, material recycling and regeneration in the ecosystem, and the principle of structure and function coordination to centrally purify rainwater collected in rainwater gardens and ecological retention ponds. Moreover, it has the characteristics of large buffer capacity, good treatment effect, simple process, low investment, low operating cost, etc., which is extremely suitable for applications in the construction of sponge cities, as shown in Fig. 5.20.

③ 蓄:削减峰值流量,调节雨洪时空分布,为雨洪资源化利用创造条件。

即把雨水留下来,要尊重自然的地形地貌,使降雨得到自然散落。现在人工建设破坏了自然地形地貌后,短时间内水汇集到一个地方,就形成了内涝。因此要想办法把降雨蓄起来,以达到调蓄和错峰。而当下海绵城市蓄水环节没有固定的标准和要求,地下蓄水形式多样,常用形式有两种:蓄水模块和地下蓄水池。

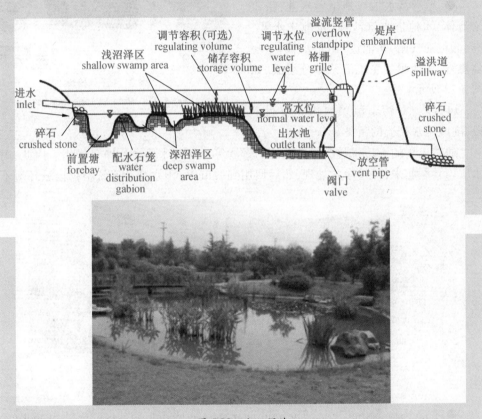

图 5.20　人工湿地
Fig. 5.20　Constructed wetland

③ Storage: Reduce peak flow, adjust the temporal and spatial distribution of rainwater, and create conditions for the utilization of rainwater resources.

That is to keep the rain water, we must respect the natural topography and landforms so that the rain can be scattered naturally. Now that artificial construction destroys the natural topography and landforms, water converges in one place within a short period of time, forming waterlogging. Therefore, it is necessary to accumulate rainfall in order to achieve regulation and storage and peak shifting. At present, there are no fixed standards and requirements for water storage in sponge cities. There are various types of underground water storage. Generally, there are two commonly used forms: plastic module water storage and underground water storage tanks.

● 蓄水模块　雨水蓄水模块是一种可以用来储存水,但不占空间的新型产品;具有超强的承压能力;95%的镂空空间可以实现更有效率的蓄水。配合防水布或者土工布可以完成蓄水、排放,但还需要在结构内设置好进水管、出水管、水泵位置和检查井,如图5.21所示。

图 5.21　蓄水模块
Fig. 5.21　Water storage module

● 地下蓄水池　地下蓄水池,由水池池体,水池进水沉沙井,水池出水井,高、低位通气帽,水池进、出水水管,水池溢流管,水池曝气系统等几部分组成,如图5.22所示。

④ 净:对污染源采取相应控制手段,削减雨水径流的污染负荷。

通过土壤的渗透,植被、绿地系统的过滤等都能对水质产生净化作用。因此,雨水应该蓄积起来,经过净化处理,然后回用到城市中。雨水净化系统根据区域环境不

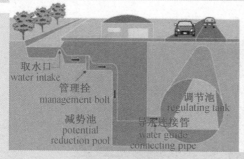

图 5.22　地下蓄水池
Fig. 5.22　Underground reservoir

● Water storage module　The rainwater storage module is a new product that can be used to store water but does not take up space; it has super pressure bearing capacity; 95% of the hollow space can achieve more efficient water storage. With tarpaulin or geotextile, water storage and discharge can be completed. At the same time, the inlet pipe, outlet pipe, water pump location and inspection well need to be set up in the structure, as shown in Figure 5.21.

● Underground reservoir　The rainwater collection tank is composed of the pool body, the water inlet settling well, the water outlet well, the high and low ventilation caps, the water inlet and outlet pipes of the pool, the overflow pipe of the pool, and the aeration system of the pool, as shown in Figure 5.22.

④ Purification: Take corresponding control measures to pollution sources to reduce the pollution load of rainwater runoff.

Through soil penetration, vegetation, green space systems, water bodies, etc., water quality can be purified. Therefore, it should be stored, purified, and then reused in the city. The rainwater purification system sets up different purification systems according to the different regional environments. According to the current situation of the city, the

同设置了不同的净化体系。根据城市现状可将区域环境大体分为三类：居住区雨水收集净化、工业区雨水收集净化和市政公共区域雨水收集净化。

⑤ 用：实现雨洪资源化，如雨水回灌、雨水灌溉及构造园林水景观等，形成雨水资源的深层次循环利用。

经过土壤渗滤净化、人工湿地净化、生物处理多层净化之后的雨水要尽可能加以利用，不论是丰水地区还是缺水地区，都应该加强对雨水资源的利用。例如，停车场上面的雨水收集净化后可用于洗车。我们应该通过"渗"涵养水源，通过"蓄"把水留在原地，再通过净化把水"用"在原地。

⑥ 排：统筹开发雨水系统、城市雨水管渠系统及超标雨水径流排放系统，构建安全的城市排水防涝体系，确保城市运行安全。

利用城市竖向与工程设施相结合，排水防涝设施与天然水系河道相结合，地面排水与地下雨水管渠相结合的方式来实现一般排放和超标雨水的排放，避免形成城市内涝等灾害。

5.7　城市园林和绿化工程

城市园林和绿化工程是美化生活环境、提高人民身心健康水平的重要工程，是生态建设的重要组成部分，是有生命的基础设施，对保持经济社会发展和改善人民

regional environment can be roughly divided into three categories: rainwater collection and purification in residential areas, rainwater collection and purification in industrial areas, and rainwater collection and purification in municipal public areas.

⑤ Usage: Realize rainwater resource utilization, rainwater recharge, rainwater irrigation and construction of garden water landscape, etc., forming a deep-level recycling of rainwater.

After the multi-layered purification of soil infiltration purification, constructed wetland purification, and biological treatment, rainwater should be used as much as possible. Whether it is in water-rich areas or water-deficient areas, the use of rainwater resources should be strengthened. Not only can it alleviate flood disasters, the collected water resources can also be used, such as collecting and purifying rainwater on parking lots for car washing. We should conserve water by "seepage", keep the water in place through "storage", and then "use" the water in place through purification.

⑥ Drainage: Coordinate the development of low-impact rainwater systems, urban rainwater pipe systems, and excessive rainwater runoff drainage systems to build a safe urban drainage and waterlogging prevention system to ensure the safety of urban operations.

Utilize the combination of urban vertical and engineering facilities, the combination of drainage and waterlogging prevention facilities with natural water system rivers, and the combination of surface drainage and underground rainwater pipes to achieve general discharge and discharge of excessive rainwater, avoiding floods and other disasters.

5.7　City garden and green engineering

The city greening engineering can not only beautify the living environment, improve people's physical and mental health, but also belongs to an important part of the

生活质量具有重要作用，已成为衡量城市文明和地区可持续发展能力的重要标志。随着社会经济的发展和人民生活水平的提高，居民对居住环境条件的要求越来越高，园林绿化工程有了更大的发展空间。

5.7.1 城市园林工程

高质量、高水平的园林工程建设，既是改善城镇生态环境和建设投资环境的需要，又是人们高质量生存、生活、工作的环境基础。通过园林工程建设，构建完整的绿地系统和优美的园林艺术景观(见图 5.23)，是净化空气、防止污染、调节气候、改善生态、美化环境的需要。

园林工程建设是集建筑科学、生物科学、社会科学于一体的综合性科学。现代园林工程建设学科已发展成为多学科边缘交叉的一门前沿科学体系。其设计主要

图 5.23 城市园林
Fig. 5.23 Urban gardens

ecological construction and life infrastructure. It plays an important role in maintaining economic and social development, and has become an important symbol to measure the sustainable development abilities of urban civilization. With the development of economy and the improvement of people's living standards, the better living conditions have been expected and landscaping work will have more development space.

5.7.1 Urban garden project

Garden constructions with high-quality and high-level cannot only improve the environment of urban ecology and the construction investment, but also belong to the basis of high-quality living and working environment. A complete green land system and beautiful landscape art are the requirements of cleaning the air, preventing pollution, regulating the climate, improving the ecology and beautifying the environment(Fig.5.23).

Garden construction is an integrated science, which combines building science, biological science, and social science. Modern landscape construction disciplines have developed into a frontier science system with the cross of many disciplines. Its design

包括如下几个部分(见表 5.4)：① 园林地形工程设计,主要是根据园林性质和规划要求,因地制宜、因情制宜地塑造地形。② 园路工程设计,即在园林中确定园路布局及园路结构设计的过程。③ 园林的给排水工程设计,主要是进行园林中的给水管网、排水系统和排水设施的设计。④ 园林植物造景工程设计,主要是进行园林中的植物种类、间距、形式的布置。⑤ 园林绿地喷灌工程设计,主要进行喷头的选型、管网的布置及灌水制度的制定等。⑥ 园林水景工程设计,主要是各种人工水体的营造设计,如湖泊、池塘、泉水等。⑦ 园林假山、置石工程设计,是综合运用力学、材料学、工程学及艺术学的知识再造自然山石的过程。⑧ 园林供电工程设计,主要是对园林输配电、照明及其他用电设备的设计和配备。⑨ 园林建筑、小品工程设计,主要是对园林中的景观建筑(如亭、廊、榭)和小型设施(如园椅、园凳、栏杆、小型雕塑)等进行的设计。

表 5.4　城市园林工程的内容
Tab.5.4　Contents in urban garden project

园林工程 urban garden project
- 地形工程 terrain engineering
- 园路工程 park road engineering
- 给排水工程 watersupply and drainage engineering
- 植物造景工程 plant landscape
- 绿地喷灌工程 green irrigation engineering
- 水景工程 waterfront engineering
- 假山置石工程 rockery,stone engineering
- 供电工程 power engineering
- 建筑、小品工程 landscape architecture

mainly includes the following parts (Tab.5.4). ① Garden terrain design is to design the garden according to the nature of the landscape, the planning requirements, and the local conditions. ② Park road design is to determine the layout and the structures of the park road. ③ Garden water supply and drainage construction design is to design the main water supply network, drainage systems and drainage facilities in the garden. ④ Plant landscape design is to arrange garden plant species, spaces and forms.⑤ Garden green irrigation is to select the nozzle, make the pipe network layout and make the irrigation system. ⑥ Garden waterfront design is to create a variety of artificial bodies, such as lakes, ponds and springs. ⑦ Design of garden rockery and stone home is to use mechanics, materials science, engineering and art learned to rebuild natural rocks. ⑧ Design of garden power supply focuses on power transmission and distribution, lighting and other designed and electrical equipment. ⑨ Landscape architecture and designed sketches focus on architecture (such as pavilions, corridors, and pavilion) on terrace and small facilities (like garden chairs, garden stools, railings, small sculptures).

园林绿化工程建设的施工顺序,一般是先整理山水、改造地形、辟筑道路、铺装场地、营造建筑、构造工程设施,然后实施绿化,如图5.24所示。

由于构成园林的要素极其复杂,既有地形、给排水、供电等工程方面的知识,又有植物、造景设计等生物方面的知识,还有各构成要素的布局、景观营造、色彩搭配等艺术方面的知识,所以园林工程的设计需要综合考虑上述各因素的影响进行详细规划。

5.7.2 城市绿化工程

城市绿化是在城市中进行的为提高城市居民生活质量、优化城市工作和生活环境、协调城市生态并创造优雅城市面貌、使其在一种健康和谐的基础上持续发展的种植植物的行为,是城市现代化建设的重要内容。城市绿化主要有人行道绿化、分车绿化、防护绿化、基础绿化、城市广场及公共建筑前的绿化、街头休息绿地、停车场绿化、立体交叉绿化、滨河路绿化、花园林荫路、建筑物绿化等,如图5.25所示。

图 5.24　园林绿化工程施工顺序
Fig. 5.24　Construction order of urban garden project

The landscape construction is arranged in the sequence of arranging landscape, transforming terrain, making roads, paving sites, building, constructing facilities and greening (Fig.5.24).

The element of landscape is extremely complex, which contains topography, drainage, electricity and other engineering aspects, the plant landscaping design and other biological engineering, and some artistic knowledge such as layout, landscape construction, and color matching. Therefore, a detailed consideration about the factors need to be made in landscape engineering design.

5.7.2　Urban greening projects

Urban greening is to improve the living quality of urban residents, optimize the urban working and living environment, coordinate urban ecology, and create an elegant appearance. Urban greening project can make the city more healthy and harmonious on the basis of the sustainable development of the growing plants, and it is an important part of urban modernization. The basic content of urban greening is sidewalk green, the drive green, green protection, basic green, city squares and public buildings green, street rest green, parking green, cloverleaf green, riverside road green, garden avenue green and building green, etc, as shown in Fig.5.25.

城市绿化工程需要定期维护,如浇水、清理垃圾、除草、施肥、修剪、病虫害防治等。在当前资源紧张,环境污染较为严重的情况下,城市园林绿化应以最少的用地、最少的用水、最少的投资,选择对周围生态环境干扰最少的绿化模式,为城市居民提供最高效的生态保障系统。

道路绿化 Road green

建筑物绿化 Building green

停车场绿化 Parking green

广场绿化 Square green

图 5.25　城市绿化
Fig. 5.25　Urban greening

Urban greening projects require regular maintenance, such as watering, cleaning up, pulling weeds, fertilization, pruning, and pest control. In the case of resource constraints and serious environmental pollution, the urban landscape should use the least amount of land, minimal water and investment, as well as disruption to the surrounding ecological environment, so as to provide the most efficient ecological protection systems for people.

5.8 市政工程发展前沿

在城市的建设中,市政工程建设居于主要地位,它代表着城市发展的主要趋势和形象,代表着城市居民的生活水平和精神面貌,因此根据社会的发展趋势对市政工程中所使用的材料、施工方法及管理措施进行更新是很有必要的。目前市政工程发展主要向以下几个方向推进。

5.8.1 新型材料在市政工程中的推广应用

随着社会的发展,传统材料在市政工程中可能会出现一些不利的影响,研发并推广新型材料是市政工程发展的首要问题,如高密度聚乙烯(HDPE)管在给排水工程中的应用(见图 5.26)、低频无极灯在城市道路照明中的应用等。

图 5.26 高密度聚乙烯管(HDPE)
Fig. 5.26 High-density polyethylene pipe

5.8 Frontier development of municipal engineering

Municipal engineering construction is a predominant aspect in the city's construction, and it represents the main trend of urban development, the standard living of urban residents and their mental outlook. It is necessary to update the materials used in municipal engineering, construction methods and management measures in accordance with the trend of social development. Currently, the development of municipal engineering contains the following aspects.

5.8.1 The promotion and application of new materials in municipal engineering

With the development of society, the traditional materials in municipal engineering have some adverse effects. Researching and promoting the new materials are the primary issues of the municipal engineering development, such as the application of high-density polyethylene (HDPE) pipe in drainage (Fig.5.26), and the induction lamp in city road lighting.

5.8.2 市政工程施工机械及方法的更新

市政工程在施工过程中会影响城市的正常运行,如给水管道的维修会暂时停止供水、城市道路的改造会增加交通的拥堵和环境污染等。通过研发新的设备使市政工程的施工过程更绿色、更高效,且尽可能实现数字化是未来市政工程发展的另一个方向。

5.8.3 市政设施管理的系统化及网络化

随着计算机和通信技术的发展,信息技术在市政工程管理过程中的作用越来越重要,通过开发市政设施管理的计算机集成应用系统并实现网络化,进而实现城市市政"无缝隙""精细化"管理,是未来市政设施管理的发展方向。

5.8.2 Update of municipal engineering construction machinery and methods

Municipal engineering in the process of construction will affect the normal operation of the city. For example, the maintenance of the water supply pipeline will temporarily cut off the water supply, the transformation of urban roads will increase traffic congestion and environmental pollution. By the research and development of new equipment, municipal engineering construction process may become more green, more efficient, and digitization as much as possible is another direction of the future development in municipal engineering.

5.8.3 Systematic management of municipal facilities and network

With the development of computer and communication technology, the information technology plays an important role in the process of municipal engineering administration. Making the computer integrated application system through the development of municipal facilities management and network are the future development direction of the municipal facilities management and making the city municipal "seamless" and "fine".

注:本章图片均来源于网络。
Note: In this chapter, all pictures are from webs.

知识拓展
Learning More

相关链接 Related Links

如果想了解市政工程的详细知识、最新发展态势及相关政策,可访问中国市政工程协会网 http://www.zgsz.org.cn/。

If you want to obtain detailed knowledge of municipal engineering, the latest development trend and related policies, please visit http://www.zgsz.org.cn/.

小贴士 Tips

市政工程的注册师制度:国内与市政工程有关的注册师有注册建造师、注册造价工程师、注册监理工程师、注册电气工程师、注册给排水工程师等,通过考试后可从事相应市政工程专业的设计、施工、管理等工作。

Registration system for municipal engineering: the registration division of municipal engineering includes registered architects, registered cost engineers, registered supervision engineers, registered electrical engineers and registered drainage engineers, etc. After passing the appropriate examinations, people can be engaged in the design construction and management of municipal engineering.

思考题 Review Questions

(1) 市政工程包括哪些内容？

What does municipal engineering include?

(2) 城市道路和桥梁可分为哪些类型？

What types can the urban roads and bridges be divided into?

(3) 城市给排水系统的特点是什么？

What are the characteristics of the urban drainage system?

 参考文献
References

［1］全国一级建造师执业资格考试用书编写委员会.市政公用工程管理与实务[M].北京:中国建筑工业出版社,2010.

［2］毛惟德.城镇排水工程[M].北京:中国建筑工业出版社,2009.

［3］蒋柱武,黄天寅.给排水管道工程[M].上海:同济大学出版社,2011.

［4］张智.城镇防洪与雨洪利用[M].北京:中国建筑工业出版社,2009.

［5］袁海龙.园林工程设计[M].2版.北京:化学工业出版社,2011.

［6］王俊安.园林绿化工程估价[M].北京:机械工业出版社,2009.

［7］范慧方.燃气供应[M].武汉:华中科技大学出版社,2011.

［8］刑丽贞.市政管道施工技术[M].北京:化学工业出版社,2004.

［9］姚时章,蒋中秋.城市绿化设计[M].重庆:重庆大学出版社,1999.

第6章 建筑环境与设备工程

Chapter 6　Building Environment and Equipment Engineering

先导案例
Guide Case

天友绿色设计中心(见图6.1 a)坐落于天津市滨海高新区,由5 600 m² 的旧电子厂房改造而成。该建筑秉承"被动技术优先,主动技术优化"的设计原则,综合应用成熟型绿色技术,包括最大限度地利用原建筑结构体系,优化围护结构热工性能,优化室内自然采光与自然通风,合理选择空调冷热源及空调末端,设置能耗监测与展示系统,设置室内空气质量检测系统等。

(a) 外观图
Appearance diagram

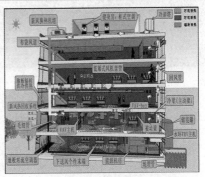

(b) 建筑节能技术应用
Building energy saving technology

图6.1　天友绿色设计中心
Fig.6.1　Tianyou Green design center

Tianyou green design center (Fig. 6.1 a) is located in Binhai New Area of Tianjin. It is an old electronic factory transformation project which covers 5 600 square meters. The building's design principle is "passive technology priority, active technology optimization". Some advanced green techniques are applied comprehensively. They include maximum utilization of the original building structure system, optimization of the building envelope's thermal performance, optimization of natural lighting and natural ventilation, the reasonable selection of air conditioning cold and heat source and air conditioning terminal device, setting energy consumption monitoring and display system, setting the indoor air quality detection system, etc.

　　其中,被动式节能技术主要包括聚碳酸酯幕墙、活动隔热墙、特朗博墙、水蓄热墙、分层拉丝垂直绿化系统;主动式节能技术包括模块式地源热泵、免费冷源、地板辐射供冷供热、能源监测与自控系统等(见图 6.1 b)。天友绿色设计中心还将种植屋面及垂直绿化引入办公建筑(见图 6.2)。

　　长久以来,人们都渴望有温暖舒适的居住和工作环境。在远古时代,人类的祖先借山洞栖息,躲避风雨严寒。随着时代的进步,科学技术的发展,人们开始有能力建造房屋,为自己搭建更安全可靠的庇护之所。但是仅有一处住所仍然是不够的,人们还希望自己的家冬暖夏凉,能方便地使用水、电等生活设施,使自己能在一个舒适的环境中度过愉快的时光,而这正是建筑环境与设备工程专业人员即公用设备工程师所从事的工作。

(a) 种植屋面 Planted roof　　　　　　(b) 垂直绿化 Vertical greening

图 6.2　种植屋面及垂直绿化

Fig. 6.2　Planted roof and vertical greening

Passive energy saving techniques include polycarbonate curtain wall, activities insula tion wall, Trumbo wall, water storage wall and stratified wire drawing vertical greening system. Active energy saving techniques include module-type ground source heat pump, free cold source, floor radiant heating, energy monitoring and control system etc.(Fig. 6.1 b). At the same time, planted roof and vertical greening are also used in this office building.(Fig. 6.2).

For a long time, people are eager to live and work in warm and comfortable environment. In ancient times, our ancestors lived in caves to escape the cold weather. With the development of science and technology, people began to have the ability to build houses which are more secure and reliable for themselves. However, only a place to stay in is not enough. People want to have a comfortable and convenient house, as well as the basic facilities like water, electricity and other living facilities. These equipment and conditions provide a convenient living environment for people, and these are what the building environment and public facility engineers should do.

如今，公用设备工程师的任务已不是简单地控制室内各个环境参数的变化范围，而是应当站在人类可持续发展的高度，从保护环境、节约能源的角度出发，合理利用室外环境，创造出低能耗的各种建筑设备系统，满足人们生活和生产的需要。

6.1　建筑环境

建筑的功能是在自然环境不能令人满意的条件下，创造一个微环境来满足居住者的安全需求及生活生产的需要。因此从建筑出现开始，"建筑"与"环境"这两个概念就不可分割。建筑环境研究包括室内外的温度、湿度、气流、空气品质、采光与照明性能、噪声与室内音质等内容，为营造一个舒适、健康的室内外环境提供理论依据。建筑环境一方面需要合理的建筑物理设计，另一方面需要高效的建筑设备。建筑环境与设备工程的主要内容如图 6.3 所示。

6.1.1　建筑热湿环境

热舒适性是人体通过自身的热平衡条件和对环境的热感觉，经综合判断后得出的主观评价或判断。除了衣着、活动方式等个人因素外，影响人体热平衡进而影响热舒适性的环境因素主要是温度、湿度、气流速度和平均辐射温度，即室内热湿环境。

Nowadays, the tasks of public facility engineers are not simply to control the variation range of each parameter in the indoor environment, they should take the sustainable development of mankind into account. From the view of environmental protection and energy conservation, they should use the outdoor environment rationally, as well as create various construction equipment systems to meet the needs of living and production.

6.1　Building environment

The function of the building is to form a local environment for its resident's living and production demand when the natural environment is not satisfying. The concept of "building" and "environment" cannot be separated. The study on building environment includes indoor and outdoor temperature, air flow, air quality, daylighting and lighting, noise and indoor acoustics, and building environment can provide the theoretical guidance for building a comfortable and healthy indoor and outdoor environment. The maintain of suitable building environment needs reasonable physical design and effective building equipment. The main contents of building environment and equipment are shown in Fig.6.3.

6.1.1　Building thermal and humidity environment

Thermal comfort is the subjective evaluation obtained by thermal balance condition, thermal feeling, and personal judgments. Besides dressing, activity and other personal factors, the main environment factors which affect the thermal comfort are temperature, humidity, air flow speed and average radiant temperature, namely, indoor thermal and humidity environment.

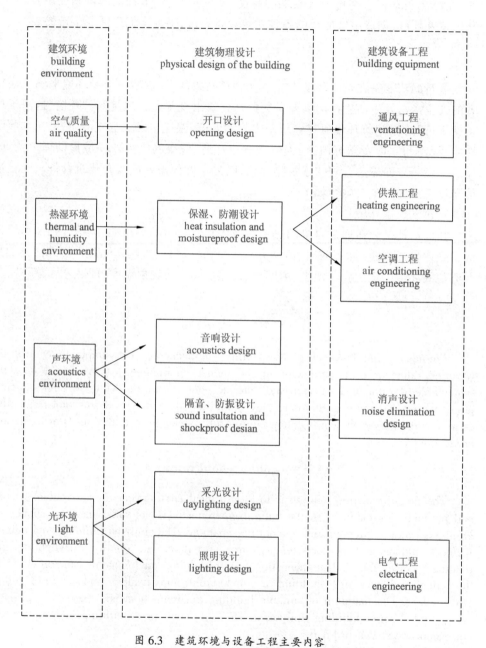

图 6.3　建筑环境与设备工程主要内容
Fig. 6.3　Content of the building environment and equipment

建筑室内热湿环境受室外气象参数与建筑围护结构的影响。通过非透明外围护结构的热传递方式为热传导,通过透明围护结构的热传递方式包括热传导与日射得热两种。通过围护结构的湿传递与室内外水蒸气分压力有关,稳定情况下,单位时间内通过单位面积围护结构的水蒸气量与两侧空气中的水蒸气压力差成正比。除围护结构外,建筑室内热湿环境受室内设备、照明及人体等室内热湿源的影响。建筑室内热湿环境的形成如图 6.4 所示。

为维持舒适的室内热湿环境及降低暖通空调系统能耗,许多国家均对围护结构的热性能指标、暖通空调系统室内温湿度设计参数做出了规定。

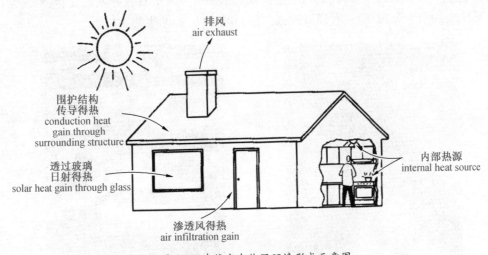

图 6.4　建筑室内热湿环境形成示意图
Fig. 6.4　Diagram of the indoor thermal and humidity environment

Indoor thermal and humidity environment are affected by weather conditions and surrounding structures. The heat transfer mode through the envelop enclosure of nontransparent surrounding structure is heat conduction, and the heat transfer mode through the envelop enclosure of transparent surrounding includes heat conduction and solar heat gain. The moisture transfer through the surrounding structure is affected by the indoor and outdoor water vapor pressure, and in the steady state, the quantity of the water vapor transferred through unit area of the surrounding structure during unit time is proportional to the difference of the water vapor pressure in the air of the two sides. Forming of the indoor thermal and humidity environment is shown in Fig.6.4.

In order to keep suitable indoor thermal and humidity environment, and decrease the energy consumption of the heating ventilating and air condition(HVAC) system, the thermal performance of the surrounding structure and the design parameters of the HVAC system for temperature and humidity are normalized in many countries.

6.1.2 室内空气品质

随着信息化的发展,越来越多的人长期在建筑内生活、学习和工作。现代化建筑的功能越来越丰富,又非常封闭,造成室内环境恶化且建筑能耗增加。典型的现代化建筑及其室内办公环境如图 6.5、图 6.6 所示。长期在现代建筑中生活和工作的人群,表现出越来越严重的病态反应,包括眼睛发红、流鼻涕、嗓子疼、困倦、头痛、恶心、头晕、皮肤瘙痒等,这种症状称为病态建筑综合征。

室内空气品质问题已引起许多国家、地区和组织的重视,各国先后制定了相关的标准。室内空气品质恶化的原因主要来自暖通空调系统和室内污染物作用两个方面。暖通空调系统方面包括通风与气流组织不好、新风量不足等。室内空气污染源主要包括来自建筑装饰材料、复合木建材及其制品所散发的有机挥发性化合物,灰尘、纤维尘和烟尘等物理污染,以及细菌、真菌和病毒引起的生物污染。

图 6.5 现代化的封闭式建筑
Fig. 6.5 Modern closed buildings

图 6.6 现代建筑中的办公环境
Fig. 6.6 Working environment in modern buildings

6.1.2 Indoor air quality

With the development of information technology, more and more people are living, studying and working in buildings. Modern buildings have more and more functions, but they are very closed, which will deteriorate the indoor environment and increase the energy consumption of these buildings. Typical modern buildings and indoor office environment are shown in Fig.6.5 and Fig.6.6. People living and working in these modern buildings have more and more serious morbidity, such as red eyes, runny noise, sore throat, sleepy, headache, nausea, dizziness, and skin pruritus. These dislocations are called sick building syndrome.

Importance has been attached to indoor air quality in many countries, regions and organizations, and its relative standard has been enacted in many countries. The reasons for poor indoor air quality come mainly from HVAC system and indoor pollutant. The reasons of the HVAC system include the unreasonable organizing of ventilation and air flow, insufficient free air supply, etc. The indoor pollutants mainly come from the volatile organic compound emitted from the construction and decorating material, compositing board and their produce, the physical pollution from the dust, fiber, soot, and the biologic pollution from the bacilli, epiphyte and virus.

　　室内污染的控制可通过以下 3 种方式实现：① 源头治理；② 通风稀释和合理组织气流；③ 空气净化。

6.1.3　建筑声环境

　　建筑声环境主要研究室内声音控制问题，包括三方面内容，即室内音质设计、建筑隔声和噪声控制。

　　音质设计问题一般只限于各类厅堂，如影剧院、音乐厅、体育馆、报告厅等。音质设计的主要内容包括房间容积设计、体型设计、噪声控制、扩声系统设计等。中国国家大剧院音乐厅的体型设计如图 6.7 所示。

　　随着城市化程度的日益加剧，隔声和噪声控制成为各类建筑面临的一个普遍性问题。噪声主要来源于建筑外部环境、建筑内部其他房间、室内设备、空调通风等。不同噪声需采取不同的控制措施，主要的措施有隔声、吸声与消声。

图 6.7　中国国家大剧院音乐厅
Fig. 6.7　The odium of Nation Centre for the Performing Arts in China

　　Three methods can be used to control the indoor pollution: ① source controlling; ② ventilation diluting and reasonable air flow organizing; ③ air purification.

6.1.3　Building acoustic environment

　　The main content of building acoustic environment is the indoor acoustic controlling, including indoor tone design, sound insulation design and noise control.

　　Indoor tone design is needed only in hall buildings, such as theater, odium, gymnasium, reporting hall. The indoor tone design mainly includes cubage confirm, shape design, noise control design, sound-reinforcement system, etc. The shape design of the odium of Nation Centre for the Performing Arts in China is shown in Fig.6.7.

　　With the development of urbanization, sound insulation and noise controlling have become common needs in all buildings. The sources of noise include the outer environment of the building, other rooms, indoor equipment, air flow, etc. Different noises need different methods to control. Main noise control methods include insulation, absorption and elimination.

6.1.4 建筑光环境

建筑光环境设计包括建筑采光设计和建筑照明设计两部分内容。建筑采光设计就是设法通过采光口使光线进入室内。建筑照明设计通过人工光源的应用,改善建筑的功能效益和环境质量,提高人们的视觉功效。随着建筑照明技术的进步和社会经济的发展,公共照明、景观照明和夜景照明正在逐步发展和完善。图6.8、图6.9为采光设计实例。图6.10为人民大会堂顶部LED照明设计。

图6.8　教室自然采光

Fig. 6.8　Daylighting of a classroom

图6.9　商场天井采光

Fig. 6.9　Daylighting of a shop through skylight

图6.10　人民大会堂顶部LED照明

Fig. 6.10　Top LED lighting in the Great Hall of the People

6.1.4　Building light environment

Building light environment includes daylighting design and light design. Building daylighting design is about lighting opening design. Building light design is to improve the function and environment of the building, enhance people's vision through artificial light source. With the advancement of lighting technology and the development of economy, public lighting and landscape lighting and night lighting have become developed and perfect. Fig.6.8 and Fig.6.9 show the examples of daylighting design. Fig.6.10 shows the LED lighting design of the Great Hall of the People.

6.2 供暖工程

供暖系统的目的是满足人们日常生活和社会生产所需要的大量热能,它是利用热媒(如热水、水蒸气或其他介质)和热力管道将热能从热源输送至各个热用户的工程技术。

6.2.1 供暖系统的分类与组成

供暖系统主要由热源、热媒输配和散热设备 3 个部分组成。根据这 3 个部分的相互位置关系,供暖系统又可分为局部供暖系统和集中供暖系统。热源、热媒输配、散热设备设置在一起的供暖系统为局部供暖系统;热源、散热设备分别设置,用热媒输送管道相连,由热源向各个部分供给热量的供暖系统为集中供暖系统。目前,集中供热已成为现代化城镇的重要基础设施之一,是城镇公共事业的重要组成部分。另外,按采用热媒方式的不同也可将供暖系统分为热水供暖系统、蒸汽供暖系统等。供暖系统的分类及组成如图 6.11 所示。

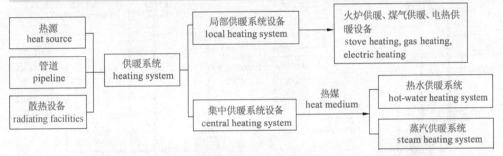

图 6.11　供暖系统的分类及组成

Fig. 6.11　Classification and composition of heating system

6.2 Heating engineering

The purpose of heating system is to meet the great requirements of heat in people's daily life and social production. It is an engineering technology using the heat medium (such as hot water, steam and other medium) to transfer the heat from the heat source to each user through the heating pipe line.

6.2.1 Classification and composition of heating system

Heating system is mainly composed of three parts: the heat source, the thermal medium distribution and the cooling device. The heating system can be divided into local heating and central heating system according to the mutual position relationships among the three parts. The heating system having heat source, heat medium conveying and cooling device on the structure is the local heating system. The heat source and the cooling device should be set respectively, then connected with the heat medium conveying pipes, so as to transmit heat to each part from the heat source. This system is the central heating system. At present, the central heating has become one of the important infrastructures of modern towns and an important part of urban public utilities. In addition, the heating system can also be divided into hot-water heating system and steam heating system according to the different using ways of heat medium. The classification and composition of heating system is shown in Fig.6.11.

6.2.2 热水供暖系统

以热水为媒介的供暖系统,称为热水供暖系统。

(1) 热水供暖系统的分类

按照热水供暖循环动力的不同, 可分为自然循环系统和机械动力系统 (见图 6.12)。

(2) 热水供暖系统的主要设备

1) 散热器

散热器是安装在房间内的一种放热设备,也是我国目前大量使用的一种散热设备(见图 6.13)。它是把来自管网的热媒的部分热量传入室内,以补偿房间散失的热量,维持室内所要求的温度,从而达到供暖目的的设备。

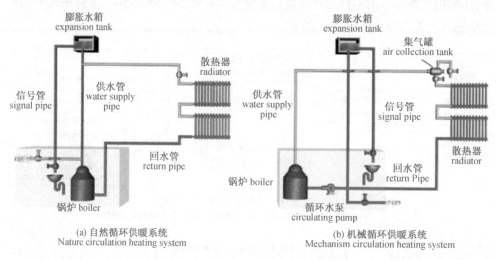

图 6.12 热水供暖系统
Fig. 6.12 Hot-water heating system

6.2.2 Hot-water heating system

The heating system using hot water as heat medium is known as the hot-water heating system.

(1) Classification of hot-water heating system

According to the different power supply for hot water cycle, the hot water cycle can be divided into natural circulation system and mechanical circulation system (Fig. 6.12).

(2) The main equipment of hot-water heating system

1) Radiator

Radiator is a heating device installed in the room, and it is also a cooling device used widely at present in China (Fig.6.13). It transfers heat from the pipe network to indoor partly, so as to compensate the heat loss in rooms and maintain the indoor temperature to achieve the purpose of heating.

2）膨胀水箱

膨胀水箱一般安装在系统的最高点，用来容纳系统加热后膨胀的体积水量，并控制系统的充水高度，保证系统压力稳定（见图 6.14）。

3）排气设备

排气设备是及时排出供暖系统中空气的重要设备。在不同的系统中可以使用不同的排气设备，如集气罐、自动排气阀（见图 6.15）、手动放气阀等。

图 6.13　散热器
Fig. 6.13　Radiator

图 6.14　膨胀水箱
Fig. 6.14　Expansion tank

图 6.15　自动排气阀
Fig. 6.15　Automatic exhaust steam valve

2）Expansion tank

Expansion tank is installed on the highest point of the system, which is used to accommodate the inflated volume of water after the system has been heated and control the water level（Fig.6.14）.

3）Exhaust equipment

Exhaust equipment is an important equipment to remove the air timely in the heating system. Different exhaust equipment can be used in different systems, such as tank, automatic exhaust steam valve（Fig.6.15）, the manual air bleed valve.

4）散热器控制阀

散热器控制阀安装在散热器入口管上,是根据室温和给定温度之差自动调节热媒流量的大小来控制散热器散热量的设备(见图 6.16)。

6.2.3 蒸汽供暖系统

以蒸汽为热媒的供暖系统称为蒸汽供暖系统,其应用极为普遍。图 6.17 为蒸汽供暖系统示意图,蒸汽从热源(锅炉)沿蒸汽管路进入散热设备(散热器),蒸汽凝结放出热量后,凝水通过疏水器靠重力流至凝结水箱,再通过凝水泵返回热源重新加热。

（1）蒸汽供暖系统的分类

按照供气压力的大小,蒸汽供暖可分为两类:供气的表压力>70 kPa 时,称为高压蒸汽供暖;供气的表压力≤70 kPa 时,称为低压蒸汽供暖。

图 6.16 温控阀
Fig. 6.16 Temperature control valve

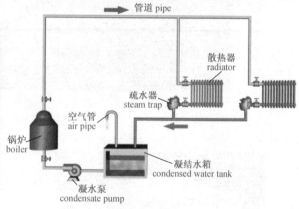

图 6.17 蒸汽供暖系统
Fig. 6.17 Steam heating system

4）Radiator control valve

Radiator control valve is the equipment installed on the inlet pipe of a radiator. According to the difference between the room temperature and the given temperature, this device can regulate the rate of the heat medium flow and control the heat released by the radiator automatically(Fig.6.16).

6.2.3 Steam heating system

Steam heating system is the heating system using steam as the heating medium, and it has a very broad application. Fig.6.17 is a schematic diagram for steam heating system. Steam from the heat source(boiler) flows along the steam pipeline into the heat radiation device (radiator). After releasing heat, steam transforms into condensed water and flows to the condensed water tank through the steam trap of gravity, and then relies on the condensate pump back to the heat source.

（1）Classification of steam heating system

According to the steam pressure, steam heating can be divided into two categories,namely supplied steam gauge pressure>70 kPa, and supplied steam gauge pressure ≤ 70 kPa. The former is also named as high pressure steam heating, the latter is called low pressure steam heating.

（2）蒸汽供暖系统主要设备

蒸汽供暖系统的主要设备如图 6.18 所示。

① 疏水器。疏水器是蒸汽供暖系统中的重要设备，其作用是自动阻止蒸汽逸漏，迅速排出热设备及管道中的凝水，排除系统中积留的空气和其他不凝性气体。

② 凝结水箱。凝结水箱是贮存凝结水的设备。

③ 管道补偿器。在供暖系统中，金属管道会因受热而伸长，又由于平直管道的两端都被固定不能自由伸缩，管道就会弯曲变形，严重时发生破裂，因此需要在管道上设管道补偿器。

6.3　通风工程

通风就是把室内被污染的空气直接或净化后排至室外，把新鲜空气补充进来，从而保证室内的空气环境符合卫生标准或满足生产工艺的需要，并在一定程度上改善室内的热湿参数。

(a) 疏水器　　　　　　　(b) 凝结水箱　　　　　　(c) 管道补偿器
Steam trap　　　　　Condensed water tank　　　Pipeline compensator

图 6.18　蒸汽供暖系统的主要设备
Fig. 6.18　Main equipment of steam heating system

（2）The main equipment of steam heating system

The main equipment of steam heating systems are shown in Fig.6.18.

① Steam trap. It is an important equipment in steam heating system. Its function is to prevent leakage of steam automatically, discharge the condensed water in the pipe rapidly, and discharge retention air and other non-condensable gas of the system.

② Condensate tank. It is the equipment to store condensed water.

③ Pipeline compensator. In heating system, the metal pipe is elongated due to heat. Two ends of the straight pipe are fixed and can not be expanded and contracted freely. The pipeline will be deformed and ruptured seriously. Therefore, pipeline compensator must be arranged on the pipe.

6.3　Ventilation engineering

Ventilation is to exhaust the indoor polluted air directly or after purification, and to supply the fresh air, so as to make the indoor environment conform to the healthy standard or production requirement, and improve the indoor thermal and humidity parameters at a certain extent.

6.3.1 通风系统的分类

通风系统按照工作动力可分自然通风和机械通风两种。

自然通风依靠室内外空气的温度差(实际是密度差)造成的热压,或者是室外风造成的风压,促使房间内外的空气进行交换,从而改善室内的空气环境。自然通风不需要另外设置动力设备,是一种经济、有效的通风方法。其缺点是无法处理进入室内的外空气质量,也难以对排出的污浊空气进行净化处理;自然通风受室外气象条件影响,通风效果不稳定。图 6.19 所示为某厂房的自然通风设计。

机械通风依靠风机作用使空气流动,对房间进行通风换气。由于风机的风量和风压可根据需要确定,这种通风方法能保证所需要的通风量,控制房间内的气流方向和速度,并可对进风和排风进行必要的处理,使房间空气质量达到要求。因此,机械通风方法得到了广泛应用。

图 6.19 某厂房的自然通风设计
Fig. 6.19 Natural ventilation design of some plant

6.3.1 Classification of ventilation system

Ventilation system can be classified into natural ventilation and mechanical ventilation according to the working power.

In natural ventilation, air exchange between indoor and outdoor depends on thermal pressure from the different temperature. Actually, it is the different density between indoor and outdoor, or wind pressure from the outdoor winding moving that improve indoor air environment. Natural ventilation does not need any power equipment, and it is an economical and effective ventilation method. One disadvantage of natural ventilation is that it can not handle the quality of air which come into indoor, nor can it cleanse the exhausted air. The other disadvantage is that natural ventilation is affected by outdoor weather, and the ventilation effect is unstable. Fig.6.19 shows the natural ventilation design of some plant.

In mechanical ventilation, air exchanging depends on fans. Because the air quantity and air pressure of fans can be decided according to demand, mechanical ventilation can meet the requirements of the air quantity and control the air flow direction and velocity in rooms. It can also handle the air intake and outtake, make the indoor environment quality parameters meet the requirements. Therefore, mechanical ventilation is applied widely.

6.3.2　通风系统的主要设备

（1）风机

风机是依靠输入的机械能提高气体的压力并排送气体的机械。按照工作原理不同，风机可分为离心式、轴流式和贯流式 3 种。离心式风机和轴流式风机如图 6.20 和图 6.21 所示。

（2）风管

风管是输送空气的管道。按其横截面形状，风管可分为圆形风管、矩形风管等，如图 6.22 和图 6.23 所示。

图 6.20　离心式风机
Fig. 6.20　Centrifugal fan

图 6.21　轴流式风机
Fig. 6.21　Axial flow fan

图 6.22　圆形风管
Fig. 6.22　Rounded air duct

图 6.23　矩形风管
Fig. 6.23　Rectangle air duct

6.3.2　Main equipment of ventilation system

（1）Fans

Fans are used to increase air pressure and transport the air depending on input mechanical energy. According to working principle, fans can be classified into three types: centrifugal, axial flow and cross flow. Centrifugal fan and axial flow fan are shown in Fig.6.20 and Fig.6.21 respectively.

（2）Air duct

Air duct is used to transport air. According to its cross section shape, air duct can be classified into rounded duct and rectangle duct, as shown in Fig.6.22 and Fig.6.23.

（3）送排风口

送排风口是用于室内送排风的装置。送排风口的位置和形式影响室内气流组织。典型的送排风口形式如图 6.24 所示.

（4）排风罩

排风罩是用于收集被污染气体的装置，按其结构形式可分为侧吸式和顶吸式。顶吸式排风罩如图 6.25 所示。

6.4 空调工程

空调工程是把特定空间内部的空气环境控制在一定的状态下，使其满足人体舒适或生产工艺要求的工程技术。它所控制的内容包括空气的温度、湿度、流速、压力、洁净度、噪声等。以生产或科学实验服务为目标的空调系统称为"工艺性空调"，而以人体舒适及健康为目标的空调系统称为"舒适性空调"。

图 6.24　送排风口形式
Fig. 6.24　Forms of supply and exhaust outlet

（3）Supply and exhaust outlet

Supply and exhaust outlet is used to supply and exhaust air. The position and form of the supply and exhaust outlet affect the organization of the indoor air. Typical forms of supply and exhaust outlet are shown in Fig.6.24.

（4）Exhaust hood

Exhaust hood is used to collect the polluted air. According to its structure, exhaust hood can be classified into side hood and top hood. Fig.6.25 shows the top exhaust hood.

图 6.25　顶吸式排风罩
Fig. 6.25　Top exhaust hood

6.4 Air-conditioning engineering

Air-condition engineering is the technology used to control the air environment in a special state, so as to satisfy the demand of the people and production technology. Air-conditioning engineering includes temperature, humidity, velocity, pressure, cleanliness, noise, etc. The air-conditioning system used for production and science experiment is called "industrial air-conditioning". The air-conditioning system used for people's comfort and health is called "comfortable air-conditioning".

6.4.1　空调系统分类

空调系统按空气处理设备的设置情况,可分为集中式系统、半集中式系统和全分散式系统;按负担室内空调负荷所用的介质,可分为全空气系统、空气–水系统、全水系统和制冷剂系统(具体分类见表 6.1)。

表 6.1　空调系统分类
Tab. 6.1　Classification of air-conditioning system

分类方法 Classification method	空调系统 Air-conditioning system	系统特征 Characteristics of the system
按空气处理设备的设置情况 according to the arrangement of air handling units	集中式系统 centralized system	空气处理设备集中在机房内,空气经处理后,由风管送入各房间 all air handling units are equipped in machine room, and air is handled and transported to each room through air duct
	半集中式系统 half-centralized system	除了有集中的空气处理设备外,在各个空调房间内还分别设有处理空气的"末端装置" besides centralized air handling unit, there are end units in each room to handle the air further
	全分散式系统 dispersed system	每个房间的空气处理分别由各自的整体式(或分体式)空调器承担 air handle is taken on by the incorporate (or separated) air conditioners in each room
按负担室内空调负荷所用的介质 according to the media taking on the air conditioning load	全空气系统 air system	全部由处理过的空气负担室内空调负荷 all air-conditioning load is taken on by handled air
	空气–水系统 air and water system	由处理过的空气和水共同负担室内空调负荷 air-conditioning load is taken on by handled air and water
	全水系统 water system	全部由水负担室内空调负荷 all air-conditioning load is taken on by water
	制冷剂系统 refrigerant system	制冷系统的蒸发器直接放在室内承担空调负荷 all air-conditioning load is taken on by the evaporator of the refrigerator directly

6.4.1　Classification of air-conditioning system

According to the arrangement of air handing units, air-conditioning system can be classified as centralized system, half-centralized system and dispersed system. According to the media taking on the air conditioning load, air-conditioning system can be classfied to air system, air and water system, water system and refrigerant system (Tab. 6.1).

6.4.2 空调系统主要设备

（1）冷热源

冷热源是空调系统冷量和热量的来源。典型的冷源包括电动压缩式冷水机组、吸收式冷水机组；典型的热源包括热泵、锅炉、城市热网。图6.26为空调系统中广泛应用的电动压缩式冷水机组。

（2）水系统相关设备

空调中的水系统包括冷冻水系统、冷却水系统和冷凝水系统。冷冻水系统承担室内的空调负荷,冷却水系统承担冷水机组的冷却负荷,冷凝水系统承担冷冻除湿形成的冷凝水的输送任务。其中冷冻水系统和冷却水系统一般需要水泵驱动。图6.27所示为冷冻水系统及其水泵。图6.28所示为冷却水系统中用的冷却塔。

图 6.26 电动压缩式冷水机组
Fig. 6.26 Electric compressing chiller

图 6.27 冷冻水系统及水泵
Fig. 6.27 Chilled water system and its water pump

图 6.28 冷却塔
Fig. 6.28 Cooling tower

6.4.2 Main equipment of air-conditioning system

（1）Heat and cold source

Heat and cold source is the source of the air-conditioning system. Typical cold sources include electric compressing chiller and absorbing chiller. Typical heat sources include a heat pump, a boiler and a heat supply network. Fig.6.26 shows the widely used electric compressing chiller.

（2）Water system and its related equipment

Water system in the air-conditioning system includes chilled water system, cooling water system and condensate water system. Chilled water system is used to take on the air load. Cooling water is used to cool the chiller, and condensate water system is used to transport the condensate water formed by the cold dehumidification. Chilled water system and cooling water system need water pumps. Fig.6.27 shows the chilled water system and its water pump. Fig.6.28 shows the cooling tower used in cooling water system.

　　(3) 风管及送排风口

　　空调中的风管、送排风口与通风工程中的设备类似。根据风管内空气的用途,风管可分为送风管道、回风管道、排风管道及新风管道等。图 6.29 所示为空调风管及送风口。

　　(4) 空气处理设备

　　空气处理设备用于对空调系统中的空气进行热湿处理及过滤净化,其中应用最广泛的热湿处理设备为喷水室和表面式换热器。将空气过滤、热湿处理单元及风机组合在一个箱体中实现空气综合处理的设备称为组合式空气处理机组(空调箱),如图 6.30 所示。将冷却盘管和风机组合在一起对房间空气直接处理的设备称为风机盘管机组,如图 6.31、图 6.32 所示。

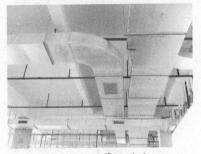

图 6.29　风管及送风口
Fig. 6.29　Air duct and supply outlet

图 6.30　空调箱
Fig. 6.30　Air box

图 6.31　卧式风机盘管
Fig. 6.31　Horizontal fan-coil

图 6.32　卡式风机盘管
Fig. 6.32　Cassette fan-coil

　　(3) Air duct, supply and exhaust outlet

　　Air duct and supply and exhaust outlet in air-conditioning system is similar to those in the ventilation engineering. According to the air function in the air duct, the duct can be classified into supply air duct, return air duct, exhaust air duct, and fresh air duct. Fig.6.29 shows the air duct and supply outlet in the air-conditioning system.

　　(4) Air handling equipment

　　Air handling equipment is used to heat, cool, humidify, dehumidify and purify the air in the air-conditioning system. The most widely used thermal and humidity handling equipment are spraying chamber and surface heat exchanger. The unit that contains air purify, thermal and humidity and fan is called combined air handling unit, as shown in Fig.6.30. The unit that is equipped with a cooling coil and a fan in one box to directly handle the air in the room is called fans coil unit, as shown in Fig.6.31 and Fig.6.32.

（5）单元式空调机

带有制冷压缩机、冷凝器、直接膨胀式空气冷却器、空气过滤器、风机和自控系统等整套装置的空气处理机组，称为单元式空调机组。典型的单元式空调机如图6.33、图6.34和图6.35所示。

6.5 建筑给水排水工程

6.5.1 建筑内部给水系统

建筑内部给水系统是将城镇给水管网或自备水源给水管网的水引入室内，选择适用、经济、合理的最佳供水方式，经配水管送至室内各种卫生器具、生产装置和消防设备，并满足用水点对水量、水压和水质要求的供应系统。

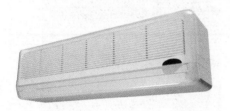

图 6.33 卧式室内机
Fig. 6.33 Horizontal indoor unit

图 6.34 室外机
Fig. 6.34 Outdoor unit

（5）Air conditioning unit

The air handling unit with a compressor, a condenser, a directly expanded air cooler, an air filter, a fan and an automatic control system is called air-conditioning unit. Typical air-conditioning unit is the air condition as shown in Fig.6.33, Fig.6.34 and Fig.6.35.

6.5 Building water supply and drainage engineering

6.5.1 Construction of internal water supply system in the building

Construction of internal water supply system is to bring the water from the urban water supply network or self-provided water network to the room, using the suitable, economical and reasonable way to supply water and deliver water to a variety of indoor sanitary ware, water nozzle, producing device and firefighting equipment through water distribution pipe. This water supply system must meet the requirements of water dosage, water pressure and water quality.

图 6.35 立式室内机
Fig. 6.35 Vertical indoor unit

（1）建筑内部给水系统的分类

建筑内部的给水系统按用途可分为供民用建筑、工业建筑内饮用、烹调、盥洗、洗涤、沐浴等用的生活用水系统，供生产设备冷却、原料产品洗涤、产品制造过程中所需生产用水的生产给水系统，供消防设备灭火用的消防给水系统。

（2）建筑内部给水系统的组成

建筑内部给水系统一般由引入管、给水管道、给水附件、给水设备、配水设施和计量仪表等组成，如图 6.36 所示。

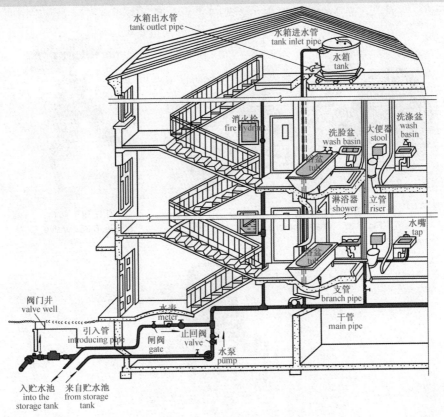

图 6.36　建筑内部给水系统
Fig. 6.36　Construction of internal water supply system

（1）Classification of internal water supply system in the building

According to different functions, the water supply system inside the building can be divided in the following parts: water supply system within the civil buildings, public buildings and industrial buildings for drinking, cooking, washing, bathing and other domestic use, water supply system for cooling production equipment, washing raw material products, supplying water in the process of product manufacturing, water supply system for fire-fighting.

（2）Composition of water supply system in the building

Internal water supply system is composed of an introducing pipe, water supply pipes, water supply accessories, water supply equipment, water distribution facilities and measurement instruments and other components, as shown in Fig.6.36.

　　① 引入管。对一幢单独建筑物而言,引入管是室外给水管网与室内给水管网之间的联络管段,也称进户管。对于一个工厂、一个建筑群体、一个学校区而言,引入管系指总进水管。

　　② 水表节点。水表节点是指引入管上装设的水表(见图 6.37)及其前后设置的闸门(见图 6.38)、泄水装置(见图 6.39)等的总称。

　　③ 管道系统。管道系统是指在建筑内部给水系统中水平或垂直设置的干管、立管、支管。

图 6.37　水表
Fig. 6.37　Water meter

图 6.38　阀门
Fig. 6.38　Valve

图 6.39　泄水阀
Fig. 6.39　Discharge valve

　　① Introducing pipe. As for a separate building, the introducing pipe is the connection part between the outdoor water supply pipe network and the indoor pipe network, and it is known as a tube into the household. As for a factory, a construction group and a school district, the introducing pipe is represented for the total inlet pipe.

　　② Water meter node. Water meter node contains a water-meter(Fig.6.37), a valve (Fig.6.38) and a discharge device(Fig.6.39) set before or after the water-meter.

　　③ Pipeline system. Pipeline system refers to the horizontal or vertical pipes, risers and manifolds inside a building.

　　④ 给水附件。给水附件指管路上的闸阀等各式阀门及各式配水龙头（见图 6.40）。

　　⑤ 增压和贮水设备。在室外给水管网压力不足或建筑内部对安全供水、水压稳定有要求时，需设置各种附属设备，如水箱、水泵、气压设备、水池等增压和贮水设备。

　　⑥ 室内消防。按照建筑物的防火要求及规定需要设置消防给水时，一般应设消火栓消防设备（见图 6.41、图 6.42）。有特殊要求时，另专门装设自动喷水灭火或水幕灭火设备等。

(a) 面盆龙头　　　　(b) 厨房水槽水龙头　　　(c) 浴缸水龙头　　　　(d) 淋浴水龙头
Basin faucet　　　　Kitchen sink faucet　　　Bathtub faucet　　　　Shower faucet

图 6.40　各式配水龙头
Fig. 6.40　Faucets

图 6.41　消火箱
Fig. 6.41　Eliminate fire box

图 6.42　消火栓
Fig. 6.42　Fire hydrant

　　④ Water attachments. Water attachments refer to the valve in the pipe system, all kinds of other valves and taps(Fig.6.40).

　　⑤ Booster and storage equipment. In the condition of low pressure of outdoor water supply network or the special requirements for water safety and steady water pressure, a variety of ancillary equipment such as tanks, pumps, pressure equipment, pools, other pressurization and storage devices are needed.

　　⑥ Indoor fire fighting. In accordance with the provisions of the building and fire safety requirements, we need to set hydrant fire fighting equipment (Fig.6.41 and Fig. 6.42). If there are special requirements, other specialized installation of automatic sprinklers or water curtain fire-fighting equipment should be installed.

6.5.2 建筑内部排水系统

建筑内部排水系统是将生活和生产过程中所产生的污、废水及房屋顶的雨水、雪水,用经济合理的方式迅速排到室外,为室外污水处理和综合利用提供条件的系统。

(1) 排水系统的分类

按照所排的污、废水性质,建筑内部排水系统可分为排除居住、公共、工业建筑生活间污、废水的生活排水系统,排除工艺生产过程中产生的污、废水的工业废水排水系统,排除多跨工业厂房、大屋面建筑、高层建筑屋面上雨雪水的屋面雨水排水系统。

(2) 排水系统的组成

建筑内部排水系统一般由卫生器具和生产设备的受水器、排水管道、清通设施、通气管道及污、废水的提升设备和局部处理构筑物等组成,如图6.43所示。

6.5.2　Internal drainage system in the building

Internal drainage system is to let the wastewater generated in the process of life, rain, and snow on the top of the house out of the room through the economic and reasonable way, and provide conditions for treatment and comprehensive utilization of outdoor wastewater.

(1) Classification of the drainage system

According to the properties of the sewage, the internal drainage system can be divided into drainage system to exclude wastewater of residential, public and industrial buildings, drainage system to exclude wastewater generated in the process of industrial production, drainage system to exclude wastewater of multi-span industrial plants, large roof buildings and high-rise building roof.

(2) Composition of the drainage system

Building internal drainage system generally consists of water heater, drain pipes, clear communication facilities, and ventilation pipes in sanitary ware and production equipment, as well as the upgrade sewage and waste treatment facilities, local structures and other components, as is shown in Fig.6.43.

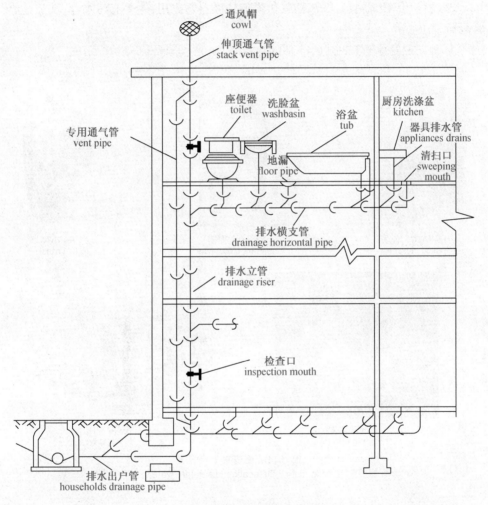

图6.43　建筑内部排水系统
Fig. 6.43　Internal drainage system in the building

1) 卫生器具和生产设备受水器

① 便溺用卫生器具。便溺用卫生器具设置在卫生间和公共厕所,用来收集生活污水。便溺用卫生器具(见图 6.44)主要包括大便器、小便器和冲洗设备。

② 盥洗、沐浴用卫生器具。盥洗、沐浴用卫生器具主要有洗脸盆、盥洗槽、浴盆、淋浴器、净身盆等。

③ 洗涤用卫生器具。洗涤器具供人们洗涤器物之用,主要有污水盆、洗涤盆、化验盆等。

④ 地漏及存水弯(见图 6.45、图 6.46)。

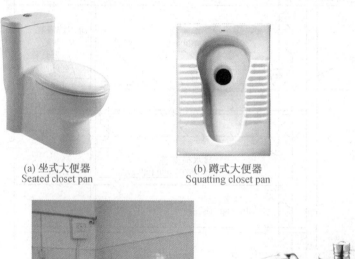

(a) 坐式大便器
Seated closet pan

(b) 蹲式大便器
Squatting closet pan

(c) 小便器
Urinal

(d) 冲洗水箱
Flushing cistern

(e) 冲洗阀
Flush valve

图 6.44　便溺用卫生器具
Fig. 6.44　Defecation in sanitary ware

1) Water heater in sanitary ware and production equipment

① Urinating and defecating sanitary ware. Urinating and defecating sanitary ware is set in the washing room, and is used to collect sewage. Urinating sanitary ware (Fig. 6.44) includes stools, urinals and washing facilities.

② Toilet, shower sanitary ware. There are washbasin, toilet tank, bathtub, shower, and bidet.

③ Washing sanitary ware. Washing appliances are used for people to wash utensils, such as sewage basins, sinks, pots and other laboratory tests.

④ Floor drain and trap are shown in Fig.6.45 and Fig.6.46.

2）排水管道

建筑内部排水管道包括器具排水管道、排水横支管、立管、埋地干管和排出管。

3）清通设施

为疏通建筑内部排水管道，保障排水畅通，需设置 3 种清通设施，即清扫口、检查口、检查口井。

4）通气管道

为防止因气压波动造成水封破坏而将排水管内臭气和有害气体排到大气中，需在建筑内部排水系统中设置通气管道，与大气相通（见图 6.47）。

5）提升设备

各种建筑地下室中的污、废水不能自流排至室外检查井，需设集水池和水泵等局部提升设备，将污水排到室外排水管道中去。

图 6.45　地漏
Fig. 6.45　Floor drain

图 6.46　存水弯
Fig. 6.46　Trap

图 6.47　通气管
Fig. 6.47　Vent

2) Drains

Internal drainage pipelines include appliances, horizontal drainage pipes, risers, buried mains and discharge pipe.

3) Clearing facilities

In order to clear the construction of internal drainage channels and guarantee the smooth drainage, we need to set three cleaning facilities: clean mouth, inspection openings, and check wells.

4) Ventilation pipe

In order to prevent the seal damage due to air pressure fluctuations, and prevent releasing harmful gases inside from venting to the atmosphere, we should use a ventilation pipe to connect the internal drainage system with atmosphere(Fig.6.47).

5) Lifting equipment

Sewage, wastewater in the basement of various buildings can not be discharged to outside inspection artesian wells, so it is necessary to set up sump pumps and other partial lifting equipment to discharge sewage drains into the outdoor drainage system.

6）污水局部处理构筑物

建筑物内部污水未经处理不允许直接排入市政排水管网或水体时，需设污水局部处理构筑物。

6.5.3　建筑内部热水供应系统

建筑内部热水供应系统主要供给生产、生活洗涤及盥洗用热水，应能保证用户限时可以得到符合设计要求的水量、水温和水质。

（1）热水供应系统的分类

按照热水供应范围的大小，建筑内部热水供应系统分为区域热水供应系统、集中热水供应系统和局部热水供应系统。

（2）热水供应系统的组成

热水供应系统的组成因建筑类型和规模、热源情况、用水要求、加热和储存设备的情况、建筑对美观和安静的要求等不同情况而异。

建筑内部热水供应系统通常由加热设备(如锅炉、太阳能热水器、直燃机、各种热交换器等)，热媒管网(蒸汽管或过热水管、凝结水管等)，热水储存水箱，热水输配水管网与循环管网，其他设备和附件组成。图 6.48 所示为一典型的局部热水供应系统。图 6.49 所示为一新型的太阳能集中热水供应系统。

6）Local sewage treatment structures

Internal untreated wastewater is not allowed to directly discharge into the municipal sewer or water system. In this case, the local sewage treatment structure needs to be set up.

6.5.3　Hot water supply system in the building

The hot water supply system inside the building mainly supplies water for production, life washing and toilet. It should ensure that users can get the amount of designed water with the right temperature and quality in time.

（1）Classification of the water supply system

In accordance with the range of hot water supply, hot water supply systems inside the building can be divided into regional water supply system, centralized hot water supply system and local water supply system.

（2）Composition of the water supply system

The composition of the hot water supply system is different due to building type and size, heat situation, water requirements, heating and storage of the device, and the demands of construction.

Hot water supply system inside the building is generally consisted of the following parts: heating equipment (such as boilers, solar water heaters, direct gas turbine, variety of heat exchangers), the heat medium pipe (steam or superheated water, condensation water, etc.), hot water storage tanks, water distribution networks and water circulation pipe network, other devices and accessories. Fig.6.48 shows a typical local water supply system. Fig.6.49 shows a new type of concentrating solar hot water supply system.

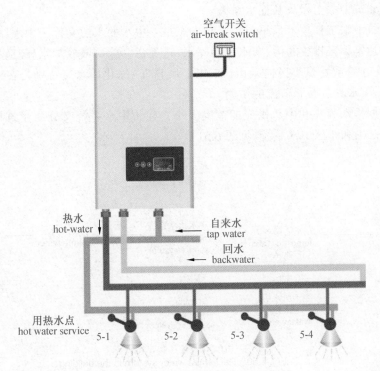

图 6.48　局部热水供应系统
Fig. 6.48　Local hot water supply system

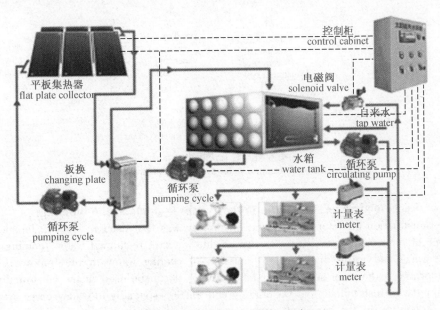

图 6.49　太阳能集中热水供应系统
Fig. 6.49　Central heating system

6.5.4 建筑中水工程及其他水系统

建筑中水工程技术最早应用于日本东京。中水的水源又称为中水原水,来自建筑物或建筑小区排放的污、废水或冷却水,这类污、废水(或冷却水)经适当水质处理后,能在建筑或建筑小区内杂用(如冲厕所、洗车、绿化用水),特别是在水资源缺乏的地区,中水具有开源节流的作用。

根据排水收集和中水供应的范围大小,建筑中水系统又分为建筑物中水系统(见图6.50)和小区中水系统(见图6.51)。

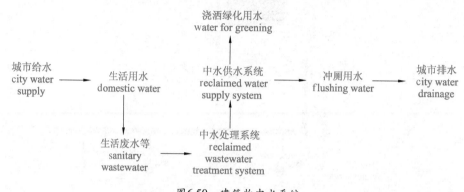

图6.50 建筑物中水系统
Fig. 6.50 Reclaimed water system in the building

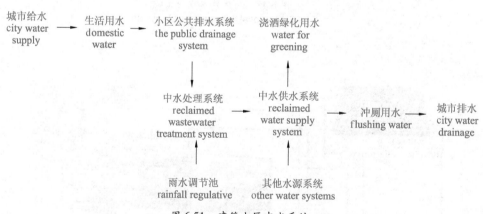

图 6.51 建筑小区中水系统
Fig. 6.51 Reclaimed water system in residential district

6.5.4 Reclaimed water system and other water systems in the building

Construction of water engineering technology was first used in Tokyo, Japan. The source of reclaimed water is also known as raw water of reclaimed water, coming from the sewage and waste discharge or emission of cooling water in the building. With appropriate treatment of sewage and waste water, it can be used in the construction or building(e.g. flush toilets, car washing, green water) especially in water-scarce areas.

According to the range of water supply and drainage collection, the reclaimed water system in the building is divided into the internal water system (Fig.6.50) and community water system(Fig.6.51).

其他的水系统有雨水系统、特殊建筑给水排水系统(如游泳池、洗衣房用水)、直饮水供应系统、喷池等景观建筑水系统等,这些水系统对完善建筑功能、改善建筑环境具有重要的作用。如在宾馆、公寓、医院等公共建筑中常设有洗衣房,用于洗涤床上用品、各类工作服等,以增加建筑的服务功能;水景不仅可以美化环境、装饰厅堂,还可以起到增加空气湿度、增加负氧离子浓度、净化空气、降低气温等改善小区气候的作用,也可以兼作消防、冷却喷水的水源。

6.6 未来展望

健康、能源、环境已成为备受人类关注的三大主题,建筑环境与设备工程和这3个方面有着密切的关系。土木工程行业将越来越关注建筑的"可持续发展"技术,关注节能环保,控制建筑设计与施工及使用过程中对自然环境造成的影响,降低室内外建筑环境控制中建筑设备的能耗。例如,国外广泛使用的被动式太阳能采暖及降温装置,为采暖、通风、空调技术提供了新型的冷源和热源;使用程序控制装置调节建筑的通风空调系统,可以使建筑物的通风量随气象参数自动调节,保证室内卫生舒适条件;使用自动温度调节器,可以保证室内采暖及空调的设计温度,并节约能源。节能建筑与绿色建筑将成为土木工程行业的重点发展方向。

Other forms of water systems include storm water system, special building water supply and drainage system such as swimming pools, laundry water, drinking water supply system, and landscape architecture system such as spray pool water system. These water systems play an important role in improving the function and environment of the building. Hotels, apartments, hospitals and other public buildings usually have laundry room for washing quilt and all kinds of clothes, etc., which can improve the service capability of architecture. Water landscape can beautify the environment, decorate hall, increase the air humidity and the concentration of negative oxygen ions, purify the air, reduce the temperature and change the climate of the area. It also can be used as the source of firewater.

6.6 Future prospects

Health, energy, and environment have become three topics recognized by people all over the world, and building environment and equipment engineering are related to these three topics closely. Civil engineering industry will pay more and more attention to the sustainable development technology. Besides, this industry will concentrate on the energy saving and environment protecting technology, so as to control the effect on nature and reduce the energy consumption of building equipment during controlling the indoor and outdoor building environment. For example, passive solar heating and cooling equipment have been widely used overseas. It provides a new type of cold and heat source for heating, ventilation and air-conditioning technology. By using the program control device regulating building ventilation and air conditioning system, we can adjust the ventilation quantity automatically along with the meteorological parameters and ensure health and comfortable indoor conditions. By using automatic temperature regulator, we can get the indoor heating and the given air-conditioning temperature and save energy. Energy saving buildings and green buildings will be the important development directions of this industry.

注:本章图片均来源于网络。
Note: In this chapter, all pictures are from webs.

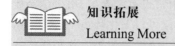

知识拓展
Learning More

上海世博会工程
绿色施工关键
技术 PPT

相关链接 Related Links

(1) 美国供热、制冷、空调工程师学会(ASHRAE)

(2) 国际室内空气品质和气候学会(ISIAQ)

(3) 中国建筑学会(ASC)

(4) 中国环境科学学会(CSES)

(5) 中国建筑业协会(CCIA)

常用应用软件介绍 Brief introduction of common application softwares

目前与本行业相关的专业软件众多,大致可以分为三类:

第一类是工程设计类,包括计算机绘图和计算软件,如AutoCAD,天正建筑设计软件,鸿业空调设计系列软件;

第二类是能耗和环境模拟软件,如FLUENT模拟软件、DOE能耗模拟软件、EnergyPlus、eQUEST快速能耗模拟软件、Ecotect生态建筑分析大师、DeST;

第三类是建筑设备诊断软件和工具等,如麦克维尔公司开发的制冷机组诊断软件等。

Currently, large numbers of professional softwares related to this industry can be divided into three categories.

The first category is engineering design categories, including computer graphics and computational software, such as AutoCAD, Tengen architecture design software, and Hongye air-conditioning design series software.

The second category is the energy consumption and environmental simulation software, such as FLUENT simulation software, DOE energy simulation software, EnergyPlus, eQUEST quick energy simulation software, Ecotect analysis of ecological building master, and DeST.

The third category is construction equipment diagnostic software and tools, such as chiller diagnostic software developed by McQuay.

专业执业资格考试 Professional qualification examination

目前,与建筑环境与设备工程专业相关的注册工程师种类主要有注册公用设备工程师、注册监理工程师和注册建造师等。

At present, the main types of registered engineers related to the building environment and equipment engineering are registered public facility engineer, registered supervision engineers and registered architect, etc.

小贴士 Tips

(1) 事故通风

在拟定工业厂房的通风方案时,对于可能突然产生大量有害气体的车间,除应根据卫生和生产要求设置一般的通风系统外,还要另设一个专用的全面机械排风系统,以便在发生上述情况时能够迅速降低有害气体的含量。这样的机械排风系统叫做"事故排风系统"。

(1) Emergency ventilation

When designing ventilation system for plant, a special mechanical ventilation system is needed which is likely to emit large amount of polluted air suddenly, besides the common ventilation system for health and production requirements. The content of the polluted air can be decreased quickly. The mechanical ventilation is called "emergency ventilation".

(2) 蓄冷空调系统

蓄冷空调系统在建筑物不需冷量或需冷量少时(如夜间),利用制冷设备将蓄冷介质中的热量移除,进行蓄冷,并将此冷量用在空调用冷的高峰期。蓄冷空调转移了制冷设备的运行时间,可以利用夜间的廉价电降低运行成本,同时减少白天的峰值电负荷,达到电力"削峰填谷"的目的。目前许多国家已将空调蓄冷技术作为重点的建筑节能技术进行推广。

(2) The cold storage air-conditioning system

The cold storage air-conditioning system removes the heat in cold storage media and accumulates cold using refrigerator when the building does not need cold or needs a little cold(for example at night). Therefore, the building can use these stored cold in the rush hour. The cold storage air-conditioning system shifts the running time of refrigerator, and reduces operating cost, because it uses the cheap electricity at night, and decreases the peak electricity load at day. Cold storage air-conditioning system has been adopted in many countries as an important energy saving technology.

(3) 地源热泵系统

地源热泵是一种利用大地能量,包括土壤、地下水、地表水等天然能源作为冬季热源和夏季冷源,然后由热泵机组向建筑物供冷供热的系统,是一种利用可再生能源的既可供暖又可制冷的新型中央空调系统。

但由于技术限制,抽取地下水水源热泵很难实现全部回灌,监督实施也比较困难,而且容易造成地下水污染。目前,国外大面积推广使用的是埋管式地源热泵技术,这是充分利用浅层地热的最佳技术途径。

目前埋管式地源热泵在欧美国家已得到普遍应用,被充分证明是成熟可行的技术。在我国,一些省市的建筑节能政策中也明确提出了要推广使用地源热泵。

(3) The ground-source heat pump system

The ground-source heat pump is a new central air-conditioning renewable energy system using the earth energy, including soil, groundwater, surface water and other natural energy as a heat source in winter and cold source in summer.

It is not easy to recharge the entire pipe with hot groundwater for the groundwater source heat pump because of the technical limitations, and the implementation is also difficult. It is likely to cause groundwater contamination. Currently, most areas in foreign countries promote the use of buried ground-source heat pump technology, and it is the best technical way of making full use of shallow geothermal.

Currently, buried ground-source heat pumps have been widely used in the United States and Europe, and have been proved to be the most mature and viable technology. In China, some building energy polices in some provinces clearly require to promote the use of ground-source heat pumps.

思考题 Review Questions

(1) 建筑、能源与环境的关系是什么？

What is the relationship among the building, the energy and the environment?

(2) 建筑环境与设备工程有哪些新技术应用？

What new technology applications have been used in the field of building environment and equipment engineering?

参考文献
References

[1] 霍达. 土木工程概论[M].北京:科学出版社,2007.

[2] 刘俊玲,庄丽. 土木工程概论[M].北京:机械工业出版社,2009.

[3] 卢军. 建筑环境与设备工程概论[M]. 重庆:重庆大学出版社,2003.

[4] 王增长. 建筑给水排水工程[M].6版.北京:中国建筑工业出版社,2010.

[5] 张国强,李志生. 建筑环境与设备工程专业导论[M]. 重庆:重庆大学出版社,2007.

[6] 朱颖心. 建筑环境学[M]. 4版. 北京:中国建筑工业出版社,2016.

[7] 柳孝图. 建筑物理环境与设计[M].北京:中国建筑工业出版社,2008.

[8] 赵荣义,范存养,钱以明. 空气调节[M].4版.北京:中国建筑工业出版社,2009.

[9] 孙一坚,沈恒根. 工业通风[M].4版.北京:北京:中国建筑工业出版社,2010.

[10] 中华人民共和国建设部.采暖通风与空气调节设计(GB 50019—2003)[S].北京:中国计划出版社,2004.

[11] 中华人民共和国公安部. 建筑设计防火规范(GB 50016—2006)[S].北京:中国计划出版社,2006.

[12] 中国建筑科学研究院,重庆大学.夏热冬冷地区居住建筑节能设计标准(JGJ 134—2001)[S]. 北京:中国建筑工业出版社,2001.

[13] 中国建筑科学研究院.夏热冬暖地区居住建筑节能设计标准(JGJ 75—2003)[S].北京:中国建筑工业出版社,2003.

[14] 中国建筑科学研究院,中国建筑业协会建筑节能专业委员会. 公共建筑节能设计标准(GB 50189—2005)[S].北京:中国建筑工业出版社,2005.

[15] 中华人民共和国建设部. 建筑节能工程施工质量验收规范(GB 50411—2007)[S].北京:中国建筑工业出版社,2007.

[16] 辽宁省建设厅. 建筑给排水及采暖工程施工质量验收规范(GB 50242—2002)[S].北京:中国建筑工业出版社,2002.

[17] 上海城乡建设和交通委员会. 建筑给水排水设计规范(GB 50015—2009)[S].北京:中国建筑工业出版社,2009.

[18] ASHRAE Handbook. American Society of Heating, Refrigerating and Air Conditioning Engineers[J].International Journal of Refrigeration,1979,2(1):56–57.

[19] CIBSE Guide B. Heating, Ventilating, Air conditioning and Refrigeration[M]. Chartered Institution of Building Services Engineers,2005.

[20] ANSI/ ASHRAE Standard 55—2004. Thermal Environmental Conditions for Human Occupancy. Atlanta: ASHRAE,2004.

第7章 土木工程防灾与减灾

Chapter 7 Civil Engineering Disaster Prevention and Mitigation

先导案例

Guide Case

日本位于亚欧板块的最东边,同时又处于环太平洋火山地震带上,一直就是全球有名的地震、海啸、火山爆发多发国家,日本平均每年有1万多次地震发生。

北京时间2011年3月11日13时46分在太平洋国际海域发生了里氏9.0级地震,震中位于北纬38.1度,东经142.6度,距日本仙台约180 km。对于此次地震,日本气象厅最初定级为7.9级,最后定级为9.0级。美国地质勘探局最初发布的震级数是8.8级,当天修改为里氏8.9级,3月14日同样定级为里氏9.0级。

由于日本的地理位置、地形状况等原因,大地震发生后经常会接连发生海啸。此次地震发生后,日本气象厅随即发布了海啸警报,称地震将会引发浪高约6 m的海啸。日本港湾空港技术研究所受政府委托,负责震后实地考察,该机构23日发布报告说,经测算,3月11日海啸最高巨浪在岩手县登陆,浪高一度达到23.6 m。

Japan lies in the east end of Eurasian plate and seismic belt in the Pacific rim of fire. It has been the famous earthquakes, tsunamis, volcanic earthquake-prone country in the world, with the national earthquake occuring more than 10 000 times per year on average.

At 13:46, March 11,2011 Beijing time, a magnitude 9.0 earthquake happened in the Pacific international waters, and the epicenter was located at 38.1 degrees north latitude, 142.6 degrees east longitude, about 180 km from Sendai, Japan. The Japan Meteorological Agency estimated this degree of the earthquake from 7.9 grading to 9.0 grading. The USGS initially released the degree of the earthquake was 8.8 grading, then changed to 8.9 grading on March 14, it was also classed as 9.0 grading on the Richter scale.

Because of Japan's geographical location, terrain conditions, after the big earthquake, tsunami is often in succession. Immediately after the earthquake, the Japan Meteorological Agency issued a tsunami warning, and said the earthquake will trigger tsunami waves of about 6 meters. Japan harbor airport technology institute, authorized by the government, is responsible for the field after the quake. The agency reported that, on the March 11 the highest tsunami waves landed in Iwate prefecture, and the waves reached 23.6 meters.

福岛核电站由于受到地震影响,冷却水无法被传送到核反应堆,导致堆芯水位下降,核燃料逐渐露出水面。由于缺少冷却水,核燃料得不到冷却,热量在空间有限的容器中聚集,高温下的水在辐射下分解为氢气和氧气,大量高浓度氢气无法正常排出,高温条件下与氧气作用,发生猛烈爆炸,导致严重的核泄漏事故。

此次灾难中共有15 844人死亡,3 450人失踪,造成经济损失1 220亿~2 350亿美元。可见,灾害对人们的人身安全、财产安全造成严重威胁。高效、正确、合理地防范和处理各种应急灾害,成为土木工程从业者必须高度重视的问题。

7.1 防灾减灾概论

7.1.1 灾害的含义与类型

灾害是指那些由于自然的、人为的或人与自然综合的原因产生的对人类生存和社会发展造成损害的各种现象。它一般具有危害性、突发性、永久性、频繁性、广泛性与区域性。土木工程灾害是指由于人们的不当活动——选址、设计、施工、使用和维护导致所建造的土木工程不能抵御突发的荷载,而致使土木工程失效和破坏,乃至倒塌而造成的损失。

Fukushima nuclear power plant was affected by the earthquake. Cooling water can't be transmitted to the reactor, resulting in a decline of core water level, fuel rods exposed above water gradually. Because of a lack of cooling water, fuel rods are not cooling; heat gathered in the container in the limited space; high temperature water is broken down into hydrogen and oxygen core radiation; a large number of high concentration hydrogen cannot be emitted normally, so under the condition of high temperature and its interaction with oxygen, a violent explosion happened and lead to a serious nuclear accident.

15 844 people died in the disaster, and 3 450 people were missing, causing economic loss of $ 122 billion to $ 235 billion. So, disasters threat people's personal safety and property security. Efficient, correct and reasonable preventing and dealing with all kinds of emergency disaster should be attached great importance by civil engineering.

7.1 Introduction to disaster prevention and mitigation

7.1.1 Definition and types of disaster

Disaster is the various phenomena which can cause damage to human beings and social development due to natural reasons, man-made reasons or comprehensive reasons. Disaster is destructive, sudden, permanent, frequent, wide and regional. The civil engineering disaster is caused by people's improper activities, such as location, design, construction, use and maintenance of civil engineering which lead to the damaged or collapse and thus cause large amount of losses.

全世界每年都会发生很多的灾害,严重的灾害会造成建筑物、构筑物的毁坏,交通通信、供水供电等工程中断,并引发次生灾害,导致大量人员伤亡,引起社会动荡,造成严重的经济损失,甚至使一个区域、一个城市在顷刻之间消失。土木工程灾害主要分为自然灾害和人为灾害。

自然灾害是自然界中物质的变化、运动造成的损害,包括地震灾害、风灾害、洪水灾害、滑坡灾害、泥石流灾害等。例如,强烈的地震可使一座上百万人口的城市在顷刻之间化为废墟。2008年5月12日,四川汶川县发生里氏8.0级地震,导致大批房屋倒塌和破坏(见图7.1),造成巨大的人员伤亡和经济损失。

人为灾害是由于人的过错或某些丧失理性的失控行为给人类自身造成的损害,包括火灾、爆炸、地陷(人为地大量开采地下水造成)及不适当的工程设施对环境造成的隐患,或者工程质量低劣造成的工程事故等。例如,2001年9月11日,美国纽约世界贸易中心大厦在飞机撞击后起火,在很短的时间内造成两栋世界标志性摩天大楼的整体倒塌,给美国造成了巨大的灾害(见图7.2)。

图 7.1 汶川地震灾害
Fig. 7.1 Wenchuan earthquake disaster

图 7.2 纽约世贸中心大厦火灾
Fig. 7.2 The New York World Trade Center Building in fire

Disasters occur every year all over the world, and the serious one will destroy the whole of building and its structure, the transportation and communication, the water supply and the electric supply. Serious disasters may trigger a secondary disaster with heavy casualties, social unrest, serious economic losses, and even the disappearance of a region or a city in an instant. Civil engineering disaster is mainly divided into natural disaster and man-made disaster.

Natural disasters are caused by physical changes and movements in nature. These disasters include earthquake, windstorm, flood, landslide, debris flow and so on. For example, a strong earthquake can make a city with millions of people into ruins in an instant. A magnitude 8 earthquake occurred in Wenchuan County of Sichuan on May 12, 2008, resulting in a large number of houses collapsed and damaged (Fig.7.1), which caused huge casualties and economic losses.

Man-made disasters are always caused by fault or irrational behaviors of human being, which include fire, explosion, subsidence (caused by excessive exploitation of ground water) and poor construction quality accidents. For example, the United States World Trade Center Building in New York was on fire after plane crash on September 11, 2001. The two world landmark skyscrapers collapsed in a short time, which caused a great damage to the United States of America (Fig.7.2).

7.1.2　土木工程防灾减灾

我国是世界上自然灾害最为严重的国家之一,灾害种类多、分布地域广、发生频率高、造成损失大。近年来,在全球气候变化和经济社会快速发展的情况下,我国自然灾害损失不断增加,重大自然灾害乃至巨灾时有发生,灾害风险进一步加剧。在这样的背景下,防灾减灾具有更为重要的意义。

防灾减灾系统工程是一个由多种防灾减灾措施组成的有机整体,主要由以下几个环节组成:① 灾害监测。灾害监测指监视测量与灾害有关的各种自然因素变化数据的工作,获取的监测资料用来认识灾害的发生规律并进行预防、预报。如监测地下岩石的运动和应力变化可以预测地震、滑坡等灾害。自然灾害的监测方式主要有:航空遥感监测、地下监测、水面和水下监测等。② 灾害预报。灾害预报是指根据灾害的周期性、重复性、灾害间的相关性、致灾因素的演变和作用、灾害发展趋势、灾源的形成、灾害载体的运移规律,以及灾害前兆信息和经验类比,对灾害未来发生的可能性做出估计或判断。③ 防灾。防灾是指在灾害发生前采取的避让性预防措施,这是最经济、最安全又十分有效的减灾措施。防灾的主要措施有规划性防灾、工程性防灾、技术性防灾、转移性防灾和非工程性防灾等。④ 抗灾。抗灾是指人类面对自然灾害的挑战做出的反应,如抗洪、抗震、抗风、抗滑坡和泥石流等,它主要包括工程结构的抗灾和工程结构灾后的检测和加固等。工程抗灾是防灾总体工作中的关键环节和重中之重。

7.1.2　Civil engineering disaster prevention and mitigation

China is one of the countries which suffers the natural disaster seriously. There are many types of disasters with wide distribution and high frequency. Under the situation of global climate change and rapid development of China's economy and society, the loss caused by natural disaster is increasing. Major natural disasters and catastrophes occur at any time, and further disasters aggravate. In this context, it has an important significance for disaster prevention and reduction. Disaster prevention and mitigation system is composed of a variety of disaster prevention and mitigation measures, which can be divided into the following parts. ① Disaster monitoring. The acquisition of monitoring data related to natural factors is always used to recognize the regulation of disaster occurrence, and offer prevention measures and predictions. For example, the monitoring of underground rock movement and the stress changes can predict earthquakes, landslides and other disasters.② Disaster prediction. The possibilities of the disaster are estimated, according to its periodicity, repeatability, relationship among disasters, its evolution and influence, its development trend, formation, movement, and precursory information and experience analogy.③ Disaster prevention. Disaster prevention measures are taken before disasters, which are the most economical, safe and effective measures. The main measures include planning of disaster prevention, engineering prevention, technology prevention, transfer of disaster prevention and other non-engineering prevention.④ Disaster resistance. Disaster resistance is the human response in front of natural disasters. It includes engineering control, such as earthquake resistance, flood resistance, landslide and debris flow resistance, wind resistance, as well as the detection of disasters and structural reinforcement. Engineering disaster resistance is a key sector in disaster pre-

⑤ 救灾。救灾是指灾害已经发生后迅速采取的减灾措施。救灾实际上是一场动员全社会甚至国际社会力量对抗自然灾害的战斗,从指挥运筹到队伍组织,从抢救到医疗,从生活到公共安全,从物资供应到维护生命线工程,构成了一个严密的系统。

⑥ 灾后重建。灾后重建是指遭受毁灭性的自然灾害,如地震、洪水、飓风等之后,在特殊情况下的建设。

土木工程防灾减灾是综合防灾减灾的重要组成部分,是防灾减灾中最有效的对策和措施。土木工程与防灾减灾的关系如图 7.3 所示。

图 7.3 还表明:

① 几乎所有的自然灾害甚至人为灾害(如战争、核泄漏)都与土木工程有关;

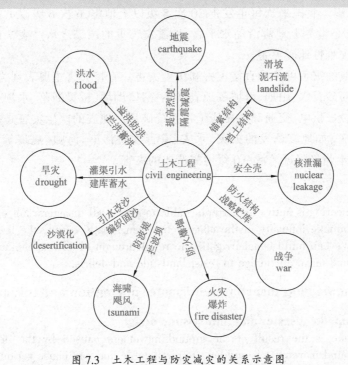

图 7.3 土木工程与防灾减灾的关系示意图
Fig. 7.3 The diagram of the civil engineering and disaster prevention and mitigation

vention. ⑤ Disaster relief. Disaster relief is the mitigation measure which is taken quickly after the disaster occurred. Disaster relief is actually an action of the whole society and even the international community power against natural disasters. From command management to team organization, from the rescue to medical treatment, from life to public security, from the material supply to maintain the lifeline engineering, it forms a tight system. ⑥ Post-disaster reconstruction. Post-disaster reconstruction is the construction in special conditions after the devastating natural disasters, such as earthquake, flood, hurricane, etc.

Civil engineering disaster prevention and mitigation is the most effective strategy and measure for disaster prevention and reduction. Fig.7.3 shows the relationship among civil engineering, disaster resistance and disaster reduction.

Fig.7.3 also shows informations as follows: ① almost all natural disasters and man-made disasters (such as war, nuclear leakage) are related to civil engineering.

　　② 土木工程几乎对所有灾害都具有极强的积极主动性和不可替代性。例如,在建设时尽可能地提高工程的地震烈度可以大大减少地震灾害时的损失。再如,筑堤、建坝可蓄洪,事先锚固可防止滑坡、泥石流;建造安全壳可防止核泄漏。几乎没有一个行业能像土木工程这样对抗灾、减灾具有如此巨大的积极主动性和不可替代性。

7.2　工程灾害与防灾减灾

7.2.1　地震灾害及抗震

　　地震是由于地壳破坏引发的地面运动,这种地面运动具有突发性和不可预测性,可能对土木工程结构造成严重破坏。全世界每年发生地震 500 万次左右,其中 1% 为有感地震。我国平均每年发生 30 次 5 级以上地震,6 次 6 级以上强震,1 次 7 级以上大震,是世界上地震活动水平高、地震灾害重的国家之一。表 7.1 列出了 21 世纪以来的灾难性地震灾害。

　　地震造成的灾害可分为直接灾害和次生灾害。直接灾害主要表现为地面裂缝、错动、塌陷、喷砂冒水、山崩、滑坡等地表破坏;房屋倒塌、桥梁断落、水坝开裂等工程结构破坏;供水、供电、交通等生命线工程系统破坏。地震的次生灾害是指由地震间接引发的灾害,如地震诱发的火灾、水灾、有毒物质污染、海啸、瘟疫等。图 7.4 为 2011 年日本本州岛附近海域 9.0 级地震造成的灾害。

② Civil engineering is active and irreplaceable to almost all disasters. For example, increasing earthquake intensity in the construction can greatly reduce the loss caused by earthquake. We can build canal irrigation to resist drought, build embankment dams to resist flood, and make anchorage to resist landslide and debris.

7.2　Engineering disaster and disaster prevention and mitigation

7.2.1　Earthquake disaster and anti-seismic design

Earthquake is the result of the ground movement caused by the destruction of crust. The ground movement is sudden, unpredictable and can cause serious damage to civil engineering structure. Earthquakes occur around the world about five million times per year, including 1% felt earthquakes. On average, 30 earthquakes above Richter magnitude 5 are monitored every year in China, in which 6 earthquakes are above Richter magnitude 6, and 1 earthquake is above Richter magnitude 7. China is one of the highest seismic active countries in the world. Tab.7.1 lists the earthquake disasters since the 21st Century.

Earthquake disasters can be divided into direct disasters and secondary disasters. Direct disasters mainly display as follows: ① The surface damage, such as the ground cracking, dislocation, collapsing, landslide, sand blasting. ② Engineering structure damage, such as houses collapsing, bridges and dams cracking off. ③ Lifeline engineering system damage, such as water supply, power supply, transportation supply disruptions. Secondary disasters are indirectly produced by earthquake, such as fire, flood, toxic pollution, tsunami, pestilence. Fig.7.4 shows the disasters caused by the Honshu island earthquake in Japan in 2011.

地震灾害不仅造成了众多结构物的倒塌、生命线工程的破坏、财产的重大损失，而且还夺去了众多的生命，对人类产生了重大的心理影响，引发了众多的社会问题。因此，抗震防灾工作具有十分重大的意义。目前，减轻地震灾害的对策从宏观上可分为三方面：地震预测预报、地震转移分散和工程抗震。

①地震预测预报主要是根据地震地质、地质活动性、地震前兆异常和环境因素等多种情况，通过多种科学手段进行预测研究，对可能发生的地震进行预报。目前，地震预报还存在着许多难以解决的问题。

表 7.1　21 世纪以来的灾难性地震灾害
Tab. 7.1　The catastrophic earthquakes happened since the 21st century

发生时间和地点 Time and place	震　级 Magnitude	造成损失 Losses
2013 年中国四川省雅安市 Ya'an, China in 2013	7.0	死亡 196 人，伤 11 470 人，200 万人受灾 196 people dead, 11 470 people injured, and two million people affected
2011 年日本本州岛海域 Honshu island, Japan in 2011	9.0	死亡 15 844 人，失踪 3 450 人 15 844 people dead, 3 450 people missing
2010 年海地太子港西部 west of port-au-prince, Haiti in 2010	7.3	死亡 27 万人，370 万人受灾 270 000 people dead, 3 700 000 people affected
2010 年智利康赛普西翁市 city of conception, Chile in 2010	8.8	死亡 802 人，近 200 万人受灾，经济损失达 200 亿美元 802 people dead, nearly two millon people affected, and with economic losses of $ 20 billion
2008 年中国四川省汶川县 Wenchuan county, China in 2008	8.0	死亡 7 万人，失踪 2 万人，伤 37 万人，损失 1300 亿美元 70 000 dead, 20 000 people missing, 370 000 people injured, and with economic losses of $ 130 billion
2005 年巴基斯坦克什米尔地区 Pakistani Kashmir in 2005	7.6	死亡 7.3 万人，数百万人无家可归 73 000 people dead and millions of people become homeless
2004 年印尼苏门答腊岛海域 the sea of Sumatra, Indonesia in 2004	7.9	死亡或失踪 20 多万人 more than two hundred thousand people dead or missing

Earthquake disasters not only cause a lot of structure collapsing, lifeline engineering damage, the loss of property, but also exert a significant psychological impact on mankind, and produce a number of social problems. Therefore, the earthquake disaster prevention is of a great significance. At present, countermeasures to mitigate earthquake disasters can be divided into: earthquake prediction, seismic shift dispersion, and anti-seismic engineering.

① Earthquake predictions are mainly based on seismic geology, geological activity, earthquake precursory anomaly, environmental factors and other conditions. Through a variety of scientific means, the potential earthquakes can be predicted. At present, there are still many problems in earthquake predictions.

    ② 地震转移分散是把可能在人口密集的大城市发生的大地震,通过能量转移,诱发至荒无人烟的山区或远离大陆的深海,或通过能量释放把一次破坏性的大地震转化为无数次非破坏性的小震。这种方法目前尚在探索研究初期。

    ③ 工程抗震是通过工程技术提高城市综合抗御地震的能力和各类建筑的耐震能力,当突发性地震发生时,将地震灾害损失降低至最轻的程度。工程抗震包括地震危险性分析和地震区划、工程结构抗震、工程结构减震控制等。

    在工程抗震方面,通过重新修订各地区的抗震设防烈度,明确提出抗震设防目标,提高了工程抗震设计和检验的标准。近十年来,结构振动控制的研究和应用成为工程抗震领域的热点。

(a) 建筑物毁坏　　　　　　　(b) 地面裂缝　　　　　　　(c) 桥梁毁坏
Buildings destroyed　　　The ground fissures　　　Bridges destroyed

(d) 地震引发海啸　　　　　(e) 炼油厂爆炸燃烧　　　　　(f) 核泄漏
Tsunami　　　　　　　Blast combustion　　　　The nuclear leakage

图 7.4　2011 年日本本州岛附近海域 9.0 级地震造成的灾害
Fig. 7.4　The disasters caused by the Honshu island earthquake in Japan in 2011

    ② The seismic dispersion is to transfer the earthquake which may occur in a dense city, through the energy transfer, to the mountains or the continental sea, or transfer from a large destructive earthquake to numerous non-destructive earthquakes through releasing its energy. Relative method is still in research stage.

    ③ Anti-seismic engineering is to improve the city's comprehensive seismic capability and the seismic capacity of buildings through engineering and technology, and reduce the damage to the lightest degree when a sudden earthquake occurs. It includes earthquake risk analysis, seismic zoning, earthquake resistance of engineering structure, and engineering structural vibration control.

    Anti-seismic engineering improves the seismic design and inspection standard by revising the seismic fortification intensity in each area. Research and application of structural vibration control in nearly ten years has become a hot topic in the earthquake engineering field.

　　传统结构抗震设计方法是依靠增加结构自身的强度和变形能力来抗震,而减震控制方法则是采用隔震、消能、施加外力、调整结构动力特性等方法来消减结构地震反应,具有安全可靠、方便有效、经济节省和适用范围广等优点,是土木工程防灾减灾积极有效的方法和技术。

　　隔震及消能减震是目前土木工程中技术较成熟且应用较广的方法。在建筑物基础与上部结构之间设置隔震装置形成隔震层,把房屋结构与基础隔离开来,利用隔震装置来隔离和耗散地震能量以避免或减小地震能量向上部结构传输,减小建筑物的地震反应,实现地震时建筑物只发生轻微运动或变形的目的,从而使建筑物在地震作用下不损坏或倒塌的抗震方法称为房屋基础隔震。传统抗震结构房屋与隔震房屋在地震中的情况对比,如图 7.5 所示。

　　隔震系统一般由隔震器、阻尼器组成,它具有竖向刚度大、水平刚度小,能提供较大阻尼的特点。结构消能减震是通过采用一定的消能装置,消耗输入主体结构的地震能量,从而减轻结构的振动和破坏。

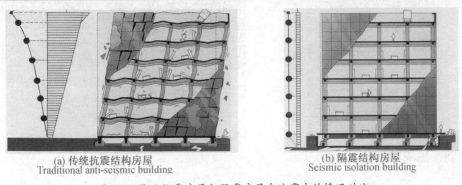

(a) 传统抗震结构房屋
Traditional anti-seismic building

(b) 隔震结构房屋
Seismic isolation building

图 7.5　传统抗震房屋与隔震房屋在地震中的情况对比
Fig. 7.5　The contrast between the traditional houses and isolation houses in the earthquake

　　Traditional anti-seismic design methods are mainly by increasing strength, deformation capacity of structure to resist earthquake. Seismic control method is mainly by seismic isolation, energy dissipation, applied force, dynamic characteristics of structural adjustment, which is safe and reliable, convenient and efficient, economical and applicable to a wide range.

　　At present, seismic isolation and energy dissipation are a more mature technology based, and have a wide application in civil engineering. Isolation devices are arranged between the building foundation and upper structure (or system) to form the isolation layer. The structure and base are isolated, and the isolation device is used to isolate and dissipate the seismic energy in order to avoid or reduce the earthquake energy transmitted to the upper structure and reduce the earthquake response of buildings. At the same time, seismic structures only have slight movement or deformation. The buildings are not damaged or collapsed in the earthquake. This anti-seismic method is called the housing base isolation, as shown in Fig.7.5. Isolation system is generally composed of isolator, damper, which has a great vertical stiffness and small lateral stiffness and can provide a larger damping.

　　We can reduce the vibration and damage of the structure by installing a certain energy dissipation device in structure.

消能装置不改变主体承载结构体系，可同时减少结构水平和竖向的地震作用，在新建和建筑抗震加固中均可采用。消能装置包括各种消能支撑、消能剪力墙、摩擦阻尼器、软钢阻尼器、黏弹性阻尼器、黏滞流体阻尼器和组合式消能减震体系等。图7.6为采用橡胶支座和黏滞阻尼器组合消能减震体系的同济大学钢框架土木大楼。

7.2.2 风灾及抗风

风是大气层中空气形成的压力作用运动。由于地球表面不同地区的大气层所吸收的太阳能量不同，造成了各地空气温度的差异，从而产生气压差。气压差驱使空气从气压高的地方向气压低的地方流动，这就形成了风。风速就是风的前进速度。相邻两地间的气压差越大，空气流动越快，风速越大，风的力量自然也就越大。因此，通常都以风力来表示风的大小。根据风速大小，国际上将风力划分为18个等级(见表7.2)。

(a) 消能减震体系位置
The location of energy dissipation system

(b) 消能减震体系施工安装
The location of energy dissipation system

图 7.6 橡胶支座和黏滞阻尼器组合消能减震体系
Fig. 7.6 Energy dissipation system combined by rubber bearing and viscous damper

Energy dissipation device does not change the main bearing structure system, but can reduce the horizontal and vertical seismic behaviors of the structure. Therefore, it can be used in the reinforcement construction and building. Energy dissipation device seismic reinforcement and new construction include various energy dissipation supports, energy dissipation shear wall, friction damper, mild steel damper, viscous elastic damper, viscous fluid damper and a combined energy dissipation system. Fig.7.6 shows the steel frame building in Tongji University, which uses energy dissipation system that combined rubber bearings and viscous dampers.

7.2.2 Wind damage and wind resistance

Wind is produced by interaction and movement of the air pressure in the atmosphere. Since the atmosphere in different regions of the surface absorbs different solar energy, the air temperature and the air pressure are also different. The pressure difference makes the air flow from the high pressure region to the low pressure region, and the air flow forms the wind. Wind speed is the forward speed of the wind.

The greater the pressure difference between two adjacent places is, the faster the air flows, the higher the wind speed is, the greater the wind power is. Therefore, the wind is usually measured by wind power. The wind is divided into 18 levels according to the wind speed (Tab.7.2).

　　风灾是全球最常见和最严重的自然灾害之一,年复一年地给人类社会带来巨大的生命和财产损失。风灾具有发生频率高、次生灾害大(如暴雨、巨浪、风暴潮、洪水、泥石流等),持续时间长等特点。

表7.2　风力等级表
Tab.7.2　Wind scale table

等级 Scale	名　称 Name	距地 10 m 高处相当风速 Equivalent wind speed at the height of 10 meters 风速/(m/s) Wind Speed	陆地地面现象 Phenomenon of the land surface	海面浪高/m Wave height
0	静风 calm	0.0~0.2	静烟直上 smoke straight up	0.0
1	软风 light air	0.3~1.5	烟示风向 smoke shows wind direction	0.1
2	轻风 light breeze	1.6~3.3	感觉有风 feel the wind	0.2
3	微风 gentle breeze	3.4~5.4	旌旗展开 flags expand	0.6
4	和风 moderate breeze	5.5~7.9	吹起尘土 blowing dust	1.0
5	清风 fresh breeze	8.0~10.7	小树摇摆 trees sway	2.0
6	强风 strong breeze	10.8~13.8	电线有声 wire audio	3.0
7	疾风 near gale	13.9~17.1	步行困难 difficulty in walking	4.0
8	大风 gale	17.2~20.7	折毁树枝 branches destroyed	5.5
9	烈风 strong gale	20.8~24.4	小损房屋 small loss of the house	7.0
10	狂风 storm	24.5~28.4	拔起树木 uprooted trees	9.0
11	暴风 violent storm	28.5~32.6	损毁重大 significant damage	11.5
12	飓风 hurricane	32.7~36.9	摧毁极大 great destroy	14.0
13~17		≥37.0		

Wind disaster is one of the most common and most serious natural disasters in human society with huge losses of life and property year after year. Wind disaster has the characteristics of high frequency, huge secondary disasters (such as rainstorm, billow, storm surge, flood, and debris flow) and long duration.

一般6级以下的风不会引起大的危害,6级及6级以上较强的风有时会造成房屋、桥梁、车辆、船舶、树木、农作物、通信系统、电力设施破坏及人员伤亡,由此造成的灾害称为风灾。

常见的导致灾害的风型主要有暴风、台风、龙卷风等。1999年5月3日,强劲龙卷风袭击美国俄克拉何马州和邻近的堪萨斯州,共造成49人丧生,摧毁了2 600间房屋,导致8 000多建筑物受损,经济损失达12亿美元(见图7.7)。2009年台风"莫拉克"造成我国500多人死亡、近200人失踪、46人受伤(见图7.8)。

从自然风所包含的成分看,风对构筑物的作用包括平均风作用和脉动风作用,从结构的响应来看,包括静态响应和风致振动响应。

平均风既可引起结构的静态响应,又可引起结构的横风向振动响应。

脉动风引起的响应则包括结构的准静态响应、顺风向和横风向的随机振动响应。当这些响应的综合结果超过结构的承受能力时,结构将发生破坏。

图7.7 龙卷风袭击美国俄克拉何马州
Fig. 7.7 Tornado hitting Oklahoma in America

图7.8 台风"莫拉克"灾害
Fig. 7.8 Losses caused by the typhoon Morakot

Generally speaking, winds under level six do not cause great damage, while others at the six level or above may cause damage to houses, bridges, vehicles, ships, trees, crops and facilities of communication systems and powers, and it may cause casualty as well. The disasters caused by the above reasons are called wind disaster.

On May 3rd, 1999, a strong tornado hit Oklahoma and its neighboring state Kansas, leading to 49 casualties, 2 600 houses destroyed, more than 8 000 buildings damaged, and the economic losses was $1.2 billion (Fig.7.7). Common wind disasters are storms, typhoons, and tornadoes. The typhoon Morakot in 2009 caused more than 500 people dead, nearly 200 people missing and 46 people injured in China (Fig.7.8).

The effect of wind on the structure can be divided into average wind effect and fluctuating wind effect according to the component of natural wind, and it can be divided into static response and wind-induced vibration response according to the response of the structure.

The average wind can result in not only the static response, but also across-wind vibration response of the structure. The fluctuating wind response includes quasi-static response of the structure, random vibration response of wind direction and across-wind direction. When the combined result of these responses exceed the bearing capacity of the structure, the structure will be destroyed.

风对构筑物的破坏主要表现在以下几个方面：

①对房屋建筑结构的破坏。这主要表现在对多高层结构的破坏，对简易房屋，尤其是轻屋盖房屋的破坏，对外墙饰面、门窗玻璃及玻璃幕墙的破坏。2005 年 8 月 29 日，飓风卡特里娜摧毁了新奥尔良凯悦酒店等许多建筑的窗户、幕墙和外墙饰面，掉落的物体砸毁了大量停在楼下的汽车及其他物品(见图 7.9)。

②对大跨结构的破坏。体育场馆、会展中心、汽车收费站等大跨结构也经常遭受风灾。2004 年河南省体育中心围护结构在 8~9 级的瞬时风袭击下严重受损(见图 7.10)。

③对桥梁结构的破坏。风对桥梁结构的破坏作用也是非常巨大的。1940 年，美国华盛顿州塔科玛海峡建造的科马悬索桥，主跨 853 m，建好后不到 4 个月，在一场风速不到 20 m/s 的风灾下，因产生上下和来回扭曲振动倒塌了(见图 7.11)。

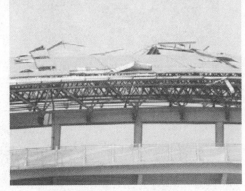

图7.9　飓风致窗户损坏
Fig. 7.9　Windows damaged by hurricane

图 7.10　体育馆遭风灾破坏
Fig. 7.10　One gymnasium destroyed by typhoon

Wind damage to structures are shown in the following forms:

① Destruction of the housing construction. This destruction mainly displays in the high-rise structure, simple houses, especially in light roof houses, external wall finishes, doors, windows, and glass curtain wall. On August 29th, 2005, hurricane Katrina ruined the windows of Hyatt Regency in New Orleans and many other buildings, curtain wall and external wall finishes, and smashed a large number of cars parked downstairs and other items (Fig.7.9).

② Destruction of long-span structures. Stadiums, convention centers, automotive toll stations and other long-span structures often suffer wind disasters. In 2004, envelope structure of Henan Sports Center building was severely damaged under the instantaneous winds of eight to nine levels (Fig.7.10).

③ Destruction of the bridge structure. The wind damaging effects on bridge are very huge. In 1940, Coma suspension bridge built on Tacoma Narrows in the state of Washington U.S., whose main span is 853 m, collapsed because of the twisting back and forth at a hurricane with the wind speed of less than 20 m/s(Fig.7.11).

④ 对电厂冷却塔等高耸结构的破坏。冷却塔也容易遭受风灾。1965 年 11 月 1 日的一场平均风速为 18~20 m/s 的大风摧毁了英国渡桥热电厂 8 个冷却塔中的 3 个(见图 7.12)。

⑤ 对输电系统等的破坏。若供电线路的电杆埋得浅,则在大风中容易被刮倒,造成停电事故,严重影响生产和生活(见图 7.13)。

⑥ 对海洋工程结构的破坏。2005 年秋季的"卡特里娜"和"丽塔"两个飓风毁坏了墨西哥湾地区 113 座石油钻井平台及 457 条油气管道(见图 7.14)。

图 7.11 美国塔科玛海峡桥遭风灾毁坏
Fig. 7.11 Tacoma narrows bridge (USA) destroyed by wind

图 7.12 冷却塔遭风灾毁坏
Fig. 7.12 Cooling tower destroyed by wind

图 7.13 高压输电塔遭风灾折断
Fig. 7.13 High-voltage transmission tower destroyed by cyclone

图 7.14 石油钻井平台遭风灾毁坏
Fig. 7.14 Oil rig destroyed by typhoon

④ Destruction of the power plant cooling towers. The cooling tower is also vulnerable to storms. On November 1st, 1965, a strong wind with average speed of 18~20 m/s destroyed three of the eight cooling towers in UK Du Bridge Thermal Power Plant (Fig.7.12).

⑤ Destruction of the power transmission system. If the burial depth of the supply line pole is shallow, the pole is easy to be blown down by strong wind, causing power outages which seriously affect the production and life (Fig.7.13).

⑥ Destruction of marine engineering structures. In the autumn of 2005, the hurricane Katrina and Rita destroyed 113 oil rigs and 457 oil and gas pipelines in Gulf of Mexico (Fig.7.14).

　　风工程研究方法有现场测试、风洞试验和理论计算 3 种。理论计算包括解析计算和数值计算。现场测试方法对风工程现象的规律性和机理的研究不适用，也无法在工程建设实施前解决相关的实际问题。现场测试方法常被作为一种有效的验证理论计算和风洞试验结果的手段。风洞试验不仅保留了直观的优点，可以节约人力、物力和时间，而且可在很大程度上人为地控制、调节和重复一些试验条件，因此它是一种很好的研究风工程现象变参数影响和机理的手段（见图 7.15 和图 7.16）。

　　目前，防止风灾害的主要措施有以下几种：

　　① 重点研究各地区的风荷载特性。例如，研究地区风压分布、地面粗糙度划分、高层建筑风效应、大跨建筑和桥梁结构风效应等，为制定和修正荷载及相关规范提供依据。

　　② 充分考虑风灾因素，加强工程结构的抗风设计。

　　③ 建造防风固沙林和防风护岸植被，以减少风力对城市和海岸的破坏。

图 7.15　大跨悬索桥风洞试验
Fig. 7.15　Long-span suspension bridge wind tunnel test

图 7.16　上海中心大厦风洞试验
Fig. 7.16　Shanghai tower wind tunnel test

　　Wind engineering research has three methods, including field test, wind tunnel test and theoretical calculation. Theoretical calculation includes analytic calculation and numerical calculation. The field test method is not applicable to the research of regularity and mechanism of wind engineering phenomena, and it cannot solve the practical problems before the implementation of engineering construction. The field test method is often used as an effective means to verify the theoretical calculations and wind tunnel test results. Wind tunnel test not only retains the advantage of intuitive, but also saves manpower, material and time relatively. Moreover, we can artificially control, adjust and repeat some test conditions to a large extent in wind tunnel test. It is a very good means to research variable parameters effect and the mechanism of the phenomenon in wind engineering (Fig.7.15 and Fig.7.16).

　　At present, the measures to prevent wind disaster include:

　　① Focus on the wind load characteristics of different regions.

　　② Take full account of the wind disaster factors and strengthen the wind-resistance design of engineering structures.

　　③ Construct windbreak, sand fixation and wind revetment vegetation to reduce wind damage to the city and coast.

④ 在经常遭受风灾害的地区,建立预报、预警机制。

⑤ 城市应编制风灾害影响区划,建立合理有效的应对策略。

传统的结构抗风对策是首先保证强度,然后验算位移,若位移过大则应通过增强结构自身刚度和抗侧力能力来抵抗风荷载作用,这是一种被动的、不经济的方法。近30多年来发展起来的结构振动控制技术开辟了结构抗风设计的新途径。结构振动控制技术就是在结构上附设控制构件和控制装置,在结构振动时通过被动或主动地施加控制力减小或抑制结构的动力反应,从而减少动力位移,以满足结构的安全性、适用性和舒适度的要求。图7.17所示为台北101大楼采用调谐质量阻尼器(TMD)进行风振舒适度控制。图7.18所示为桥梁黏滞阻尼器风振控制。

图 7.17   台北 101 大楼调谐质量阻尼器风振
舒适度控制
Fig. 7.17   TMD wind vibration comfort control in Taipei
101 building

图 7.18   桥梁斜拉索黏滞阻尼器风振控制
Fig. 7.18   Wind vibration control of bridge with
viscous damper

④ Establish the mechanisms of forecast and early warning in the regions which often suffer the wind disaster.

⑤ Compile regionalization affected by wind and establish reasonable and effective coping strategies.

The traditional structural countermeasure against wind is to ensure the strength and then check the displacement. However, when displacement is too large, we will strengthen the structure stiffness and lateral resistance to resist the effect of wind loads. This is a passive and uneconomical way. The structural vibration control technology developed in nearly 30 years has offered a new approach to structural wind-resistance design. The structural vibration control technology is used to attach the control members and control devices to the structure, by applying passively or actively control to reduce or inhibit the dynamic response of structures, when the structure vibrates thereby reducing the dynamic displacement in order to meet the requirements of structural safety, applicability and comfort. Fig.7.17 is Taipei 101 building which adopts tuned mass damper (TMD) for wind-induced comfort control. Fig.7.18 is the wind vibration control of bridge with viscous damper.

7.2.3　冰雪灾害及抗灾

在冬季,各类土木工程还会受到雪荷载的作用,大的雪荷载及由此引起的冰雨还会引起建筑结构、道路、桥梁及输电线塔等结构产生破坏。2008 年 1 月期间,我国南方大部分地区相继出现了持续的大范围灾害性冰雪天气,此次雨雪冰冻天气过程影响范围大、持续时间长、涉及面广、危害程度大,给当地的交通、电力、通信和人民生活带来严重影响(见图 7.19 和图 7.20)。

《建筑结构荷载规范》(GB 5009—2012)中所采用的雪荷载,是根据历史记录,经统计分析,结合结构设计原则与方法确定的。一方面,随着极端灾害的发生,需要对所采用的标准及具体数值进行不断地修正和调整。另一方面,随着人们对工程安全性、适用性与耐久性等功能要求的不断提高,工程抗灾能力的设防标准也要不断提高。

7.2.4　火灾及防火

火灾是指时间和空间上失去控制的火,在其蔓延发展过程中将给人类的生命财产造成损失的一种灾害性的燃烧现象。

图 7.19　雪荷载引起的厂房倒塌
Fig. 7.19　The collapse of factory building caused by snow load

图 7.20　雪荷载及冻雨引起的输电线塔破坏
Fig. 7.20　The damage of transmission tower caused by snow load

7.2.3　Disaster of snow and resisting disaster

All kinds of civil engineering are affected by the snow load in winter, and the sleet caused by the large snow load may destroy the building structures, roads, bridges and transmission line towers. In January 2008, most regions in southern China continuously suffered a wide range of severe snow and ice. The results of the freezing rain and snow weather had a wide influence, long duration, high harm degree and serious impact on transportation, electricity, communication and people's life (Fig. 7.19 and Fig. 7.20).

According to the historical data and the statistical analysis, snow load used in *Building Structures Load Specification* is determined by combining with structural design principles and methods. On the one hand, with the occurrence of extreme disasters, the adopted standards and the specific values need to be continuously corrected and adjusted. On the other hand, people's needs for the functional requirements of the engineering safety, suitability and durability are continuously improved, and the fortification standards of engineering resilience must be continuously improved.

7.2.4　Fire disaster and fire prevention

Fire disaster is a disastrous combustion phenomenon. In its spreading process, the fire is out of control with the loss of life and property.

　　火灾是各种灾害中发生最频繁、影响面最广的灾种之一。火灾可以分为自然火灾和建筑火灾两大类。随着城市化的发展,建筑火灾及其危害越来越严重。这些火灾不仅带来了重大的人员伤亡和财产损失,也严重影响了建筑结构的安全。例如,上海静安区胶州路公寓大楼特大火灾(见图 7.21)以及中央电视台新址配楼火灾(见图7.22)。

　　火灾是一个燃烧过程,要经过发生、蔓延和充分燃烧几个阶段。火灾的严重性取决于持续时间和温度,而这两者又受建筑类型、燃烧荷载等诸多因素的影响。控制和改善影响燃烧的各种因素是建筑防火设计首先要考虑的问题。

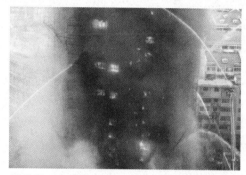

图 7.21　上海静安区胶州路公寓大楼特大火灾
Fig. 7.21　Catastrophic fire in Shanghai

图 7.22　央视新址配楼大火
Fig. 7.22　The new CCTV side building fire disaster

　　Fire disaster is the most frequent disaster and has the greatest impact. Fire disaster can be divided into two major categories: natural fire and building fire. With the development of urbanization, building fire is more and more serious. These fires not only bring heavy casualties and property losses, but also seriously affect the structural safety. For example, the catastrophic fire in Jiaozhou Road, Jing'an District, Shanghai Condominium (Fig.7.21) and the new CCTV side building fire disaster (Fig.7.22).

　　Fire disaster is a combustion process which has to go through the stages of occurence, spreading and full combustion. The seriousness of the fire depends on the duration and temperature, both of which are affected by building types, combustion load, ect. Controlling and improving various factors that affect the burning are the first thing to be considered when designing the building fire safety.

对于建筑结构构件,在受火时,随着温度的升高和持续时间的增加,构件的力学性能下降到不足以承受设计规定的荷载,此时该构件将部分或全部失去正常工作的能力。为避免火灾对建筑结构安全性的影响,防止结构在火灾中发生破坏或坍塌,必须对结构进行抗火设计;火灾发生后,应当及时、有效地进行扑救,并在灾后对建筑结构进行损伤鉴定和加固修复,降低火灾的危害。

①建筑防火,包括建筑火灾基础科学、建筑总体布局、建筑内部防火隔断、防火装修及消防扑救、安全疏散路线、自动防排烟系统的设计和研究。在建筑防火方面,我国已制定《建筑设计防火规范》(GB 50016—2014)。

②建筑结构抗火性能,主要包括结构材料的抗火性能,结构在火灾高温下的强度、刚度、变形、承载能力,建筑结构耐火时间及结构抗火构造等内容。目前,欧洲各国颁布了钢结构耐火设计技术规范,法国还制定了混凝土结构的耐火强度计算方法,日本建立了建筑结构耐火设计数据库,而我国至今还没有建筑结构抗火设计规范。

③火灾后建筑结构的损伤鉴定和加固修复。火灾后建筑物的损伤鉴定主要包括现场调查及火灾温度判断;火灾后建筑材料及结构性能的检测;受损分析和剩余承载力的计算;结构受损综合评定。火灾后的结构加固修复方法,目前常用的有喷射混凝土法、粘钢加固法、碳纤维增强复合材料加固法等。

With the increase of temperature and duration in the fire, the mechanical properties of building structural member will not be able to hold the design loads, and the member will partially or totally loss the ability to work at that time. In order to avoid the fire influence on the building structure and prevent the damage or collapse of the structure in the fire, we need to make fire resistance design of structures. When the fire disaster happens, we should fight the blaze promptly and effectively, we also need to make a damage identification, the reinforcement and restoration of the structure so as to reduce the fire hazard.

① Building fire protection includes the basic science of building fire, the overall building layout, the cut-off of building interior fire, fire decoration and firefighting, safe evacuation routes, automatic anti-exhaust system design and research. In building fire protection, we have *Code for Fire Protection Design of Buildings* (GB 50016—2014).

② Fire-resistance performance of building structures includes fire-resistance performance of structural materials, the strength, stiffness, deformation and bearing capacity of the structure under the high temperature in the fire, fire resistance time of building structure, etc. European countries have enacted steel fire design and technical specifications, and France has also develop a fire-resistant strength calculation method on the concrete structure. Japan has established a database about fire resistance design of building structures.

③ Damage identifying, reinforcement and restoration of building structure after fire disaster. Damage identifying of building structures includes field investigation and determination of fire temperature, detection of building materials and structural properties, the damage analysis and residual bearing capacity calculation, and the integrated assessment of structural damage. The structural reinforcement method includes sprayed concrete, bonded steel reinforcement, carbon fiber reinforced composites reinforcement.

7.2.5 地质灾害及防治

自然的变化和人为的作用都可能导致地质环境或地质体发生变化,当这种变化达到一定程度,其产生的后果便给人类和社会造成危害,称为地质灾害。由于我国处于特殊的地质构造部位,且全国 2/3 的地区属于山地,地质灾害分布广、类型多、频率高、强度大,每年都造成重大人员伤亡和严重经济损失,已成为影响我国城乡建设和人民生存环境的重大问题。地质灾害包括滑坡、泥石流、崩塌、地面沉降等。

滑坡是指斜坡上大量土体或岩体由于种种原因在重力作用下,沿一定的滑动面整体向下滑动的现象(见图 7.23)。滑坡的防治途径主要有以下几种:① 终止或减轻诱发滑坡的外部环境条件,如截流排水、卸荷减载和坡面防护。② 改善边坡内部力学特征和物质结构,如土质改良。③ 设置抗滑工程直接阻止滑坡的发展,如抗滑桩、挡土墙和预应力锚固等。

图 7.23　山体滑坡对建筑物造成破坏

Fig. 7.23　Damages caused by landslide

7.2.5 Geological disaster and prevention

Natural variations and human activities may lead to changes of the geological environment or geological body. When such changes reach a certain level, their consequences will cause harm to humans and society, which are called geological disasters. Since China is in a special geological structure part, and 2/3 of the area belongs to the mountain, geological disasters in China are widely distributed with multitype, high frequency and intensity, which caused many casualties and serious economic losses, and it has become the major issues affecting the urban and rural construction and people's living environment. Geological disaster includes landslide, debris flow, collapse and ground subsidence.

Landslide is the phenomenon that the overall soil or rock mass on the slopes slide downward under gravity along the sliding surface due to various reasons (Fig.7.23). The main ways of landslide prevention and treatment are as follows: ① Terminate and reduce the external environmental conditions which caused the landslide, such as closure and drainage, unloading and slope protection. ② Improve the internal mechanical characteristics and the structure of the material in a slope, such as soil improvement. ③ Set anti-sliding engineering which prevents the development of the landslide directly, such as pile, retaining wall and prestressed anchor.

泥石流是一种工程动力地质现象,它是一种水与泥沙、石块混合在一起流动的特殊洪流,具有突然爆发、流速快、流量大、物质容量大和破坏力强的特点。泥石流爆发时,大量泥沙石块沿山沟奔腾而下,在很短的时间内,冲进乡村和城镇,冲毁铁路、公路和航道等交通设施,淹没农田,造成灾害(见图7.24)。

泥石流的防治措施主要有:① 跨越。处于泥石流地段的铁路、公路线路,可采用桥梁、隧道等方式跨越泥石流。② 排导。修筑排导沟、急流槽、导流堤等工程,将泥石流顺利排走。③ 拦挡。在泥石流沟中修筑一系列的低矮小坝。④ 拦截。修筑拦淤库和储淤场,减弱下泄物质总量和洪峰流量。⑤ 水土保护,包括生物措施和工程措施两类防护方法。

图 7.24　2010 年甘肃舟曲特大泥石流灾害
Fig. 7.24　Extremely heavy debris flow disaster in Zhouqu, Gansu

Debris flow is an engineering geological phenomenon, and it also belongs to a special torrent mixed with water, mud and rock. It's features include sudden outbreak, high flow velocities, and strong destruction. When debris flow happens, large amount of sands and stones swoop down along the ravine, rush into the villages and towns, destroy rails, roads and waterway transportation facilities, and flood farmland in a very short time (Fig.7.24).

The prevention and control measures of debris flow are as follows: ① Cross. Railway and highway lines can step across debris flow by bridge and tunnel. ② Exhaust. We can drain away debris flow smoothly by building exhaust ditch, rapid slot and diversion dike.③ Resist. We can build a series of low dams.④ Intercept. We can build silt library and silt storage field, which can reduce the total discharged substances and peak flow. ⑤ Soil and water conservation, including biological and engineering measures.

崩塌是指较陡斜坡上的岩土体在重力作用下突然脱离母体崩落、滚动、堆积在坡脚的地质现象(见图 7.25)。崩塌是多山地区及黄土高原常见的自然灾害之一。崩塌可以单独发生,也可能与滑坡一起发生,甚至可能和泥石流同时发生。防治崩塌的常用措施有:① 清除坡面危岩。② 加固坡面。在易风化剥落的边坡地段,修建护墙、挡墙。③ 拦截、遮挡。通过设置拦石网、拦石墙、落石槽等方法进行拦截。④ 修筑排水构筑物,防止水流大量渗入掩体而恶化斜坡的稳定性。

地面沉降是指由于自然动力因素,如地壳的下降运动、地震、火山活动等,或受地下开采、地下施工或灌溉等人工活动的影响,造成地下空洞或使地下松散土压缩固结,导致地面标高下降的现象(见图 7.26)。地面沉降可能引起地面建筑物、市政管道和交通设施等城市基础设施的损坏。

图 7.25　重庆彭水山体崩塌毁坏道路

Fig.7.25　The road destroyed by collapse of the mountain

Collapse is a geological phenomenon that the rocks and soil on the steeper slopes separated from the body, collapsed, scrolled and piled up at the foot of the slope (Fig.7.25). Collapse is one of the common natural disasters in the mountainous area and loess plateau. Collapse can occur separately, and it may also occur related with landslide, and even occur with debris flow. Common measures about prevention and treatment of collapse are as follows: ① Clearing unstable rocks on slope. ② Reinforcing slope. ③ Intercepting rocks through setting block stone network and retaining wall.④ Building draining structures.

Ground subsidence is a phenomenon of ground elevation drop caused by downward movement of the earth's crust, earthquakes, volcanic activity, or by underground mining, underground construction or irrigation and other human activities (Fig.7.26). Ground subsidence can damage the ground buildings, municipal pipelines, transport facilities and other urban infrastructures.

　　例如,上海市由于过度开采地下水及高楼密度过大,自 20 世纪 90 年代起,地面平均每年下降 1cm。

　　当前对地面沉降的控制和治理措施分为表面治理和根本治理两大类。表面治理是对已有地面沉降的地区,通过地面整治和改善环境的手段,减弱地面沉降造成的损失。根本治理是从研究引起地面沉降的根本因素入手,谋求缓和,直至控制或终止地面沉降的措施。其主要方法有人工补给地下水;限制地下水开采,调整开采层次,以地面水源代替地下水源;限制或停止开采固体矿物。

7.2.6　工程事故灾害

　　工程事故灾害是指由于勘察、设计、施工和使用过程中存在重大失误造成工程倒塌(或失效)而引起的人为灾害。1999 年 1 月 4 日,重庆市綦江县旧彩虹桥发生整体垮塌(该桥建成不足 3 年),造成 40 人死、14 人伤的惨剧和 600 多万元的重大经济损失(见图 7.27)。

图 7.26　浙江温岭保障房小区地面沉降
Fig.7.26　Ground subsidence in a residential district

　　Shanghai, for example, due to over-exploitation of groundwater and excessive highrise building, the average annual decline is 1cm since the 1990s.

　　Currently, the control measures of ground subsidence can be divided into surface control and fundamental control. Surface control can reduce the losses by ground remediation and environmental improvement. Fundamental control is used to ease the subsidence until control or end it by researching how to eliminate the fundamental factors. The main methods include three aspects: artificial recharge of groundwater; restrict groundwater extraction, and adjust the level of exploitation of ground water instead of groundwater sources; restrict or stop the mining of solid minerals.

7.2.6　Engineering accident disaster

　　Engineering accident disaster is the man-made disaster caused by the collapse of the project or failure due to the significant errors in the process of survey, design, construction and use. On January 4th, 1999, the overall collapse of the old Rainbow Bridge in Qijiang county, Chongqing caused the death of 40 people, the injuring of 14 people and the major economic loss about more than 6 million yuan (Fig.7.27).

事故发生后,专家对彩虹桥工程进行质量鉴定指出,该工程存在多处致命的质量问题:① 拱架钢管焊接存在严重缺陷,焊接质量不合格;② 钢管混凝土抗压强度不足,所用混凝土平均强度低于设计强度的 1/3;③ 桥梁构造设计不合理,致使连接拱架与桥梁、桥面的钢绞线拉索、锚具、锚片严重锈蚀;④ 压力灌浆不密实。

从技术方面分析,发生工程事故灾难的原因有:① 地质资料勘察严重失误或根本没有进行勘察;② 地基承载力不够,同时基础设计又严重失误;③ 结构方案、结构计算或结构施工图有重大错误或凭"经验"设计,无图施工;④ 材料和半成品的质量严重低劣,甚至采用假冒伪劣的产品和半成品;⑤ 施工和安装过程中偷工减料,粗制滥造;⑥ 施工的技术方案和措施有重大失误;⑦ 使用过程中盲目增加使用荷载,随意变更使用环境和使用状态;⑧ 任意对已建成工程打洞、拆墙、移柱、改扩建等。

(a) 垮塌前Before the collapse

(b) 垮塌后After the collapse

图 7.27　重庆市綦江县旧彩虹桥整体垮塌

Fig.7.27　Overall collapse of the old Rainbow Bridge in Qijiang county, Chongqing

This bridge was built in less than 3 years. After the accident, the experts made the quality identification for Rainbow Bridge and found many fatal quality problems: ① The arch steel pipe welding had serious defects, and the welding was unqualified. ② The strength of steel pipe concrete was insufficient and the strength of the used concrete was less than one-third of the design strength.③ The bridge structural design was unreasonable, which resulted in the severe corrosion of the strand cable, the anchor and the anchor piece connecting the arch, the bridge and the deck. ④ The pressure grouting was not dense.

Analyzing from the technical aspects, the reasons of the engineering accident disasters are as follows: ① The geological exploration data are seriously wrong or no exploration at all. ② Bearing capacity of the foundation is not enough, and the basis design is also seriously wrong at the same time. ③ Structural program, structural calculation or structural drawing have major mistakes, and the design is made by experience. ④ The used materials and semi-finished products are of seriously poor quality or even belong to fake and shoddy product.⑤ There are cutting corners and shoddy in the construction and installation process. ⑥ There are serious mistakes in the technology programs and measures. ⑦ The load is blindly increased during the production, and the environment and the using state are freely changed. ⑧ The completed project is freely changed by the holes, tearing down walls, column shift and expansion.

从管理方面分析,发生工程事故灾难的原因有:① 由非相应资质的设计、施工单位进行设计、施工;② 建筑市场混乱无序,出现"六无"工程项目(无报建程序、无设计图纸、无勘察资料、无招投标、无资质施工、无质量监督);③ "层层分包"使设计、施工的管理处于严重失控状态;④ 企业经营思想不正,片面追求利润、产值,没有建立可靠的质量保障制度;⑤ 无固定技工队伍,技术工人和管理人员素质不高。

7.3 工程结构灾害检测与加固

7.3.1 结构灾害检测与鉴定

工程结构检测与鉴定是采用各种检测方法对工程结构进行耐久性检测,并对其安全性、可靠性进行鉴定,得出鉴定等级和是否需要加固的结论的整个过程,其程序为:检测任务委托—调查—编制检测方案—现场检测—发出检测报告—进行鉴定评级—出具鉴定报告。例如混凝土和砌体结构的检测过程,通常应先根据图纸数据进行理想结构的受力分析,找出关键构件,如关键的梁、柱或墙;然后在建筑现场对关键构件的关键部位去除表面装饰层和粉刷层,采用适当的检测方法确定混凝土碳化深度和弹性模量,或确定砌体结构砂浆和砌块的实际强度和弹性模量(图 7.28 所示为砌体原位压力机检测砌体强度)。

Analyzing from the management aspects, the reasons of the engineering accident disasters are as follows: ① The construction and the design are not finished by the corresponding construction units and the design units. ② Due to the disorder of building market, there exists "six none" projects, including no reported procedure, no design drawings, no survey data, no bidding, no qualified construction and no quality supervision. ③ "Mulit-level sub-contracting" makes the management of design and construction out of control. ④ The enterprise pursues one-side profit and production value, and has no reliable quality assurance system. ⑤ The construction company has no fixed team of skilled workers, and the quality of the skilled workers and managers are not high.

7.3 Disaster detection and reinforcement of engineering structures

7.3.1 Structure disaster detection and identification

Engineering structure detection and identification is a complete process by using a variety of detection methods for durability testing of the engineering structure and identification of the security and reliability, then the identified level and the conclusion about whether the reinforcement is needed will be reached. The program includes detection commissioned, investigation, making detection plan, field testing, test report, rate identification and issuing reports. For example, in the testing process of concrete and masonry structure, firstly you should analyze the force on ideal structure and find out the key components such as beam, column or wall, according to the drawing data. Secondly, at the scene of the construction, the surface decoration layer and paint layer are cleaned in the key components of the key parts, and concrete carbonation depth and elastic modulus, or the actual strength of the mortar and brick masonry structure and elastic modulus are determined by using the appropriate test method(Fig.7.28).

对部分关键混凝土构件可采用钻孔取芯得到真实的混凝土块,在实验室测得其真实混凝土强度(见图 7.29)。将实测材料强度和弹性模量代入结构计算模型进行受力分析,分析结果可以评估结构是否安全并指出不安全的构件,作为结构加固的依据。

7.3.2 工程结构改造及加固

工程结构改造加固学是一门研究使受损的工程结构重新恢复使用功能或适应新的使用功能的学科。由于土木工程和现代科学技术的迅猛发展,这门学科的发展也异常迅速。工程结构需要加固和改造的原因通常有以下几种:① 使用荷载增加。如桥梁车辆吨位增加、对建筑物进行加层或二次装修等,使原有结构负担增加。② 抗震加固。现有结构达不到抗震设防指标,对结构的基础、柱、砌体墙、梁板进行加固改造,增强抗震能力(见图 7.30)。③ 结构损害。建筑物年久失修或由于各种灾害导致结构损伤或破坏,不能满足目前的使用要求或安全度不足。

图 7.28 原位压力机检测砌体强度
Fig.7.28 Press testing in masonry strength

图 7.29 钻孔取芯机取样
Fig.7.29 Sampling by core drilling machine

As to some key concrete components, concrete blocks can be obtained by drilling core, then the real strength of concrete can be measured in the laboratory (Fig.7.29). The measured material strength and modulus of elasticity are brought into structure calculation model. As the basis of structure reinforcement, the results of the analysis can evaluate whether the structure is safe and detect unsafe components.

7.3.2 Engineering structural transformation and reinforcement

Engineering structure reinforcement learning is the study that restoring the using function of the damaged engineering structure or adapting to the new function. Due to the rapid development of civil engineering and the modern science and technology, this subject also developed rapidly. The reasons that buildings need to be reinforced and reconstructed usually fall into the following kinds: ① The increase of using load. Adding tonnage of vehicles on the bridge layer or secondary decoration of buildings can increase the burden of the original structure.② Seismic strengthening. The existing structures can not meet the seismic fortification target, then seismic capability should be increased for the structure of the foundations, columns, masonry walls, beams and slabs reinforcement renovation (Fig.7.30).③ The structural damage. The dilapidated and damaged structures caused by various disasters cannot meet current demand or insufficient degree of safety.

④ 纠正设计或施工失误。如纠正设计中配筋不足、构件截面太小,或施工中偷工减料等问题。⑤ 历史性建筑的保护(见图 7.31)。

在发达国家和地区,工程改造加固已成为建筑业的重要组成部分。目前我国既有建筑物总量达 400 多亿平方米,建筑业也已开始由大规模新建时期迈向新建与维护并重时期。土木工程结构的加固方法主要有加大截面法、外包钢加固法、预应力加固法、钢绞线网聚合物砂浆加固法、改变传力途径加固法、粘钢加固法、化学灌浆法、基础托换、基础纠偏等,如图 7.32 所示。

7.4　工程防灾减灾的新成就与发展趋势

7.4.1　工程防灾减灾的新成就

城市化的加速,使得生命线工程系统防灾减灾成为研究的热点。

图 7.30　中学教学楼基础隔震抗震加固
Fig.7.30　Building seismic strengthening

图 7.31　南京博物院老大殿整体顶升与隔震加固
Fig.7.31　The Integrally jack-up and isolation reinforcement

④ Correcting the design or construction mistakes. ⑤ The protection of the historic buildings(Fig.7.31).

In the developed countries and regions, engineering reinforcement has become an important part of the construction. Currently, the total quantity of existing buildings in China is more than 40 billion square meters, and the construction in China has also started from a largescale construction period towards a building and maintenance period. There are many methods of the reinforcement applied to different engineering structures respectively (Fig.7.32). For example, enlarging section method, the steel-encased reinforcement method, the prestressed strengthening method, the steel wire mesh-polymer mortar strengthening method, changing force's transfer road, stick steel strengthening method, chemical grouting method, and foundation underpinning, etc.

7.4　New achievements and trends of engineering disaster prevention and mitigation

7.4.1　New achievements of disaster prevention and mitigation

With the acceleration of urbanization, the lifeline systems engineering disaster prevention and mitigation of lifeline system engineering has become a research hotspot.

　　近年来,我国学者对生命线工程系统减灾开展了基础研究并取得了重大的研究成果。

　　① 建立了生命线工程系统的场地危险性评估方法。

　　② 建立了地下管网等生命线工程系统在地震作用下的反应分析方法。

图 7.32　土木工程结构主要加固方法
Fig.7.32　Strengthening methods of civil engineering

　　In recent years, Chinese scholars carried out many basic researches and made significant achievements in the lifeline engineering system mitigation.

　　① Established the site risk assessment method of lifeline engineering systems.

　　② Established the earthquake response analysis method of the underground pipe network system.

③ 建立了城市多灾害损失的评估模型。

④ 提出了基于性能的结构抗震设计方法。

⑤ 研究了城市地震触发滑坡、岩溶塌陷、采空区塌陷、地震火灾及渗水引发滑坡等灾害链现象,并提出了相应的评估方法。

⑥ 提出了大跨度建筑与桥梁结构的抗震抗风分析和隔震减振控制方法。

7.4.2　工程防灾减灾的发展趋势

自然灾害的预测、预报、预防和救助涉及自然科学领域的众多方面,而且大多数都是当前世界性的高科技前沿课题。防灾减灾科学研究处于自然科学、技术科学和社会科学的交汇点,体现了三者的相互渗透和结合。防灾减灾科学技术具备跨学科性,其成果具有广泛的社会应用性。例如,防灾减灾工作中的综合减灾、灾害系统论、减灾模型、失误控制论、灾害哲学及文化、人为灾害、数字减灾系统等概念及评估方法都体现了上述特点。实践还表明,防灾减灾预测预防基础研究具有鲜明的超前性、突破途径的非常规性、某些重大发现的偶然性及科学创新的艰巨性。由于灾害涉及的学科较多,灾种间还有交叉,各地区所面临的重点有所不同,所以土木工程防灾减灾学的内容可以扩展,并不断地发展和完善。

① 开展自然灾害危险性评价和风险评估。美国和俄罗斯等国家重点对实际风险和可承受风险进行评估, 其成果已成为减灾应用基础研究的重要前沿领域之一,并为城市生命线工程的抗灾能力评估提供了依据。

③ Established loss assessment model for city multi-hazard.

④ Put forward performance based seismic design method.

⑤ Studied the disaster chain phenomena including landslides, collapse caused by urban earthquakes, and landslides caused by seepage, then put forward the corresponding evaluation method.

⑥ Put forward the seismic and wind resistance analysis method as well as the vibration control method of long-span buildings and bridges.

7.4.2　New trends of disaster prevention and mitigation

Natural disaster prediction, forecasting, prevention and rescue involve many aspects of the natural science, and most of them are the current worldwide high-tech frontier. Disaster prevention and mitigation science research is the intersection of natural science, technology science and social science. It embodies the mutual infiltration and combination of all three. Disaster prevention and mitigation is interdisciplinary science and technology, and its achievements with extensive social application is its important characteristic. For example, the comprehensive disaster reduction, disaster system theory, disaster reduction model, disaster philosophy and culture, man-made disasters, digital disaster reduction system concept and evaluation method in the disaster prevention and mitigation work reflect the above characteristics.

① Carrying out the risk assessment for natural disasters. The United States, Russia and other countries are focusing on practical and affordable risk assessment. The achievements have become one of the important frontier fields in the basic research, and provided a basis for the evaluation of urban lifeline engineering disaster ability.

② 开展承灾体易损性研究。美国等国家积极开展大城市的底层易损性研究、自然灾害社会易损性研究、积极易损性与社区易损性研究,这些研究在理论和方法上促进了承灾体易损性研究的深入,也表明承灾体易损性研究是综合科技减灾的前沿领域。

③ 灾害信息系统建设。美国、日本、加拿大和欧盟等,为进行灾害和紧急事物的管理,更好地沟通灾害信息,减轻灾害损失,均建立了灾害信息系统,实现灾害信息共享,以达到在灾害面前各方快速应急反应的要求。

④ 新材料、新技术的应用。传统的结构加固方法所用到的材料,如焊接、螺栓连接等会对原结构产生新的损伤和应力,而纤维增强复合材料是由高性能纤维材料与基体按一定比例并经过一定工艺复合形成的一种高性能新型材料,具有高强、轻质、耐腐蚀及施工方便等优点,并开始以各种形式应用于土木与建筑结构加固工程中。

② Researching on vulnerability of hazard-affected bodies. The United States and other countries actively develop the underlying vulnerability research in big cities, social vulnerability research, positive vulnerability and community vulnerability research.

③ Disaster information system construction. America, Japan, Canada and some countries in the European Union have established the disaster information system, for the purpose of disaster management, better communication of disaster information and reducing disaster losses.

④ The application of new material and technology. Traditional structure reinforcement methods, such as welding, and bolt connection on the original structure produce new trauma and stress. The fiber reinforced polymer (FRP) is a high-performance new material which is composed of a certain proportion of high-performance fiber and matrix and through a certain process to form. Due to the advantages of high strength, light weight, corrosion resistance, and it begins to be applied in civil reinforcing engineering structures.

注:本章图片均来源于网络。
Note: In this chapter, all pictures are from webs.

知识拓展
Learning More

相关链接 Related Links

(1) 中国国家减灾网

(2) 中国地震局

(3) 太平洋地震工程研究中心

小贴士 Tips

我国《建筑抗震设计规范》(GB 50011—2010)明确提出设防要求：“小震不坏，中震可修，大震不倒。”

The design requirements were put forward in *Code for Seismic Design of Buildings* (GB 50011—2010)in China. No damage in small earthquake, repairable under moderate earthquake, no collapse under rare earthquake.

思考题 Review Questions

(1) 土木工程灾害有哪些主要类型？

What are the main types of civil engineering disasters?

(2) 减灾系统工程包括哪些内容？

What contents do disaster reduction system engineering include?

(3) 如何减轻地震灾害？

How can earthquake disasters be mitigated?

(4) 如何防止风灾？

How to prevent wind disaster?

(5) 如何防治地质灾害？

How to prevent and control geological disaster?

参考文献
References

[1] 中华人民共和国住房和城乡建设部. 建筑抗震设计规范(GB 50011—2010)[S].北京:中国建筑工业出版社,2011.

[2] 中华人民共和国住房和城乡建设部. 建筑结构荷载规范(GB 50009—2012)[S].北京:中国建筑工业出版社,2012.

[3] 中华人民共和国公安部. 建筑防火设计规范(GB 50016—2014)[S].北京:中国计划出版社,2006.

[4] 张相庭. 结构风工程:理论·规范·实践[M].北京:中国建筑工业出版社,2006.

［5］江见鲸,徐志胜,等.防灾减灾工程学[M].北京:机械工业出版社,2005.

［6］李爱群.工程结构减振控制[M].北京:机械工业出版社,2007.

［7］叶志明.土木工程概论[M].3版.北京:高等教育出版社,2009.

［8］周新刚.土木工程概论[M].北京:中国建筑工业出版社,2011.

［9］周云.土木工程防灾减灾学[M].广州:华南理工大出版社,2002.

［10］董羡,黄林青.土木工程概论[M].北京:中国水利水电出版社,2011.

［11］袁海军,姜红.建筑结构检测鉴定与加固手册[M].北京:中国建筑工业出版社,2003.

［12］任建喜.土木工程概论[M].北京:机械工业出版社,2011.

［13］王茹.土木工程防灾减灾学[M].北京:中国建材工业出版社,2008.

［14］李文虎,代国忠.土木工程概论[M].北京:化学工业出版社,2011.

［15］Palanichamy M S. Basic Civil Engineering[M].北京:机械工业出版社,2005.

［16］Shen Zuyan. Introduction of Civil Engineering[M].北京:中国建筑工业出版社,2010.

［17］Simiu , Scanlan R H. Wind Effects on Structures: Fundamentals and Applications to Design[M]. 3rd ed. Hoboken: John Willey & Sons, INC., 1996.

［18］Williams A. Seismic Design of Buildings and Bridges[M]. 北京:中国水利水电出版社,2002.

第8章 道路与渡河工程

Chapter 8 Road and River Crossing Engineering

先导案例
Guide Case

四川雅西高速公路(雅安—西昌),堪称世界上最困难的工程项目(见图8.1)。它是一个双重示范项目,设计时速80 km/h,全线长240 km,总投资206亿元,从四川盆地一直延伸到青藏高原,横穿横断山脉的深谷,跨越青衣江、大渡河、安宁河等水系,全线共有桥梁270座,隧道25座,整条公路的桥隧比达到55%,实现了桥、隧道、路的完美结合。这条高速公路跨越12条地震断裂带,海拔高,环境恶

图 8.1 四川雅西高速公路
Fig. 8.1 Yaxi Expressway in Sichuan

Sichuan Yaxi Expressway (Ya'an–Xichang) is the most difficult engineering project in the world (Fig.8.1). It is a dual demonstration project with a design speed of 80 km/h, a total length of 240 km and a total investment of 20.6 billion yuan. It extends from Sichuan Basin to the Qinghai-Tibet Plateau, crosses the deep valleys of the Hengduan Mountains and crosses the Qingyi River, Dadu River, Anning River and other water systems. There are 270 Bridges and 25 tunnels in the whole road, and the bridge-tunnel ratio of the whole highway reaches 55%. The perfect combination of bridge, tunnel and road has been realized. Crossing 12 seismic fault zones, high altitude and harsh environment, such a complex road is rare in the world. In order not to destroy the landform, vegetation and to protect the environment,

劣,这样复杂的路段设计全球少见。为了不破坏地貌、植被,保护环境,雅西高速公路有20多公里建在安宁河里。它的建成真的是一个奇迹,展示了我国无与伦比的基建水平和实力。

本章主要介绍道路、桥梁与隧道工程的基本概念、特点及技术要求。通过本章的学习,可以加深对道路、桥梁及隧道工程的性质、组成、功能、特点、技术要求的认识理解,了解道路与渡河工程行业对于人类生产、生活及国民经济的重要作用,了解未来的发展趋势,激发起对道路与渡河工程专业的学习兴趣,明确作为交通人的责任。

8.1 道路工程

8.1.1 道路的定义及分类

道路包括公路和城市道路,其中公路按其使用的功能和性质可分为国家干线公路(简称国道)、省级干线公路(简称省道)、县级公路(简称县道)及专用公路;城市道路按其在城市道路系统中的地位、交通功能分为快速路、主干路、次干路和支路四类,根据城市规模、规划交通量和地形等因素,除快速路外,各类道路划分为Ⅰ,Ⅱ,Ⅲ级。

more than 20 kilometers of the Yaxi expressway has been built in the Anning River. It is truly a miracle that it is built at all, demonstrating the unparalleled level and strength of our infrastructure.

This chapter mainly introduces the basic concepts, characteristics and technical requirements of road, bridge and tunnel engineering. Through the learning of this chapter, we can deepen the understanding of the nature of road, bridge and tunnel engineering, composition, function, characteristics and technical requirements, understand the roads and river crossing engineering industry for human production, life and the important role of the national economy, understand the development trend of the future, set up learning interest in road and river crossing engineering, make clear responsibility in traffic.

8.1 Road engineering

8.1.1 Definition and classification of roads

Roads include roads and urban roads. According to their functions and properties, roads can be divided into national trunk roads (national highways), provincial trunk roads (provincial highways), county roads (county roads) and special roads. Urban road according to its position in the urban road system, traffic function is divided into the following four categories: expressway, trunk road, secondary trunk road and branch road, according to the city size, planning traffic volume and terrain and other factors, in addition to expressway, all kinds of roads are divided into grade Ⅰ, Ⅱ, or Ⅲ.

8.1.2　道路设计的基本要素

（1）平面线形设计

公路平面线形是指道路中心线的空间线形在水平面上的投影（见图 8.2）。公路的平面线形主要包括直线、圆曲线及缓和曲线，也称为公路平面线形的三要素。

（2）纵断面设计

通过公路中线的竖向剖面展开的图称为路线纵断面图。纵断面设计主要是指纵坡和竖曲线设计，采用直角坐标，横坐标表示里程桩号，纵坐标表示高程。为了清楚地反映路中心线上地面起伏情况，通常公路纵断面图横坐标的比例尺采用 1:2000，

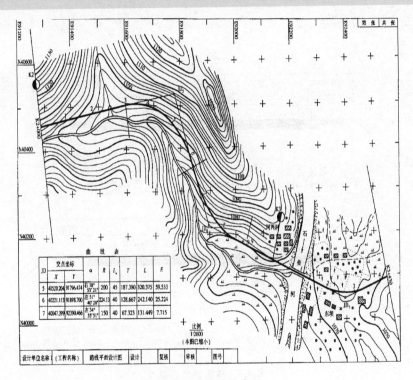

图 8.2　公路平面线形
Fig.8.2　Highway plane linear form

8.1.2　Basic elements of rood design

（1）Plane alignment design

The highway plane alignment is the projection of the spatial alignment of the road centerline on the horizontal plane (Fig. 8.2). Highway plane alignment mainly includes straight line, round curve and ease curve, also known as three elements of highway plane alignment.

（2）Vertical section design

A vertical section spread through the highway center line is called a route profile. The vertical section design mainly refers to the vertical slope and vertical curve design. Cartesian coordinates are adopted. The abscissa represents the mileage pile number and the ordinate represents the elevation. In order to clearly reflect the ground fluctuation on the road center line, the scale of abscissa of the highway profile is usually 1:2000, and

纵坐标采用 1:200。

(3) 横断面设计

① 公路横断面的组成及布置

沿着公路平面中心线的法线方向作一垂直剖面,这个剖面称为公路横断面。各等级公路的路基横断面如图 8.3 所示。

② 横断面设计图

比例尺一般用 1:200,在横断面图上要标注桩号、填(挖)高度、填(挖)面积、边坡坡度,在有超高、加宽的横断面上还要标明其相应数值。

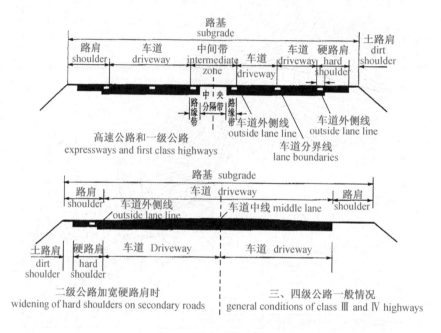

图 8.3　路基横断面组成
Fig. 8.3　Cross section composition of subgrade

the scale of ordinate is 1:200.

(3) Cross section design

① Composition and layout of highway cross-section

Make a vertical section along the normal direction of the highway plane center line, this section is called the highway cross-section. The cross-section of roadbed of each grade highway is shown in Fig. 8.3.

② Cross-sectional design drawing

The scale is generally 1:200. Pile number, filling (digging) height, filling (digging) area and slope gradient should be marked on the cross-sectional drawing, and corresponding values should be marked on the over-high and widened cross-sections.

8.2　桥　梁

桥梁是指跨越江河、沟谷或其他交通线路等各种障碍并供铁路、公路、渠道、管线及行人使用的承重结构物,它是架空的路,在道路交通及城市建设中起着控制性作用。

桥梁是随着人类社会的进步而逐渐发展起来的。随着陆上新的交通工具的出现和不断变化,以及人们对交通快捷、舒适的期望越来越高,桥梁在载重、跨度、美观等方面的性能要求也逐渐提高,而新材料、新设备、新工艺的出现与发展,电子计算机的普及和计算技术的不断发展为桥梁工程的发展提供了巨大的动力。

现代桥梁已不纯粹以满足功能为目的,而是成为具有强烈的时代特征的一种空间艺术。桥梁巨大的跨度、强烈的形体表现力、超凡的尺度均对城市或大地景观产生显著的影响,成为人类文明和城市发展的重要标志之一。

8.2.1　桥梁的基本组成及分类

（1）桥梁的基本组成

桥梁通常可以划分为上部结构、下部结构和附属结构,典型的桥梁概貌和基本组成如图 8.4 所示。

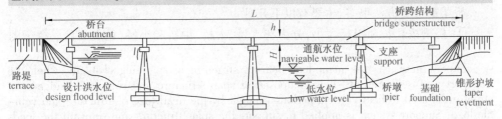

图 8.4　桥梁的基本组成
Fig. 8.4　Compositions of a typical bridge

8.2　Bridges

A bridge is a load-bearing structure that comes across rivers, gorges or other barriers in transport line, and it is built for the use of railway, highway, channel, pipeline and pedestrian. It plays a crucial role in the development of road transport and urban construction.

Bridges have been gradually developed with the progress of society. As the advent of new inland vehicles and the improvement of people's requirements on convenience, comfortable of transport, the performance requirement of bridge is improved in load-bearing, span capacity and artistic design. Simultaneously, a huge source power for the progress of bridge is provided by the development of high performance materials, new devices and technology, along with the great improvement of computer and computational capacity.

Nowadays, bridge is not only a key part of transport line, but also a spacial art with distinct characteristics of the times. Its huge span, strong body expressive force and extraordinary scale affect remarkably the city appearance and the landscape, and has become one of important marks of human civilization and urban development.

8.2.1　Basic compositions and classification of bridge

（1）Basic compositions of bridge

Usually, a bridge can be divided into superstructure, substructure and accessory structures. The compositions of a typical bridge are shown in Fig.8.4.

上部结构是线路中断时跨越障碍的主要承载结构,由桥跨结构和支座系统组成。桥跨结构用于承受恒载和车辆等竖向荷载的作用;支座系统则用于支承桥跨结构并传递荷载到桥梁墩台上,应满足上部结构在荷载、温度或其他因素作用下预计的位移要求。桥梁跨越的幅度(即跨径)越大,结构构造要求越复杂,施工难度也相应增大。

下部结构设置在地基上用以支撑桥跨结构,并将其荷载传递至地基的结构物,如桥墩、桥台及墩台基础等,是桥跨结构极为重要的组成部分。

① 桥墩,多跨桥梁中处于相邻桥跨之间支撑上部结构的构造物。

② 桥台,位于桥梁两端,一端与路堤相接并防止路堤滑塌,另一端支承桥跨上部结构的构造物。

③ 墩台基础,保证墩台安全并将荷载传至地基的结构部分。墩台基础是桥梁的根基,对保证整个结构的安全具有举足轻重的作用,同时也是桥梁施工中最复杂、难度最大的部分。

附属结构包括桥面铺装(或称行车道铺装)、排水防水系统、栏杆(或防撞护栏)、伸缩缝和灯光照明等,均为与桥梁服务功能有关的部件,总称为桥面构造。

(2) 桥梁的分类

桥梁的分类方法很多,主要有以下几种:

① 桥梁按总长和跨径大小,分为特大桥、大桥、中桥和小桥(见表 8.1)。桥梁的

Superstructure is the main load-bearing structure of spanning obstacles when the transport line is broken off. It is composed of span structure and supports. Bridge span structure is used to bear vertical actions brought by the dead load of structure and vehicle load. Support system is applied to support the span structure and transmit the loads to the piers, so it should be suitable for the displacements restrictions under the actions of loads, temperature and other factors. The bigger the span of bridge is, the more complicated the structure is and the more difficult the construction is.

Substructure refers to the structure that is laid on the ground to support the bridge span structure and transmit loads to the ground, including piers, abutment, pier foundation and so on. It is a very important part of bridge structure.

① Bridge pier is located between the adjacent spans in the multi span structure to support superstructure.

② Abutment is located on both ends of the bridge. One side of abutment connects with the embankment to prevent its slumping and the other one supports the superstructure of the bridge.

③ Pier foundation protects the pier and transmits loads to the ground. It is the base of bridge and has vital effect on the safety of the whole structure. Besides, it is the most complicated and difficult part during the bridge construction.

Accessory structure includes bridge deck pavement(or called carriageway pavement), drainage and waterproof system, railing (or crash barrier), expansion joints and lighting. All of the above are related to the service functions of bridge, called bridge deck structure.

(2) Classification of bridges

There are many methods for classifying bridges, and the common ones are listed as follows:

① According to the total length and span length, bridges can be classfied as super-large bridge, large bridge, medium bridge and small bridge (Tab.8.1). The span reflects

跨径反映了桥梁的建设规模。

②桥梁按主体结构材料,分为钢桥、混凝土桥、石桥、木桥等。

③桥梁按用途,分为铁路桥、公路桥、公铁两用桥、人行桥、输水桥等。

④桥梁按桥面系和上部结构的相对位置,分为上承式桥、中承式桥和下承式桥。

⑤桥梁按平面布置,分为正桥、斜桥、弯桥等。

⑥桥梁按结构体系,分为梁桥、拱桥、缆索体系桥和组合体系桥等。

8.2.2　主要的桥梁类型

（1）梁式桥

梁式桥是中国古代最早出现的桥型且应用普遍,它以梁作为承重构件,在竖向恒载和交通荷载作用下主要承受剪力和弯矩,无水平反力。结构形式包括简支梁桥、

表 8.1　桥梁按总长和跨径大小分类
Tab. 8.1　Classification of bridges according to the total length and span length

桥梁分类 Kinds	多孔跨径全长 L/m The total length of multi-span bridge L/m	单孔跨径 l/m The length of single-span bridge l/m
特大桥 Super-large bridge	$L \geqslant 1000$	$l \geqslant 150$
大桥 Large bridge	$100 \leqslant L < 1000$	$40 \leqslant l < 150$
中桥 Medium bridge	$30 \leqslant L < 100$	$20 \leqslant l < 40$
小桥 Small bridge	$8 \leqslant L < 30$	$5 \leqslant l < 20$

数据来源:《公路工程技术标准》(JTG B01—2014)。
Data Sources：*Technical standard of highway engineering* (JTG B01—2014).

the construction scale of bridge.

② According to the structural material, bridges can be classfied as steel bridge, concrete bridge, stone bridge, wooden bridge and so on.

③ According to the usage, bridges can be classfied as railway bridge, highway bridge, combined railway and highway bridge, footbridge, water bridge and so on.

④ According to the bridge deck system and the position of superstructure, bridges can be classfied as deck bridge, half through bridge and through bridge.

⑤ According to the plane layout, bridges can be classfied as main bridge, skew bridge and curved bridge.

⑥ According to the structural system, bridges can be classfied as beam bridge, arch bridge, cable system bridge and combination system bridge.

8.2.2　Main kinds of bridge

（1）Beam Bridge

Beam bridge is one of the ancient structural styles of bridge. Its beam is the main load-carrying element of the structure to bear the shear force and bend moment under the action of vertical dead loads and traffic loads. The outstanding feature is no horizontal inner force existing in the structure. The structural forms include simply supported beam bridge, con-

连续梁桥和悬臂梁桥等,如图8.4所示。

(2)拱桥

拱桥也是桥梁工程中使用广泛且历史悠久的一种桥梁结构类型,其桥跨结构由主拱圈、传力结构和桥面系组成(见图8.5),与梁桥的受力性能有着本质区别。具有

(a) 简支梁桥
Simply supported beam bridge

(b) 连续梁桥
Continuous beam bridge

(c) 悬臂梁桥
Cantilever beam bridge

图8.4 梁式桥
Fig. 8.4 Beam Bridge

(a) 重庆朝天门大桥
Chaotianmen Bridge in Chungking

(b) 上海卢浦大桥
Lupu Bridge in Shanghai

(c) 韩国傍花大桥
Bangwha Bridge in Korea

(d) 美国新河谷大桥
New River Gorge Bridge in America

图8.5 拱桥
Fig. 8.5 Arch Bridges

tinuous beam bridge, cantilever beam bridge and so on, which are shown in Fig.8.4.

(2) Arch bridge

Arch bridge is also a widely-used ancient type of bridge. It is composed of main arch, transmit structure and deck system, as shown in Fig.8.5, which is distinctively

曲线外形的拱圈是拱桥的主要承重结构,承受轴向压力,但也具有较大的水平反力,这能在很大程度上抵消拱圈内的弯矩作用。该类桥梁具有拱弯矩小、外形优美、经济性好的优点,但对地基条件要求较高。

(3) 悬索桥

现代悬索桥,是由 19 世纪的吊桥演变而来的,由桥塔、主缆索、吊杆、加劲梁、锚定及鞍座等主要部件组成的承载结构体系(见图 8.6)。其中,桥塔又称塔架、主塔,是支承主缆的重要构件,悬索桥的全部活载和恒载及加劲梁支承塔上的反力,都将通过桥塔传递至下部的桥墩和地基;主缆是悬索桥的主要构件,除承受自身荷载外,还通过吊杆来承担加劲梁的恒载、桥面上的活载及横向风荷载,并将这些荷载传至桥塔;锚锭是用来锚固主缆的重要构件,锚锭将主缆中的拉力传递给地基;吊杆又称吊索,将活载和加筋梁等恒载传至主缆;加劲梁则主要防止桥面发生过大挠曲变形及扭转变形。由于这一结构形式能充分利用悬索桥各方面材料的强度,具有用料省、自重轻等特点,故其跨越能力比其他桥型更强。

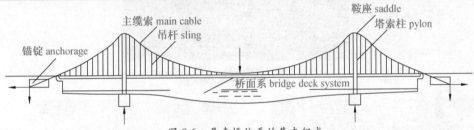

图 8.6　悬索桥体系的基本组成
Fig. 8.6　Basic composites of suspension bridge

different from the loading state of beam bridges. The main arch is the principle load-carrying element to bear axial pressure. At the same time, it also bears a big horizontal force which offsets partly the bend moment acting on the arch. Therefore, this kind of bridge has advantages of lower bend moment, attractive appearance and cost-saving characteristic. However, its application is restricted to the high requirements on foundation.

(3) Suspension bridge

Modern suspension bridge is evolved from hanging bridge in the 19th century. Its load-bearing system is consisted mainly of pylon, main cable, suspension rod, stiffening girder, anchorage and saddle, as shown in Fig.8.6. The pylon, also known as main tower, is an important element to support the main cable. The live load acting on the bridge, dead load and the counter force of stiffening girder are transmitted to the piers and foundation by the pylon. The main cable, which is the main structural element of suspension bridge, bears not only the dead load of itself, but also the dead load of stiffening girder, the live load acting on the deck of bridge and transverse wind load by means of suspension rods, and transmits these loads to the pylon. The anchorage is used to anchor the main cable and deliver the tension in the cable to the ground. The suspension rod delivers the loads from the deck to main cable. Finally, the stiffening girder plays the role of avoiding big flexural deflection and torsion deformation of deck. As this kind of bridge can make full use of the strength of materials, it has many merits, such as saving materials and lightening the structure system. As a result, its spanning capacity is bigger than other kinds of bridges.

近年来,随着悬索桥设计理论和计算方法的发展和完善,以及施工技术的进步,悬索桥成为发展较快的桥型之一,尤其是在所需跨度大,或跨越大河、深谷等不易修筑桥墩的工程中具有较突出的优势。

悬索桥的典型工程有主跨 1 991 m 的日本明石海峡大桥(见图 8.7 a)、主跨为 1 490 m 的江苏润扬大桥南汉大桥、主跨 1 385 m 的江阴长江大桥、主跨 1 650 m 的舟山西堠门跨海大桥(见图 8.7 b)、主跨 1 377 m 的香港青马大桥等(见图 8.7 c)。但与其他桥型相比,悬索桥的刚度和振动的固有频率较低,因此在桥梁设计时需要重点

(a) 日本明石海峡大桥 (主跨 1991 m)
Akashi Kaikyo bridge in Japan

(b) 中国西堠门跨海大桥
Xihoumen Bridge in China

(c) 香港青马大桥
Tsing Ma Bridge in Hongkong

图 8.7　悬索桥典型工程
Fig. 8.7　The typical suspension bridges

In recent years, as the development and improvement of the bridge design theory and calculation method, suspension bridge becomes more and more popular. It has outstanding advantages especially in the cases where the requirement span of bridge is too long, or it is difficult to construct piers.

The typical suspension bridges are the Akashi Kaikyo bridge in Japan with the main span of 1991 m, the south branch of the Runyang Bridge with the main span of 1490 m, Jiangyin Yangtze River Bridge in China with the main span of 1385 m, Xihoumen Bridge in China with the main span of 1650 m, and Tsing Ma Bridge in Hongkong with the main span of 1377 m (Fig.8.7). However, compared with the other kinds of bridges, suspension bridge has lower stiffness and natural frequency, so its

考虑结构的抗风稳定性。同时,作为传统缆索材料的钢材自重大,且在侵蚀性环境中容易腐蚀,导致钢缆索承载效率随着跨度的增大而降低,运营中的维修养护费用较高等问题。

(4) 斜拉桥

斜拉桥是我国大跨径桥梁的优选桥型之一,由主梁、斜索和塔柱 3 部分组成,其主要承载结构体系如图 8.8 所示。与连续梁桥相似,桥面恒载和交通荷载仍由主梁承担,而主梁则由高强材料制成的斜索多点吊起,斜索的另一端锚固在索塔上,将主梁的恒载、活载及风载等传递至索塔,再通过索塔传至地基。

在这样的承载体系中,每根斜索分别取代连续梁桥中桥墩的弹性支撑,其受力特征如同多跨弹性支承的连续梁一样,故梁中的弯矩得以大大降低,从而使主梁截面尺寸大大减小,自重减轻,既节省了结构材料,又能大幅度地增强桥梁的跨越能

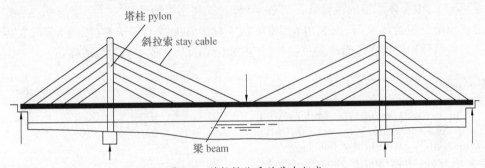

塔柱 pylon

斜拉索 stay cable

梁 beam

图 8.8　斜拉桥体系的基本组成
Fig. 8.8　Basic composites of cable-stayed bridge systems

wind resistant stability should be stressed in the bridge design. Meanwhile, steel which has big dead load and tends to be corroded is often used as the cable material. Consequently, the capacity efficiency is reduced with the increase of span and maintenance cost becomes high in the service of bridge under the severe environment.

(4) Cable-stayed bridge

Cable-stayed bridge is one of the preferred type of long-span bridge in China. Its main load-bearing structure is composed of three parts: main beam, cables and tower, as shown in Fig.8.8. Similar to continuous beam bridge, dead load of bridge deck and traffic load are still carried by the main beam. On the other hand, the main beam made of high strength materials is hung up by several cables and the cables are anchored to the pylon at their other ends, so that the constant load, live load and wind load can be transmitted to the pylon, and then to the ground.

In such bearing system, each cable replaces the elastic supports of the bridge pier in continuous beam bridge, and the mechanical characteristics are very similar to the continuous beam with multiple elastic supports. In such way, the bending moment of the beam is decreased greatly and consequently the bending moment is also reduced, the structural material is saved and the spanning capacity is improved. At the same time, horizontal component of the tension in inclined cables prestress the beam, which enhances the crack resistance of the main girder and reduces the consumption of high-strength steel. The Su Tong

力。同时,斜索拉力产生的水平分力可以对梁产生预压力,从而增强主梁的抗裂性能,节约高强钢材的用量。中国苏通大桥是典型的斜拉桥结构,如图 8.9 所示。

一般来说,对于主跨 200~700 m 的桥梁,斜拉桥在技术上和经济上具有相当优越的竞争力。德国著名桥梁专家 F.leonhardt 认为,即使跨径 1 400 m 的斜拉桥也比同等跨径悬索桥节省一半的高强钢丝用量,造价更是低 30%左右。然而,随着斜拉桥跨度的增大,将会面临桥塔过高和斜索过长等一系列技术问题,如增加了索–梁或索–塔的连接构造的复杂性和施工中高空作业的难度。另外,斜拉索的防腐问题及长索在风、雨激励下的振动和疲劳控制等问题仍是当今桥梁界的重要研究课题。

（5）组合体系桥梁

组合体系桥梁指承重结构采用两种或两种以上基本体系, 或者一种基本体系与某些受力构件组合在一起,同时利用不同结构体系或构件的力学优势、共同承载的桥梁结构体系。从 1858 年奥地利人兰格尔(Josef Langer)获得刚性梁柔性拱的系杆拱桥专利至今,组合体系桥的设计或设想逐渐增多,如利用梁的受弯特性和拱的承压特性组成的梁拱组合体系(见图 8.10 a)、将梁与立柱组合而成的钢架桥(见图 8.10 b)、斜拉索与吊杆均作用在桥面的斜拉拱桥(见图 8.10 c,d)及其他类型的组成体系。

Bridge in China (Fig.8.9) is the typical structure of cable-stayed bridge.

In general, among the bridges with the main span in the scope of 200 m to 700 m, cable-stayed bridge has superiorities in technology and economic aspects. F.leonhardt, a famous German bridge engineer, even believes that when the span reaches 1400 m, the consumption of the high strength steel in cable-stayed bridge can also be saved 1/2 of that in the suspension bridge with the same span, and the cost is lower about 30%. However, with the increase of the span, there will

图 8.9 中国苏通大桥(主跨 1 088 m)
Fig. 8.9 Su Tong Bridge in China(span is 1 088 m)

be a series of technical difficulties caused by too high pylon and too long cables. For example, the complexity of the connection structure of the cable-girder or cable-pylon and the difficulty in the overhead working are increased. In addition, the corrosion prevention, the vibration control and fatigue of stayed cables under the wind and rain are still some important research subjects in bridge.

（5）Composite bridge

Composite bridge means the load-bearing structure of bridge which is composed of two or more than two basic systems or combined with one basic system and some structural components to utilize the advantages of different structure systems and components simultaneously for carrying load. Since the tied-arch bridge system has been invented by an Austrian engineer Josef Langer in 1858, which is combined with rigid beam and flexible arch, the conception and design of combination system bridges is increasing. For example, the beam-arch combined system utilizes the outstanding performances of the beam in bending resistance and the arch in compression resistance(Fig.8.10 a). A steel frame bridge is combined with beam and column (Fig.8.10 b). A cable-stayed arch composite bridge has stay cables and suspension rods acting on the bridge deck

　　组合体系桥梁并不是简单地组合基本结构体系。当我们将不同的结构体系组合在一起时,首先要考虑这种组合是否能够获得某种优势,这种优势是否具有价值;其次,还需考虑到各种基本体系桥各自的特点和适应性,当它们组合在一起并获得某些优势时,是否会带来新的问题,这些新的问题应采取何种措施加以解决。

　　随着科学技术的发展,生产力水平空前提高,桥梁在跨越能力提高的速度、结构形式的组合化趋势、各种新型材料的应用和合理配置等多方面都得到长足的发展。通过各种桥梁方案的优化比较和组合实践,桥梁结构将更趋于多样、适用、合理和美观。

(a) 梁拱组合体系
Beam-arch combination system

(b) 钢架桥
Rigid frame bridge

(c) 马来西亚的 Seri Saujana 桥
Seri Saujana bridge in Malaysia

(d) 中国湘潭莲城大桥
Xiangtan city bridge in China

图 8.10　组合体系桥梁
Fig.8.10　Composite bridge

(Fig.8.10 c,d). And there are many other kinds of combination system bridges.

　　A composite bridge is not a simple combination of basic systems. Firstly, the advantages of the combination and its value must be considered before combining different structures together. Secondly, the features and adaptabilities of each basic system should be considered. For example, whether there are new problems occurring when some advantages are obtained by the combination and how to resolve these problems.

　　With the progress in science and technology, bridge has great development in the spanning capacity, the trend of structural combination, the application and the reasonable allocation of various new materials. The bridge structures tend to be more diverse, suitable, reasonable and artistic by optimizing and comparing with the various bridge programs and the practice of composite bridge.

8.2.3 桥梁的设计要求

桥梁的设计需根据其使用任务、性质和将来发展的需要,实现安全、经济、适用和美观的目标。

一般而言,需满足下列要求:

① 使用上的要求。桥梁的设计需保证车辆和行人安全畅通,满足所在线路将来交通发展的需要。长期超负荷运营往往是很多公路桥梁出现损伤病害甚至结构破坏的主要因素之一,因此,设计时应对桥梁的远景交通流量和组成有一定的前瞻性,充分考虑现阶段的施工和管理水平以及材料工艺水平,以保证桥梁的使用性能。同时,桥梁的类型、跨度大小和桥下净空还应满足泄洪、安全通航和通车的要求。

② 经济上的要求。桥梁的建造应经济合理,在满足功能要求的前提下,不仅需要考虑工程建设成本和完成工期,使之尽快通车运营,还需考虑桥梁在运营期的养护维修费用,使其全寿命周期的成本最低。

③ 结构上的要求。整个桥梁结构及其部件应具有足够的安全度, 即足够的刚度、强度、稳定性和耐久性,这需要通过结构计算确定和校核。

④ 美观上的要求。桥梁的设计应与周围环境相协调,体现力学与美学结合的技术美学原则、功能优先原则、环境与生态保护原则和景观创新原则。桥梁设计既要对其功能、构造技术、形态美学、材料肌理进行研究,还应与社会对环境品质的更高要求相适应,使其结构性与美观性和谐统一。

8.2.3 Overall design requirements of bridges

The bridge design should achieve the goals of safety, economy, application and aesthetic appearance based on its task, the nature and future development requirements.

Generally speaking, requirements for the design of bridge are as follows:

① Requirements in usage. Bridge design must ensure the safety and convenience of vehicles and pedestrians, and meet the need of future traffic growth in the traffic line. Long-term overload operation is one of the main reasons that many roads and bridges are damaged and even destroyed. Therefore, designers should have foresight to predict the traffic flow and composition in the future and fully take into account the level of construction, management and materials technological level at the stage to ensure the performance of the bridge. At the same time, the type of bridge, span scale, bridge clearance should meet the requirements prevention of flood, navigation and traffic.

② Economic requirements. The construction of bridge should be economical and reasonable. On the basis of the functional requirements, the construction costs and duration should be considered. Besides, the cost of maintenance and repairation should be taken into account during the operation of the bridge to achieve the lowest cost of the whole life cycle.

③ Structural requirements. Entire bridge structure and its components should have adequate safety including sufficient stiffness, strength, stability and durability, which are determined by calculation and checking the structures.

④ Aesthetic requirements. Bridges should be designed in harmony with the surrounding environment, embodying the technological aesthetics principle, function priority principle, environmental and ecological protection principle, and landscape innovation principle. Bridge design not only need to study the function, construction technology, aesthetic form and material texture, but also need to meet with higher and higher demands of the environment quality so as to achieve the harmony of structural and aesthetic properties.

8.2.4　桥梁的质量控制与管理养护

建造一座现代大桥,除需大量的建设资金(用来支付造桥所需付出的工资、材料费、机具使用费等)外,还需要具备 2 种能力:① 设计能力:提供各种设计图纸、提出施工建议、编制预算;② 施工能力:组织工人施工、使用机具,将材料加工成构件,再将构件安装成桥梁结构,将设计图纸变成实桥。以上两方面中任何一方的质量出现问题,都可能出现严重的工程问题。国内外桥梁建设中曾发生过很多这样的事故,并为此付出了惨重的代价。

加拿大圣劳伦斯河之上的 Quebec Bridge(见图 8.11)是由著名设计师 Theodore Cooper 设计的世界上最长的跨度钢悬臂桥。大桥的主跨由 490 m 延伸至 550 m,以此节省建造桥墩基础的成本。然而,该桥在 1907 年第一次建造时,因设计师对桥梁重量计算不精确导致大桥杆件失稳而突然倒塌,75 名施工工人落水溺亡。

<table>
<tr><td>(a) Quebec Bridge 在第一次施工中倒塌
The Quebec Bridge collapsed during the first construction</td><td>(b) 建成后的 Quebec Bridge
The Quebec Bridge in service</td></tr>
</table>

图 8.11　加拿大 Quebec Bridge
Fig. 8.11　Quebec Bridge in Canada

8.2.4　Quality control and maintenance management of bridge

When we construct a modern bridge, there are two components needed except massive construction funds for the salary of workers, material expenses and the machine fees: ① design skills for providing design drawings, putting forward the construction suggestions and budget proposal; ② the construction capacity for organizing workers to construct, using the machines, processing the material to get the structural elements, installing the elements into a bridge structure and finally getting a real bridge from design drawings. If any quality problem appears in the above aspects, serious problems may occur in engineering. There have been many accidents in the bridge construction both in native and abroad, and people have paid a heavy price for them.

The Quebec Bridge, over the Laurent Canada Su River, is the longest span steel cantilever bridge (Fig.8.11) in the world designed by Theodore Cooper, a famous bridge designer. In order to reduce the expense of the pier foundations, the main span of the bridge is extended from 490 m to 550 m. However, during the first construction in 1907, the rods of the bridge lost stability and collapsed suddenly because the designer took an imprecise calculation of the bridge weight. The accident caused 75 workers falling into the river and drowned.

1913—1916 年间,该桥再次重建,却因施工过程中一个支撑点的材料指标不足而使桥身塌陷,13 名工人死亡。直到 1917 年,经历了两次惨痛的教训后,Quebec Bridge 终于竣工通车。

湖南省凤凰堤溪沱江大桥是圬工(没有钢筋,由石头、水泥、砂等构成)拱桥,属于湖南省的重点工程,设计桥长 328 m。由于施工过程中工程管理混乱、施工单位擅自变更原施工方案、施工质量差等原因,2007 年 8 月施工中的桥身突然整体垮塌(见图 8.12),致 64 人死亡,直接经济损失近 4 000 万元。

此外,桥梁运营中的安全养护管理工作也十分重要。尤其是旧桥,受建造时的设计、材料、施工等方面的影响和局限,加上交通量的逐年增长,检测和加固显得非常迫切。为了保证桥梁的耐久性,必须定期对桥梁进行安全养护及检查,掌握桥梁状况,及时消除安全隐患,有效延缓因疲劳导致的桥梁功能退化,确保桥梁安全运营。

图 8.12　沱江大桥垮塌现场
Fig. 8.12　Collapsed scene of the Tuojiang Bridge

The bridge was reconstructed from 1913 to 1916, ended by the collapse of bridge deck. The reason is that material strength at one supporting point wasn't high enough. 13 workers died in this accident. In 1917, the Quebec Bridge was completed finally and started its operation after experiencing two painful lessons.

Tuojiang Bridge in Di Xi, Feng Huang city, Hunan Province in China was a masonry arch bridge. The bridge was made up of stone, cement, sand, and so on, without steel bars. It is a key project in Hunan Province. Design length of the bridge is 328 m. The bridge entirely collapsed suddenly during the construction in August, 2007, because the project management was in disorder(Fig.8.12). The construction units changed the original construction scheme without authorization and construction quality was poor. The accident killed 64 people and caused a direct economic loss nearly 40 million yuan.

In addition, maintenance and management of bridge on site is also very important. Special attention should be paid to old bridges. The inspection and reinforcement technologies become very urgent for old bridges which were affected and restricted by the design, materials and construction at the design stage. Besides, the rapid growth of the current traffic flow aggravates the damage of bridges. In order to ensure the durability of the bridge, the conservation and security checks of bridges must be regularly carried out, the bridge status should be understood and potential hazard should be eliminated timely. Therefore the degradation of bridge caused by fatigue can be delayed effectively and the safe operation of the bridge can be guaranteed.

8.3 隧 道

8.3.1 隧道的分类及发展

(1) 隧道的定义及其分类

隧道,是修筑在岩体、土体或水底的,两端有出入口的通道,它是线路穿越山岭、丘陵、土层、水域等天然障碍最有效的途径。1970 年 OECD(世界经济合作与发展组织)会议定义:以任何方法修建,最终使用于地表面以下的条形建筑物,其空洞内部净主断面大于 2 m^2 以上者均为隧道。从这个定义出发,隧道包括的范围很大。隧道的分类方法也很多,通常可按用途对隧道进行分类,如图 8.13 所示。

(2) 隧道的发展

近代隧道兴起于运河时期,从 17 世纪起,欧洲陆续修建了许多运河隧道。法国的兰葵达克运河隧道建于 1666—1681 年,长 157 m,它可能是最早用火药开凿的隧道。1830 年前后, 铁路成为新的运输手段, 铁路隧道也随之出现并逐年增多,至 1906 年已出现了穿越阿尔卑斯山脉长达 19.73 km 的铁路隧道。目前世界上的长大道路隧道(2 km 以上)和长大水底隧道(0.5~2.0 km)已超过百条,如图 8.14 所示,最长的铁路隧道已达 53.85 km。

8.3 Tunnel

8.3.1 Classification and development of tunnel

(1) Definition and classification of the tunnel

The tunnel is built in soil, rock masses or submarine and has inlet and outlet at the both ends. It is an effective path when traffic line goes through the mountains, hills, soil, water area and other natural obstacles. In 1970, OECD (Organization for Economic Cooperation and Development) defined the tunnel in the conference. That is, the tunnel is a strip-type structure constructed by any method and finally used below the ground surface whose net main section is greater than 2 m^2. According to this definition, the tunnel has a large range. There are many classification methods for tunnels. They are usually classified according to the usage, as shown in Fig.8.13.

(2) The development of the tunnel

Modern tunnel originated during the Canal Period. From the 17th century, Europe has built gradually many canal tunnels. Languedoc Canal Tunnel in France was built in 1666—1681 with the length of 157 m. It is probably the earliest tunnel dug with gunpowder. Around 1830, the railway had become a new means of transport, with which railway tunnel also appeared and increased year by year. In 1906, a 19.73 km-length railway tunnel appeared which went through the Alps. Nowadays, the number of long road tunnel (more than 2 km) and long and large underwater tunnel (0.5 ~2.0 km) in the world is more than one hundred (Fig.8.14). The length of the longest railway tunnel has reached up to 53.85 km.

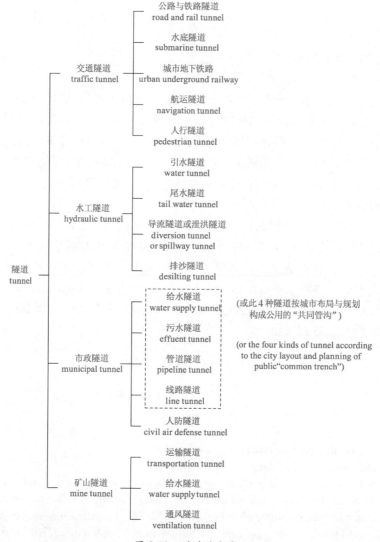

图 8.13　隧道的分类
Figure 8.13　Classification of the tunnel

(a) 中国终南山隧道 (18 km)　　　　　　(b) 挪威洛达尔隧道 (24.5 km)
Zhongnan Mountain Tunnel in China (18 km)　　Los dahl tunnel in Norway (24.5 km)

图 8.14　长大道路隧道
Fig.8.14　Long and large road tunnel

　　隧道工程要在地下挖掘所需要的空间，并修建能长期承受外部压力的衬砌结构。工程进行时由于要承受周围岩土(山岭隧道)或水体(水底隧道)等产生的压力，不但要防止可能发生的坍塌，而且还要避免由于地下水涌出所产生的不良影响。因此，为了适应多种多样的条件，隧道技术也是复杂多样的。

　　隧道技术与地质学和水文学、岩土力学、应用力学和材料力学等有关学科有着密切的联系。同时，它涉及测量、施工机械、炸药、照明、通风、通信等诸方面技术，以及对金属、水泥、混凝土、压注药剂之类化学制品等的有效利用，因而与许多领域有着广泛的联系。

8.3.2　隧道的组成与设计

　　隧道的结构包括主体构造物和附属设备两部分。主体构造物由洞身衬砌和洞门组成，附属设备包括避车洞、消防设施、应急通信和防排水设施，长大隧道还有专门的通风和照明设备。高速铁路隧道内不设置供养护维修人员待避的洞室，但应考虑设置存放维修工具和其他业务部门需要的专用洞室。

　　相比地上结构，隧道工程更易受水文地质条件和施工方法的影响。由于地层内结构及地质环境的复杂性，施工场地空间有限、光线暗等因素，隧道的设计与施工有很多特殊之处，也极其困难。

Tunnel engineering needs enough space under the ground, and the lining structure should be constructed to withstand long-term external pressure. During the construction, pressure affected on the structure produced by the surrounding soil (mountain tunnels) or water (underwater tunnel), so the possible collapse and sometimes some bad effect generated by the groundwater gushing should be avoided. Therefore, the tunnel technology is complex and diverse in order to adapt to a wide variety of conditions.

Tunnel technology is closely linked with geology and hydrology, soil mechanics, applied mechanics, material mechanics and other related disciplines. At the same time, it involves measuring, construction machinery, explosives, lighting, ventilation, communications and other aspects of technology, as well as the effective use of metal, cement, concrete and some chemical products such as pressure injection agent and the like. Thus it is related extensively to many areas.

8.3.2　Composition and design of tunnel

The structure of tunnel includes the principal structures and ancillary equipment. The principal structure is composed of the hole body lining and portal. Ancillary equipment includes refuge holes, fire-fighting facilities, emergency communications and drainage facilities. In long and large tunnels, there are some specialized ventilation and lighting equipment. In high-speed railway tunnels, chambers are not often provided for maintenance personnel, but special chambers should be set up for the storage of maintenance tools and the needs of other operating departments.

Compared with overground structures, the tunnel project is vulnerable to be affected by hydrogeological conditions and construction methods. Due to many restrictions such as the complexity of the stratum structure and geological environment, limited operating space for construction and low-light situation, the design and construction of the tunnel is quite special and extremely difficult.

　　隧道设计首先需确定隧道的空间位置,这是通过隧道几何设计完成的,包括隧道平面线形、纵断面线形、与平行隧道或其他结构物的间距、引线、隧道横断面设计等。几何设计要综合考虑地形、地质等工程因素和行车安全因素。

　　隧道设计的另一个主要方面是隧道结构设计,即计算确定隧道主体构造物和附属构造物。主体构造物是为了保持岩体稳定和行车安全而修建的。洞身衬砌的平、纵、横断面的形状由道路隧道的几何设计确定,衬砌断面的轴线形状和厚度需通过结构计算确定。在山体坡面有发生崩坍和落石可能时,往往还需要接长洞身或修筑明洞。下面以山岭隧道为例介绍其主要结构物。

　　(1) 洞身衬砌

　　山岭隧道的衬砌结构形式,主要是根据隧道所处的地质地形条件,考虑其结构受力的合理性、施工方法和施工技术水平等因素确定的。地质条件不同,结构类型也随着变化,如垂直围岩压力大但水平围岩压力很小时采用直墙式衬砌(图 8.15 a),围岩中存在较大水平压力时采用曲墙式衬砌(图 8.15 b),洞口相对洞身的围岩条件差,常采用复合式衬砌等加强形式。

The spatial position of the tunnel needs to be determined firstly in tunnel design, and it is completed by the geometric design of tunnel, including the plane alignment of tunnel, alignment of longitudinal section, spacing between the designed tunnel with the other parallel ones and structures, lead line of tunnel and design of cross section. The topography, geology and other engineering factors should be taken into consideration in geometric design, as well as the traffic safety.

Another major aspect of the tunnel design is the structure design, namely, to calculate and determine the principle structures and subsidiary structures. Principle structures are built to keep the stability of rock mass and traffic safety. The shape of hole body lining is determined by the geometric design, while the axial shape and thickness of the lining section are determined by structural calculation. We need to prolong the cave body of tunnel or build an open cut tunnel when it is possible for landslides and rockfall to occur on the slope of mountain. The mountain tunnel is taken as an example to explain the main structures of tunnel below.

(1) Lining of cave body

The structural form of the lining of mountain tunnel is determined by many factors, such as reasonability of loading status, construction methods and technology, based on its special geological and topographical conditions. The structural type will change with the different geological conditions. For example, the upright wall lining (Fig.8.15 a) is applied in the cases where high vertical pressure and small horizontal pressure exist in surrounding rock. Once the horizontal pressure gets high, the curved wall lining is needed (Fig.8.15 b). Rock conditions of the entrance are worse than those of the cave inner, so composite lining and other reinforcement forms are often adopted.

（2）洞门

洞门是隧道两端的外露部分，其作用在于保证洞口边坡的安全和仰坡的稳定，引离地表流水，以稳固洞口，保证线路行车安全。洞门的构造形式由岩体的稳定性、通风方式、照明状况、地形地貌及环境条件等多方面因素决定，主要形式如图8.15至图8.21所示。

(a) 直墙式衬砌 Upright wall lining　　　(b) 曲墙式衬砌 Curved wall lining

图 8.15　洞身衬砌类型
Fig.8.15　Lining of cave body

图 8.16　环框式洞门　　　　　　　图 8.17　端墙式洞门
Fig. 8.16　Ring frame of the tunnel portal　　Fig. 8.17　Side wall tunnel portal

图 8.18　翼墙式洞门　　　　　　　图 8.19　台阶式洞门
Fig. 8.18　Wing wall tunnel portal　　Fig. 8.19　Sidestep tunnel portal

（2）Tunnel portal

Tunnel portal is the exposed part of both ends of tunnel. Its function is to ensure the safety of opening slope and the upward slope, lead off the surface water, stabilize the cave opening, and ensure the traffic safety. The structural form of portal is determined by the stability, ventilation, lighting conditions, topography and environmental conditions and other factors of rock mass, and the main forms are shown in Fig.8.15 to Fig.8.21.

道路隧道在照明上有相当高的要求。因隧道出入口区域亮度悬殊,为了处理好司机在通过隧道时一系列视觉上的变化,有时需要在入口一侧采取减光措施,形成一个光过渡带,优化行车环境,如设置植被减光或用减光建筑将衬砌接长,构成新的入口(见图8.22)。

(3) 洞内附属设备

附属构造物是为了运营管理、维修养护、给水排水、供蓄发电、通风、照明、通信、安全等而修建的构造物。不同用途的隧道有不同的附属设施。如公路与铁路隧道常在两侧边墙上交错均匀修建人员躲避和设备存放的避车洞,以保障洞内行人、维修人员及维修设备的安全(见图8.23)。

图 8.20　柱式洞门
Fig. 8.20　Column-type tunnel portal

图 8.21　削竹式洞门
Fig. 8.21　Bamboo-truncating tunnel portal

图 8.22　设置减光棚
Fig. 8.22　Setting light-reduce shed

图 8.23　隧道两侧设避车洞
Fig. 8.23　Setting refuge chamber on both sidewalls

Road tunnels have very high requirements on lighting. Because of the great disparity of brightness at the area of tunnel entrance, the drivers have to experience a series of changes on vision when they drive through the tunnel. Sometimes several measures are taken to reduce brightness at the entrance in order to form a transition zone of brightness so as to optimize the driving environment, such as setting vegetal dimming or using dimming buildings to prolong the lining for getting a new entrance(Fig.8.22).

(3) Ancillary service inside the tunnel

Subsidiary structure is the structure constructed for operation management, repair and maintenance, water supply and drainage, power supply storage, ventilation, lighting, communications and security. Tunnels for different purposes have different ancillary facilities. For example, highway and rail tunnels often have some staggered refuge chambers evenly located on both sidewalls for hiding personnel and storing equipment in order to protect the safety of pedestrian, maintenance personnel and maintenance equipment in the tunnel(Fig.8.23).

8.3.3　水底隧道工程

时下,水下隧道的研究和建设日益成为行业和社会关注的焦点。水底隧道指修建在江河、湖泊、海港或海峡底下的隧道。和跨江跨海大桥一样,水底隧道将成为人类跨越大自然的阻隔和屏障的重要通道。

人类对于修建水底隧道的尝试与实践始于 19 世纪法国人对贯通英法之间海峡的隧道修建构想。20 世纪 40 年代,日本建成世界上第一条海底隧道——长 6.3 km 的关门海峡隧道,1985 年又修建完成世界上最长的水下公路隧道——全长 53.85 km 的青函海底隧道(见图 8.24)。至今,世界各国修建的水底交通隧道已有数百座。我国水底隧道从 20 世纪 70 年代开始迅速发展,截至 2011 年,已经建成和正在建设的水下公路隧道有 30 余条,成为水底隧道建设大国(南京长江隧道见图 8.25)。

水底隧道与桥梁相比,具有运营不受恶劣天气影响、不影响航运、对生态环境影响小、抗震性能优越、比建造长大桥的工程造价和维修保养费用低等优势。

图 8.24　日本青函海底隧道
Fig. 8.24　Japanese Seikan Tunnel

图 8.25　南京长江隧道
Fig. 8.25　Nanjing Yangtze River Tunnel

8.3.3　Underwater Tunnel

Nowadays, research and construction of underwater tunnel is becoming the focus of the industry and society. It refers to the tunnels constructed under rivers, lakes, harbors or straits. Just like the river-crossing and sea-crossing bridges, the underwater tunnel will become an important path for human to overcome the natural hinders and barriers.

The first attempt and practice of the underwater tunnel construction began in the 19th century when French tried to construct a tunnel crossing the strait between England and France. In the 1940s, the first subsea tunnel in the world, Guanmen Straits Tunnel with the length of 6.3 km, was completed in Japan. After that, the longest underwater highway tunnel in the world, Japanese Seikan Tunnel with the length of 53.85 km, was finished in 1985 (Fig.8.24). So far, there have been hundreds of underwater traffic tunnels around the world. The Chinese underwater tunnels started and developed rapidly in the 1970s. By 2011, there are more than 30 underwater highway tunnels which have been completed or under construction, which indicates that China has been an important power of underwater tunnel construction. Fig.8.25 is Nanjing Yangtze River Tunnel.

Compared with bridge, underwater tunnel has some outstanding advantages. For example, its operation will not be affected by the weather conditions and won't influence shipping. It has little impact on the ecological environment too.

但由于其环境的特殊性,在建造和运营方面存在诸多困难,如深水下地质勘查精确度较低、围岩的高孔隙水压力和地下水渗流梯度大,使结构稳定性降低且易发生突涌水、防水排水困难、施工难度大、海底隧道环境的腐蚀性强、施工风险高等。

总体而言,对于水底隧道的建造,桥、隧方案的比选和优化是前期方案和工程可行性设计中的重中之重,而做好工程地质、水文地质勘查和超前预报工作,提高遇险应变能力,解决施工探水、治水和防塌三大技术难点,是水底隧道施工成败的关键。

And it has superior seismic performance, lower construction and maintenance cost than that of long-span bridges.

On the other hand, due to the special environment nature, there are many difficulties in the construction and operation of tunnel: geological exploration has low accuracy under deep water; high pore water pressure of rock mass and large seepage gradient of groundwater lower the structural stability and become prone to sudden gushing; water-resistance, water-drainage and construction are very difficult; corrosive feature of subsea tunnel environment gets strong and high risk exists in the construction of underwater tunnel.

As a whole, for the construction of underwater tunnel, the comparison and optimization of bridge scheme and underwater tunnel scheme is the emphasis of preliminary design and engineering feasibility study. Before the construction of underwater tunnel, a good engineering geology, hydrogeology survey and advanced forecasting should been carried out and the anti-risk capacity has to be improved gradually. It is crucial for success in the underwater tunnel construction to solve the three key technical aspects of exploring water, flood control and anti-collapse.

知识拓展
Learning More

相关网站　Related Links
(1) 中国桥梁网
(2) 中国隧道网
(3) 中国公路网

小贴士　Tips
(1) 国内与桥梁、隧道工程有关的注册师有注册土木工程师、注册建造师、注册造价工程师、注册监理工程师等,通过考试后可从事相应工程专业的设计、施工、管理等工作。

(2) 想了解更多国内隧道的实际工程与建设成就,可参阅中国公路学会隧道工程分会 2009 年 9 月编辑出版的《中国公路隧道建设成就画册》。

(1) The qualifications related to bridge and tunnel engineering are certified civil engineer, certified national architect, certified cost engineer, certified supervision engineer and so on. People can be engaged in relevant municipal engineering design, construction and management after passing the corresponding exams.

（2）If you want to know more about actual engineering and construction achievements in domestic tunnel engineering, you can refer to *the Album of the Achievement in Chinese Highway Tunnel Construction* edited and published by Tunnel Engineering Branch, Chinese Highway Society in September 2009.

思考题　Review Questions

1. 公路是如何分级的？

How to grade the roads?

2. 城市道路是如何分类的？

How to classify the urban roads?

3. 道路平面线形"三要素"是指什么？

What are the "three elements" of road plane alignment?

4. 简述公路横断面的组成。

Briefly describe the composition of highway cross section.

5. 请简述桥梁总体设计的主要原则。

Please describe the main principles in bridge design briefly.

6. 桥梁的基本组成有哪几部分？各自的作用是什么？

What are the basic components of a bridge? What are their roles?

7. 桥梁有哪些主要结构体系？你见过的桥梁结构体系有哪些？请列出 4~5 种，并说明其主要特点和优缺点。

What are the main structural systems of the bridge? How many kinds of bridge structures have you seen? Please list 4–5 kinds and explain what the main features and advantages are.

8. 隧道是由哪几部分组成的？

What parts are tunnels made of?

9. 隧道衬砌的作用是什么？衬砌是由哪几部分组成？

What is the function of tunnel lining? What are the components of the lining?

10. 隧道设计的主要内容是什么？要考虑哪些因素？

What is the main content in tunnel design? What factors should be considered?

11. 请比较跨海大桥与海底隧道孰优孰劣？为什么？

Please compare the sea crossing bridges with the undersea tunnel, and explain which one is btter. Why?

参考文献
References

[1] 中华人民共和国交通运输部公路局,中交第一公路勘察设计研究院有限公司.公路工程技术标准(JTG B01—2014)[S].北京:人民交通出版社,2014.

[2] 中华人民共和国住房和城乡建设部.城市道路工程设计规范(CJJ 37—2012)[S].北京:建筑工业出版社,2012.

[3] 中交第一公路勘察设计研究院有限公司.公路路基设计规范(JTG D30—2015) [S].北京:人民交通出版社,2015.

[4] 中华人民共和国交通部公路科学研究所.公路沥青路面施工技术规范 (JTG F40—2004)[S].北京:人民交通出版社,2004.

[5] 中交路桥技术有限公司.公路沥青路面设计规范(JTG D50—2017) [S].北京:人民交通出版社,2017.

[6] 中华人民共和国交通运输部公路科学研究院.公路路面基层施工技术细则(JTG/T F20—2015)[S].北京:人民交通出版社,2015.

[7] 中交公路规划设计院.公路水泥混凝土路面设计规范(JTG D40—2011) [S].北京:人民交通出版社,2011.

[8] 中华人民共和国交通运输部公路科学研究院. 公路水泥混凝土路面施工技术细则(JGT/T F30—2014) [S].北京:人民交通出版社,2014.

[9] 中交公路规划设计院.公路桥涵设计通用规范 (JTG D60—2015) [S]. 北京:人民交通出版社,2015.

[10] 朱彦鹏,王秀丽. 土木工程导论[M].2版. 北京:化学工业出版社,2021.

[11] 许金良. 道路勘测设计[M].北京:人民交通出版社,2019.

[12] 孙家驷. 道路勘测设计[M]. 3版. 北京:人民交通出版社,2012.

[13] 杨少伟.公路勘测设计[M].3版. 北京:人民交通出版社,2009.

[14] 何景华.公路勘测设计[M].北京:人民交通出版社,1985.

[15] 姚玲森.桥梁工程[M].3版. 北京:人民交通出版社,2021.

[16] 范立础.桥梁工程(上册)[M]. 3版. 北京:人民交通出版社,2017.

[17] 王毅才. 隧道工程 [M]. 北京:人民交通出版社,2006.

[18] 黄成光. 公路隧道施工[M]. 北京:人民交通出版社,2001.

[19] 何修美.中国公路(1979—1989) [M].北京:中国画报出版社,1989.

[20] 万明坤,程庆国,项海帆,等.桥梁漫笔 [M].北京:中国铁道出版社,1997.

[21] 吉姆辛.缆索承重桥梁:构思与设计[M] .姚玲森,林长川译.北京:人民交通出版社,1992.

[22] Leonhardt F Z.Bridges Aesthetics and Design [M]. Cambridge: MIT Press,1984.

第9章　岩土工程

Chapter 9　Geotechnical Engineering

先导案例
Guide Case

　　比萨斜塔是一个钟塔,修建于 1173 年,位于意大利罗马式大教堂右后侧,因倾斜而闻名于世。比萨斜塔从基底到塔顶高 58.36 m,从地面到塔顶的高度为 55.86~56.67 m,钟楼墙体在地面上的宽度为 4.09 m,在塔顶宽 2.48 m,总重约 14 500 t。塔体呈圆形,底层共有 15 根巨柱支撑,二层至七层各有石柱 30 根,顶层略小,有石柱12 根,大钟置于顶层。塔内有 294 阶楼梯,可盘旋登上。计算可知,塔体的重心位于地基上方 22.6 m 处,圆形地基面积为 285 m²,对地面的平均压强为 497 kPa。

　　比萨塔自 1173 年开始施工后不久便开始倾斜。直到 1178 年,第三层主体结构才完工,塔却继续由北侧向南侧倾斜。此后由于与周边国家的战争而停工 94年,1273 年继续施工,1350 年建成此塔。塔建成后,塔顶偏离垂直中心 2.1 m,成了斜塔。据测定从 1829 年至 1910 年,比萨塔平均每年倾斜 3.8 mm。

The Leaning Tower of Pisa was built in 1173. It is known worldwide for its unintended tilt. It is situated behind the Cathedral and is the third oldest structure in Pisa's Cathedral Square after the Cathedral and the Baptistery. The height of the tower is 58.36 m, It is 55.86 m from the ground on the low side and 56.67 m on the high side. The width of the walls at the base is 4.09 m and the top is 2.48 m. Its weight is estimated at 14 500 metric tons. The tower has a cylinder body and there are 15 giant marble columns to support the upper loads at the first floor. 30 pillars were set from the second to the seventh floor and 12 pillars with small size were at the top floor. The big clock was placed on the top floor. Tourists can move spirally through the 294 steps to the top floor. Based on the calculation, the center of gravity of the tower is located 22.6 m above the foundation. The area of the circular foundation is 285 m², and the average pressure on the ground is 497 kPa.

The tower has been leaning since its initial construction in 1173. By 1178, the third floor was completed. It was observed at this stage to continue leaning on the north side. Then the construction was on hold for the next 94 years because of neighboring wars. The construction was restarted in the year of 1273 and established, when the tower shifted to the one side about 2.1 m from the vertical center in 1350. According to the testing results from 1829 to 1910, around 3.8 millimeters displacement was found from the vertical center at the top floor every year.

为什么比萨塔会一直发生倾斜呢？这个问题的答案就与岩土工程有关。首先，比萨塔的地基土由砂土和黏土组成。不稳定的混合土层导致地基土以不同的速度压缩，最终导致塔下沉30~40 cm，并由北向南倾斜。其次，塔体自身重量太大，地基负荷超重，地基土在南侧沉降速率大于北侧。第三，地基埋深太浅，只有3 m，地下水层在深约1 m处，塔身倾斜与地下水位的波动也存在关联。

比萨斜塔的纠偏修复工作开展于1990年到2001年，修复工作需要用到岩土工程相关知识。1992年，采用涂塑钢索环绕斜塔南侧的二楼，防止塔身墙体发生"屈曲"破坏，但是效果不明显，塔体仍继续向南侧倾斜。1993年，在北侧施加了750 t的铅锭作为支持，使得北侧的重量等于南侧的重量，即使得施加于地基土上的荷载南北侧一致，最终南侧地基土的下沉速度得到控制。通过现场检测发现，塔身已经停止继续向南侧倾斜。1995年，开始对地基土进行处理，一个简单的纠偏方法得以实施，即从斜塔的北侧地基下抽出部分土，使斜塔的倾斜自然北移。41条抽土管深深地插入20 m深的北侧塔基地下，工程以日抽土100 kg的速度进行。

图9.1　比萨斜塔
Fig. 9.1　The Leaning Tower of Pisa

Why the tower was continuing leaning? The answers to this question might be related to the geotechnical engineering. First, the tower's foundation comprised of layers of sand and clay. The unstable mixture of soil material caused the soil to compact at a different rate, causing the tower to sink 30~40 cm and sink to the south side. Second, soil at south side compressed faster than the north side and the weight of the tower was the main factor of tilt. Third, the embedded depth of tower foundation is only 3 m and the underwater table depth is about 1 m. Tilting was also due to the fluctuations of the water levels.

The restoration work was performed between 1990 and 2001, which also needed the geotechnical engineering related knowledge. In 1992, plastic coated steel wires were wrapped around the south side of the second floor to prevent a type of failure called "Buckling". Buckling occurs when overly stressed walls suddenly burst outwards. But the tower continued to incline to the south side. In 1993, the incline was halted by stacking lead ingots on the north side of the tower. There were 750 metric tons of leads acted as support. This caused the weight of north side equal to the weight of south side. The ground below the south side was no longer compressing faster than the north side. Monitoring equipment was placed and indicated that the tower's incline stopped due to the counterweights. In 1995, the work on the foundation started. The ground was frosted and preparation to add cables and weights went underway. As the ground unthawed some unseen plates cracked, the tower began slowly falling south, so a crane stacked a few more lead ingots on the north side. Contractors removed soil from the north side with drilling equipment.

与此同时,几条钢索环绕捆住斜塔的二层,向北拉住塔身,120 部精密仪器从各个角度和高度严密监测斜塔的动静。斜塔的倾斜开始出现了回复的迹象,最终塔倾斜角度由 5.5°减小为 3.99°。2001 年 4 月斜塔得以重新对外开放。

岩土工程是土木工程的分支,是以岩体力学、土力学与基础工程、工程地质学为基础理论,研究和解决工程建设中与岩土有关技术问题的一门新兴的应用科学。按照工程建设阶段划分,可以分为岩土工程勘察、岩土工程设计、岩土工程治理、岩土工程监测与检测。土木工程建设中遇到的岩土工程问题促进了该学科的发展,本章主要介绍岩土工程勘察,基础工程,基坑与地下工程,地基处理,边坡工程的相关内容,并对岩土工程的发展作了展望。

9.1　岩土工程概述

人类早就将岩土材料用于防洪筑堤,修建水利工程、墓穴、建筑物基础。例如,公元前 20 世纪,在古埃及古美索不达米亚和新月沃土地带,以及印度河流域文明中心的摩亨佐·达罗和哈拉帕古城,人们通过修建堤防、坝址、运河,筑堤防洪,引洪灌溉。随着城市的扩大,出现了成型的地基基础用以支撑上部结构,如古希腊人建造的垫板基础和筏板基础。

The tower started to sink on the north side, therefore reducing some of the stress that was built up on the south side. Suspension cables were loosely fitted to the tower so that it could pull back the tower in case it started leaning. The tower reopened in April, 2001.

Geotechnical Engineering (GE) is a branch of Civil Engineering (CE), which is concerned with the engineering behavior of earth materials such as rocks and soils. It is a new application discipline aiming at studying and solving the technology problems in the engineering construction related to rocks and soils based on the theories of rock mechanics, soil mechanics and foundation engineering, engineering geology. According to the engineering construction stage, a GE project can be divided into four stages, named as geotechnical investigation, geotechnical design, geotechnical treatment, geotechnical inspection and monitoring respectively. Geotechnical problems in civil engineering construction promote the development of this discipline. This chapter mainly introduces the contents of geotechnical investigation, foundation engineering, foundation excavation and underground engineering, slope engineering, and presents the prospects for the development of geotechnical engineering.

9.1　Overview of geotechnical engineering

Humans have used soil as material for flood control, irrigation purposes, burial sites, building foundations, and construction material for buildings. First activities were linked to irrigation and flood control, as demonstrated by traces of dykes, dams, and canals dating back to at least 2000 BC that were found in ancient Egypt, ancient Mesopotamia and the Fertile Crescent, as well as around the early settlements of Mohenjo Daro and Harappa in the Indus valley. As the cities expanded, structures were erected supported by formalized foundations.

直到 18 世纪,仍未出现岩土设计的相关理论,此时更多是依靠过去的经验。岩土工程理论源于挡土墙结构的压力计算理论的提出。1717 年,法国工程师 Henri Gautier 提出了不同土壤的天然边坡静态休止角的概念。1773 年,Charles Coulomb(库伦,物理学家、工程师、陆军上尉)最早将力学基本原理应用于土壤,为确定军事堡垒的土压力提出改进的方法。库伦理论与 Christian Otto Mohr(莫尔)二维应力状态,被称为莫尔–库仑理论,至今仍然在实践中使用。

19 世纪,Henry Darcy(达西)提出了达西渗透定律,以描述流体在多孔介质中的流动;Joseph Boussinesq(布辛涅斯克,数学家和物理学家)提出了弹性固体中应力分布理论,可以有效地估计地面以下不同深度的应力分布;William Rankine(郎肯,工程师和物理学家)提出了一种可替代库仑土压力理论的郎肯土压力理论。Albert Atterberg(阿太堡)提出了黏性土的一致性指数,至今仍被用于土的分类。

现代岩土工程开始于 1925 年,以 Karl Terzaghi(太沙基,工程师和地质学家)出版土力学(*Erdbaumechanik*)一书为标志,因此太沙基被公认为现代土力学和岩土工程之父。太沙基不仅提出了有效应力原理,土体的抗剪强度是由有效应力控制的,而且为地基承载力提出了理论框架,并提出固结理论用于黏性土层的沉降速率计算。

Ancient Greeks notably constructed pad footings and strip-and-raft foundations. Until the 18th century, however, no theoretical basis for soil design had been developed and the discipline was more of an art than a science, relying on past experience. The earliest advances occurred in the development of earth pressure theories for the construction of retaining walls. Henri Gautier, a French royal engineer, recognized the "natural slope" of different soils in 1717, an idea later known as the soil's angle of repose. The application of the principles of mechanics to soils was documented as early as 1773 when Charles Coulomb (a physicist, engineer, and army Captain) developed improved methods to determine the earth pressures against military ramparts. By combining Coulomb's theory with Christian Otto Mohr's 2D stress state, the theory became known as Mohr-Coulomb theory, which is still used in practice today.

In the 19th century, Henry Darcy developed Darcy's Law which describing the flow of fluids in porous media. Joseph Boussinesq (a mathematician and physicist) developed theories of stress distribution in elastic solids that proved useful for estimating stresses at depth in the ground. William Rankine, an engineer and physicist, developed an alternative to Coulomb's earth pressure theory. Albert Atterberg developed the clay consistency indices that are still used for soil classification.

Modern geotechnical engineering began in 1925 with the publication of *Erdbaumechanik* by Karl Terzaghi (a civil engineer and geologist), who is recognized as the father of modern soil mechanics and geotechnical engineering. He not only developed the principle of effective stress, and demonstrated that the shear strength of soil is controlled by effective stress, but also developed the framework for theories of bearing capacity of foundations, and the theory for prediction of the rate of settlement of clay layers due to consolidation.

20 世纪 60 年代末至 70 年代初,岩体力学、土力学及基础工程、工程地质学三者逐渐结合为一体并应用于土木工程实际而形成新学科——岩土工程。岩土工程运用土力学和岩石力学原理研究地下条件和材料,确定岩土材料的物理、力学和化学性质,评价天然边坡和人造土场地稳定性、风险及对岩土工程结构与基础的设计、场地条件进行施工监测等。岩土工程是土木工程的一个重要组成部分。在房屋、市政、能源、水利、道路、航运、矿山、国防等各种建设中,都有十分重要的意义。

9.2 岩土工程勘察

岩土工程勘察工作是设计和施工的基础,其目的主要是查明工程地质条件,分析存在的地质问题,对建筑物区域做出工程地质评价。

根据勘察对象的不同可分为水利水电工程(主要指水电站、水工构造物)、铁路工程、公路工程、港口码头、大型桥梁及工业、民用建筑勘察等。水利水电工程、铁路工程、公路工程、港口码头等工程一般比较重大、投资造价及重要性高,国家分别对这些类别的工程勘察进行了专门的分类,编制了相应的勘察规范、规程和技术标准等,这些工程的勘察被称为工程地质勘察。

In the late 1960s to 1970s, a new application discipline namely Geotechnical Engineering was formed based on the combination of rock mechanics, soil mechanics and foundation engineering as well as engineering geology. Geotechnical engineering uses principles of soil mechanics and rock mechanics to investigate subsurface conditions and materials; determine the relevant physical/mechanical and chemical properties of these materials; evaluate stability of natural slopes and man-made soil deposits; assess risks posed by site conditions; design earthworks and structure foundations; and monitor site conditions, earthwork and foundation construction. Geotechnical engineering is not only an important part in civil engineering, but also used by municipal, highway, military, mining, petroleum, or any other engineering concerned with construction on or under the ground.

9.2 Geotechnical investigation

Geotechnical investigation work is the basis for the design and construction, and the main purpose is to find out the engineering geological conditions, analyze the geological problems, assess the engineering geological conditions of building district.

According to the different objects, geotechnical investigation can be divided into different categories such as hydraulic and hydroelectric engineering(mainly refers to the hydropower station and hydrotechnical construction), railway engineering, highway engineering, wharfs, large bridges, industrial and civil construction geotechmical investigation. Investment cost and significance for hydraulic and hydroelectric engineering, railway engineering, highway engineering, port engineering, some special classifications were given by our country, and the corresponding investigation standards, regulations and technical standards were also formulated. These engineering investigations are usually called the engineering geological investigation.

因此,通常所说的"岩土工程勘察"主要指工业、民用建筑工程的勘察,勘察对象主体包括房屋楼宇、工业厂房、学校楼舍、医院建筑、市政工程、管线及架空线路、岸边工程、边坡工程、基坑工程、地基处理等。岩土工程勘察的内容主要有工程地质调查和测绘、勘探及采土试样、原位测试、室内试验、现场检验和检测,最终根据以上几种或全部手段,对场地工程地质条件进行定性或定量分析评价,编制满足不同阶段所需工程勘察的成果报告文件。

9.2.1 工程地质调查与测绘

工程地质调查与测绘是岩土工程勘察的基础工作,一般在勘察的初期阶段进行。这一工作的本质是运用地质、工程地质理论,对地面的地质现象进行观察和描述,分析其性质和规律,并以此推断地下地质情况,为勘探、测试工作等其他勘察方法提供依据。

在地形地貌和地质条件较复杂的场地,必须进行工程地质测绘。但对地形平坦、地质条件简单且较狭小的场地,则可采用调查代替工程地质测绘。通过调查和测绘,借助计算机软件或其他诸如遥感等测试技术,可以建立场地的地质地貌模型(见图9.2)或者地质地貌实际图像(见图9.3)等。

Therefore, commonly called "geotechnical investigation" mainly refers to industrial, civil construction engineering investigation, and the investigation object includes the residential buildings, industrial buildings, school buildings, hospital construction, municipal engineering, pipeline and overhead line, shore engineering, slope engineering, foundation engineering, and ground improvement, etc. Geotechnical investigation mainly includes the engineering geological survey and mapping, exploration and sampling, in-situ test, laboratory test, field test and detection. Finally according to several or all of the above methods, qualitative or quantitative analysis evaluation of site engineering geological conditions are developed and reported of engineering investigation to meet different stages.

9.2.1 Engineering geological survey and mapping

Engineering geological survey and mapping is a basic work of geotechnical investigation, which is usually carried out in the early stages. According to the application of geology, engineering geological theory, geological phenomenon on the ground were observed and described. At the same time the nature and change laws were analyzed and followed by the underground geological conditions, which can provide the basis for exploration, testing and other survey methods.

In the topography and complex geological site conditions, the engineering geological survey and mapping should be carried out. However, on flat terrain, simple geological conditions and relatively narrow filed space, filed survey can be adopted instead of engineering geological mapping. Through the survey and mapping, using computer software or other technology such as remote testing, geological and geomorphic model can be built (Fig.9.2), or actual images of geology and geomorphology can be obtained (Fig. 9.3).

9.2.2　勘探与取样

勘探工作包括物探(见图 9.4)、钻探(见图 9.5)和坑探(见图 9.6)等各种方法。通过勘探,可以调查地下地质情况,并且可利用勘探工程取样进行原位测试和监测。实际工程中,应根据勘察目的及岩土的特性选用不同的勘探方法。

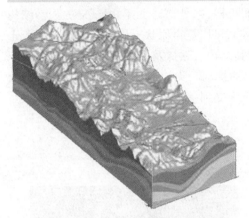

图 9.2　GeoView 软件建立的三维地质模型
Fig. 9.2　Three-dimensional geological model by GeoView

图 9.3　江苏大学遥感影像图
Fig. 9.3　Remote sensing image of Jiangsu University

图 9.4　物探
Fig. 9.4　Geophysical survey

图 9.5　钻探
Fig. 9.5　Drilling survey

图 9.6　坑探
Fig. 9.6　Pitting survey

9.2.2　Exploration and sampling

Exploration work includes geophysical survey (Fig.9.4), drilling survey (Fig.9.5), pitting survey (Fig.9.6) and other methods. It is used to investigate the underground geological conditions and carried out in-situ testing and monitoring by exploration engineering sampling. Different kinds of above exploration methods should be chosen carefully according to the characteristics of rocks and soils as well as geotechnical investigation purposes.

物探是一种间接的勘探手段,它的优点是较钻探和坑探轻便、经济而迅速,常常与测绘工作配合进行。钻探和坑探也称勘探工程,均是直接勘探手段,能可靠地了解地下地质情况,在岩土工程勘察中是必不可少的。当钻探方法难以查明地下地质情况时,可采用坑探方法。

现场取样(见图9.7)伴随着勘探过程,根据取土样质量等级划分为两种:一是原状土样,即保持原有的天然结构未受破坏的土样;二是扰动土样,即试样的天然结构已遭受破坏。通过一定的取样仪器(见图9.8),可以获得原状土样。但是需要注意,土样脱离母体后,应力状态发生了变化,影响其结构。钻探及采样过程中,钻具在钻压过程中必然要对周围土体产生一定程度上的扰动。无论何种取土器都有一定的壁厚、长度和面积,它在压入过程中,也会使土样受到一定的扰动。

图9.7 现场取样
Fig. 9.7 Field sampling

图 9.8 地表取样钻机
Fig. 9.8 Drilling equipment for sampling

Geophysical survey is an indirect exploration means with the advantages of portability, economy and speediness compared with that of drilling and pitting survey. It is often used in conjunction with the mapping work. Drilling and pitting survey are two direct exploration means which can reliably provide underground geological conditions information. Therefore, it is indispensable in geotechnical investigation. Pitting survey method is chosen when drilling method is difficult to ascertain the underground geological conditions.

Field sampling (Fig.9.7) is accompanied with the exploration process. According to the quality level, soil samples can be divided into two categories namely undisturbed soil sample and disturbed soil sample, respectively. Undisturbed soil sample keeps its original natural unspoiled and the disturbed soil specimens have been destroyed. Undisturbed soil sample can be obtained by certain sampling instrument (Fig.9.8). It should be noticed that soil stress state had been changed when soil specimens were taken out from the filed soil layers, the soil structure will be affected. Drilling tool is bound to have a certain degree of disturbance on the surrounding soil in the process of drilling. It will also make the soil samples under certain disturbance in the process of penetration due to different wall thickness, length and area of the sampler.

9.2.3　原位测试与室内试验

原位测试与室内试验的主要目的是为岩土工程问题分析评价提供所需的技术参数,包括岩土的物理指标、强度参数、固结变形特性参数、渗透性参数和应力、应变及时间关系的参数等。原位测试一般都借助于勘探工程进行,是详细勘察阶段一种主要的勘察方法。原位测试的优点是试样不脱离原来的环境,基本上在原位应力条件下进行试验,所测定的岩土体尺寸大,能反映宏观结构对岩土性质的影响,代表性好;缺点是试验时的应力路径难以控制、边界条件也较复杂,有些试验耗费人力、物力较多,难以大量进行。原位测试的方法比较多,图 9.9 和图 9.10 分别为原位静力触探试验和原位十字板剪切试验的现场。

图 9.9　原位静力触探试验　　　　　　　图 9.10　原位十字板剪切试验
Fig. 9.9　In-situ static cone penetration test　　Fig. 9.10　In-situ vane shear test

9.2.3　In-situ and laboratory tests

The main objective of in-situ tests and laboratory tests is to provide technical parameters for the analysis and evaluation of geotechnical engineering problems. These param eters include physical properties of rocks and soils, strength parameters, consolidation deformation characteristics parameter, permeability parameter and stress, strain and time related parameters, etc. In-situ test is usually carried out by exploration engineering. It is an investigation method in the process of detailed exploration. The advantages of in-situ tests suggest that soil specimen is not separated from the original environment and tests are basically conducted in the in-situ stress condition. Meanwhile, test size of rocks and soils is large, which can reflect the macroscopic structure of the properties. The disadvantages of in-situ tests indicate that stress path is difficult to control and the boundary conditions are more complex. Meanwhile, some tests consume more manpower and material resources than those of the corresponding laboratory tests. Fig. 9.9 and Fig. 9.10 show the in-situ static cone penetration test and in-situ vane shear test site, respectively.

室内土工试验是研究土特性的试验,主要优点是试验条件比较容易控制,边界条件明确,应力应变条件可以控制。室内土工实验主要包括:① 土的物理性质试验;② 土的水理性质试验;③ 土的力学性质试验(静态);④ 土的动力性质室内试验。室内试验仪器有很多,包括常规物理性质试验仪器、三轴压缩试验仪器(见图9.11 和图9.12)和固结试验仪器(见图 9.13)等。

9.2.4 现场检验与监测

现场检验与监测的主要目的在于保证工程质量和安全,提高工程效益。现场检验包括施工阶段对先前岩土工程勘察成果的验证核查,以及对岩土工程的施工监理和质量控制。现场监测则主要包括施工作用和各类荷载对岩土反应性状的监测、施工和运营中的结构物监测和对环境影响的监测等方面。检验与监测所获取的资料,可以反求出某些工程技术参数, 并以此为依据及时修正设计, 使之在技术和经济方面得以优化。

图 9.11　GDS 三轴仪	图 9.12　普通三轴仪	图 9.13　固结仪
Fig. 9.11　GDS triaxial instrument	Fig. 9.12　Normal triaxial instrument	Fig. 9.13　Consolidmeter

Laboratory test is ready for testing soil properties. Main advantages of laboratory tests suggest that it is easy to control the test conditions and the boundary conditions are explicit, stress and strain conditions can be easily controlled, etc. The indoor experiments mainly include: ① soil physical properties test; ② soil water physical properties test; ③ soil mechanical properties test (static) ④ soil dynamic properties test. There are many indoor test equipments such as conventional physical properties test instruments, triaxial compression test instrument(Fig.9.11 and Fig.9.12), consolidmeter test instrument(Fig.9.13) and so on.

9.2.4 Field inspection and monitoring

The main purpose of the field inspection and monitoring is to ensure the project quality and safety as well as improve the engineering benefit. The field inspection includes both the verification and the validation of previous geotechnical engineering investigation results in construction stage and the supervision and quality control of geotechnical engineering construction. The field monitoring mainly includes the geotechnical behavior monitoring under construction and all kinds of loads on the rocks and soils, structure monitoring during construction and operation and the monitoring of influence on environment, etc. Data obtained from inspection and monitoring can recalculate some engineering parameters. A modified design could be given according to the tests data and the optimization can be achieved in the aspects of technology and economy.

图 9.14 为地基自动化监测系统示意图。

9.3　基础工程

建筑物向地基传递荷载的下部结构称为基础。其作用是将上部结构的荷载安全可靠地传给地基。岩土工程师设计基础时，应考虑结构的荷载特性、场地土壤和/或基岩的性质。基础按埋置深度分为浅基础和深基础。

9.3.1　浅基础

通常把埋置深度不大，只需经过挖槽、排水等普通施工程序就可以建造起来的基础称为浅基础。它可扩大建筑物与地基的接触面积，使上部荷载扩散。

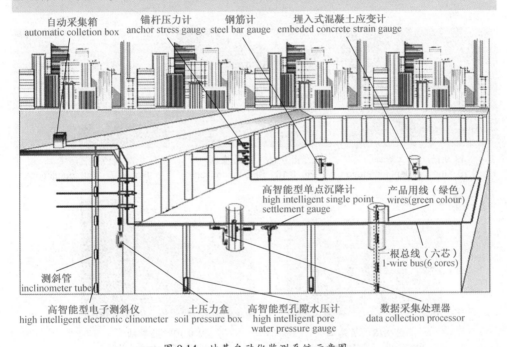

| 自动采集箱 automatic colletion box | 锚杆压力计 anchor stress gauge | 钢筋计 steel bar gauge | 埋入式混凝土应变计 embedded concrete strain gauge |

高智能型单点沉降计 high intelligent single point settlement gauge

产品用线（绿色） wires(green colour)

一根总线（六芯） 1-wire bus(6 cores)

测斜管 inclinometer tube

高智能型电子测斜仪 high intelligent electronic clinometer

土压力盒 soil pressure box

高智能型孔隙水压计 high intelligent pore water pressure gauge

数据采集处理器 data collection processor

图 9.14　地基自动化监测系统示意图
Fig. 9.14　Schematic diagram of the foundation automation detection system.

Fig.9.14 shows the schematic diagram of the foundation automation detection system.

9.3　Foundation engineering

A building's foundation transmits loads from buildings and other structures to the earth. Geotechnical engineers design foundations based on the load characteristics of the structure and the properties of the soils and/or bedrock at the site. On the basis of buried depth, foundation can be divided into two categories—shallow foundation and deep foundation, respectively.

9.3.1　Shallow foundation

Shallow foundations, often called footings, are usually embedded about a meter or so into soil. It can be constructed only after dredging, drainage and other common construction programs, which can enlarge the contact area of the building and the ground. Therefore, the upper load is diffused.

浅基础主要有:① 独立基础(如大部分柱基),如图9.15所示;② 条形基础(如墙基),如图9.16所示;③ 筏板基础(如水闸底板),如图9.17所示。

9.3.2 深基础

深基础是埋深较大,以下部坚实土层或岩层作为持力层的基础,其作用是把所承受的荷载相对集中地传递到地基的深层,而不像浅基础那样,通过基础底面把所承受的荷载扩散分布于地基的浅层。因此,当建筑场地的浅层土质不能满足建筑物对地基承载力和变形的要求,而又不适宜采用地基处理措施时,就要考虑采用深基础方案了。深基础包括桩基础(见图9.18)、墩基础(见图9.19)等类型。

图9.15 独立基础
Fig. 9.15 Independent foundation

图9.16 条形基础
Fig. 9.16 Strip foundation

图9.17 筏板基础
Fig. 9.17 Raft foundation

图9.18 桩基础
Fig. 9.18 Pile foundation

图9.19 墩基础
Fig. 9.19 Pier foundation

Shallow foundations can be divided into: ① independent foundation (such as the pillar foundation), as shown in Fig.9.15; ② strip foundation(such as a wall foundation), as shown in Fig.9.16; ③ raft foundation(such as sluice plate), as shown in Fig.9.17.

9.3.2 Deep foundation

Deep foundation is used to transfer loads down through the upper weak layer of topsoil to the stronger layer of subsoil below. There are different types of deep footings including impact driven piles(Fig.9.18), drilled shafts, caissons, helical piles, geo-piers and earth stabilized columns (Fig.9.19). Deep foundations are used for structures or heavy loads when shallow foundations cannot provide adequate capacity, due to size and structural limitations. They may also be used to transfer building loads to weak or compressible soil layers.

9.4 基坑与地下工程

9.4.1 基坑工程

随着基坑开挖越来越深、面积越来越大,基坑围护结构的设计和施工也越来越复杂,对理论和技术的要求也越来越高,远远超出了作为施工辅助措施的范畴,基础施工单位没有足够的技术力量来解决复杂的基坑稳定、变形和环境保护问题,而研究和设计单位的介入解决了基坑工程的理论计算和设计问题,由此逐步形成了一门独立的学科分支——基坑工程。基坑工程涉及岩土工程、结构工程和环境工程等众多学科领域,综合性高,影响因素多,设计计算理论目前还不成熟,在一定程度上还依赖于工程实践经验。

基坑土方开挖的施工工艺一般有两种:放坡开挖(无支护开挖)和在支护体系保护下开挖(有支护开挖),分别如图 9.20 和图 9.21 所示。

图 9.20 无支护开挖基坑
Fig. 9.20 Slope excavation (without support)

图 9.21 有支护开挖基坑
Fig. 9.21 Excavation with support system

9.4 Foundation excavation and underground engineering

9.4.1 Foundation excavation engineering

With the increase in depth and area of excavation, the design and construction of building foundation excavations become more and more complex. More and more advanced theories and technologies are also needed. It is not just an auxiliary construction measures. Since the construction units do not have enough technical support to solve the complex problems related to foundation stability, deformation and environmental protection, research and design units participate in and solve the corresponding theoretical calculation and design problems. An independent branch of discipline namely foundation excavation engineering is gradually formed, which covers many fields such as geotechnical engineering, structure engineering and environmental engineering. Currently the design calculation theory is not mature due to its high comprehensiveness and large amount of influencing factors. Therefore, the design is also subjected to the experience of engineering practice to some extent.

Construction technology of soil excavation in foundation excavation engineering can be generally divided into two categories: slope excavation without support and excavation with support system, as shown in Fig.9.20 and Fig.9.21, respectively.

　　放坡开挖既简单又经济,但应具备放坡开挖的条件,即基坑不太深而且基坑平面之外有足够的空间供放坡。当不具备放坡开挖条件时,应选择有支护开挖,常见的支护开挖方法见9.4.2节。

9.4.2　基坑支护方法

　　① 板桩式主要有钢板桩、钢管桩、钢筋混凝土板桩等,适用于黏性土、砂性土和粒径不大于100 mm的砂卵石地层。由于施工时打桩噪声大,宜用于远离居民区的施工当中。常见的钢板桩主要是H型钢(见图9.22),间距在1.2~1.5 m。钢板桩的特点是有成品制作,可反复使用;施工简便,但施工有噪声。钢管桩(见图9.23)的截面刚度大于钢板桩,在软弱土层中开挖深度可增大,日本工程案例显示最大开挖深度达30 m,同时需有防水措施相配合。

图 9.22　H 型钢钢板桩

Fig. 9.22　H type steel sheet pile

图 9.23　钢管桩

Fig. 9.23　Steel pipe pile

　　The slope excavation is simple and economical, but excavation conditions should be guaranteed in case that the foundation is not too deep and there is enough space for the slope outside of the foundation plane. Otherwise, excavation with support system should be selected. Common support methods for foundation excavation are summarized in the following section.

9.4.2　Foundation excavation support methods

　　① Panel pile type support methods mainly include steel sheet pile, steel pipe pile, reinforced concrete sheet-pile, etc. It is suitable for cohesive soil, sandy soil and grain size not greater than 100 mm. Construction should be away from residential areas due to the piling noise of construction. Furthermore, a common type of steel sheet pile is called H type steel with spacing between 1.2 m and 1.5 m, as shown in Fig.9.22. It is pre-produced in the factory and can be used repeatedly in the construction. Although the construction process is simple, the construction noises can not be avoided. Steel pipe pile (Fig.9.23) has higher section stiffness than that of the steel sheet pile, where excavation depth can be deeper in the soft soils matched with waterproof measures. Engineering projects show that excavation depth could be maximized to 30 m in Japan for instant.

② 柱列式灌注桩主要有钻孔灌注桩和挖孔灌注桩,见图 9.24 和图 9.25。其特点是施工噪声小,适于城区施工,对周边地层、环境影响小;刚度大,可用于深大基坑;需降水或和止水措施配合使用。

③ 地下连续墙(见图9.26),可兼做永久结构,适于逆筑法、半逆筑法。施工时振动小、噪声低,墙体刚度大,对周边地层扰动小;开挖深度大,适用于所有土层;强度大,变位小,隔水性好,可兼做主体结构的一部分,但造价高。

图 9.24 钻孔灌注桩
Fig. 9.24 Cast-in-place bored pile

图 9.25 挖孔灌注桩
Fig. 9.25 Artificial drill-pouring pile

图 9.26 地下连续墙施工
Fig. 9.26 Construction of underground diaphragm wall

② Soldier type cast-in-place pile support methods mainly include cast-in-place bored pile and artificial drill-pouring pile, as shown in Fig.9.24 and Fig.9.25, respectively. Little construction noise is the main advantage for this type. So it is suitable for urban construction with small environmental impact and small effect on the surrounding soil layers. It can also be used for deep foundation excavation engineering due to its high stiffness. However, some dewatering or watertight measures should be operated in coordination.

③ Underground diaphragm wall (Fig.9.26) can be used as a permanent structure and suitable for inverse construction method and semi-inverse construction method. Low noise is a merit for this type due to small vibration in construction. Big wall stiffness results in small disturbance to the surrounding soil layers and big excavation depth. It is suitable for all soil layers with big strength, small deflection and good waterproof. How-

④ 沉井(箱)法(见图9.27和图9.28)是指在垂直方向上,将各种形状的井筒(沉井)或箱体(沉箱)边排土边沉入地下,最后固定在地层中,形成地下建筑物或构筑物的施工方法。事先在地面上用钢筋混凝土制成的井筒(沉井)或箱体(沉箱)结构作为基坑坑壁的支撑,在井壁的保护下,通过机械和人工在井内挖土,并在其自重作用下沉入土中。该方法施工占地面积小,不需要板桩维护,与大开挖相比,挖土量小,对邻近建筑的影响较小,操作简便,无须特殊的专用设备。

⑤ SMW法(原状取土灌注桩技术)(见图9.29),即在水泥土桩内插入H型钢或钢板、钢管等(可拔出反复使用,经济性好),将承受荷载与防渗挡水结合起来,使之成为同时具有受力与抗渗两种功能支护结构的围护墙。

图 9.27　沉井法
Fig. 9.27　Sunk-well method

图 9.28　沉箱法
Fig. 9.28　Caisson method

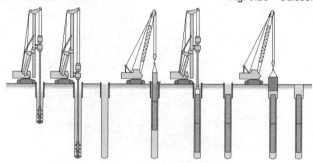

图 9.29　SMW 法示意图
Fig. 9.29　Schematic of SMW method

ever, the cost for this type is high.

④ Sunk-well method (Fig.9.27) and caisson method (Fig.9.28) are two foundation excavation support methods. In vertical direction, various shapes of wellbore or box are set to sink into the ground and finally fixed in soil layers. Different shapes of wellbore or box were made of reinforced concrete as the support system for foundation pit wall in advance. Mechanical or artificial digging in the well or box under the protection of the borehole wall enable the well or box to sink into the soil layers under self-weight. These two methods cover a small area in construction without the support of panel piles. Compared with the large excavation, it has small amount of digging and less effect on the adjacent buildings as well as easy operation with no special equipments in construction.

⑤ SMW method refers to soil mixing wall, as shown in Fig.9.29. Cement-soil pile was first inserted by the H type steel, steel sheet or steel pipes and so on, which can be pulled up and repeatedly used with small cost.

SMW 支护结构的特点是:施工时基本无噪声,对周围环境影响小,结构强度可靠,特别适合于以黏土和粉细砂为主的松软地层,挡水防渗性能好,不必另设挡水帷幕,可以配合多道支撑应用于较深的基坑,具有较好的发展前景。

⑥ 自立式水泥挡土墙(见图 9.30),分为深层搅拌桩挡土墙和高压旋喷桩挡土墙,将水泥材料作为固化剂,采用特殊机械,如深层搅拌机(见图 9.31)和高压旋喷机(见图 9.32),将其与原状土强制拌和,形成具有一定强度、整体性和稳定性的圆柱体(柔性桩),桩体相互搭接,组成整体结构性的水泥土墙或形成格栅状墙体,以保证基坑边坡的稳定。其优点是自立式,无须支撑,开挖方便,有良好的隔水性能,充分利用原状土,节省材料,造价低;缺点是水泥土墙体的材料强度比较低,不适于支撑作用,墙体变形和位移较大。

图 9.30 自立式水泥土墙
Fig. 9.30 Self-standing cement-soil retaining wall

图 9.31 深层搅拌机
Fig. 9.31 Machine for deep mixing pile

图 9.32 高压旋喷机
Fig. 9.32 Machine high-pressure jet grouting pile

This method combines two functions of bearing loads and seepage prevention, which has many advantages including little construction noise, small impact on the environment and enough strength reliability. Therefore, it is suitable for soft soil layers mainly composed of clay and silty sand. Blocking water curtain can not have to be set due to its well seepage prevention. It can be applied in the deep foundation pit cooperated with multi-layer support, which has good development prospect.

⑥ Self-standing cement-soil retaining wall (Fig.9.30) can be classified as deep mixing pile retaining wall and high-pressure jet grouting pile retaining wall, respectively. With the special machinery, such as deep mixing machine (Fig.9.31) and high-pressure jet grouting machine(Fig.9.32), intact soil is forced to mix with solidified agent of cement to form a cylinder body (flexible pile) possessing a certain strength, integrity and stability. These flexible piles are lapped to form cement-soil retaining wall or grille wall, which can ensure the slope stability of foundation pit. It has many advantages including self-standing, no horizontal support, easy digging, well seepage prevention, low cost and using intact soil to save material. However, the deformation and displacement of the cement-soil retaining wall is relatively large since the strength for materials of the wall is relatively low and without horizontal support.

⑦ 土钉墙是由土钉和喷射混凝土面层(含钢筋网)及原位土体组成的基坑支护结构,如图9.33和图9.34所示。通过形成土钉复合体,显著提高基坑整体稳定性。土钉墙的特点是施工设备简单,钉长一般比锚杆的长度小得多,且不需要施加预应力;成本低,施工噪声、振动小,不影响环境;本身变形很小,对相邻建筑物影响不大。

⑧ 锚杆支护(见图9.35)是一种岩土主动加固的基坑支护方法,其作用原理是将锚杆一端锚入稳定的土体中,另一端与其他形式的支护结构(如腰梁、冠梁等)连接,同时施加预应力,通过杆体受拉作用,调动深部地层潜能,从而维持基坑稳定。锚杆支护具有成本低、支护效果好、操作简便、使用灵活、占用施工空间少等优点。

图 9.33　土钉施工
Fig. 9.33　Soil nail construction

图 9.34　喷射混凝土面层
Fig. 9.34　Shotcrete surface layer

⑦ Soil-nailed wall is one of the foundation excavation support methods, which are made up of soil nails and shotcrete surface layer(including steel mesh) together with the in situ soil, as shown Fig.9.33 and Fig.9.34. Stability of foundation pit can be improved by composite structure. Construction equipment for soil-nailed wall is simple. The length of each nail is generally much smaller than that of the anchor bolt and it does not need to be pre-stressed. It has many advantages including little construction noise, small impact on the environment, small vibration and low cost. It has small impact on adjacent buildings due to its small deformation.

图 9.35　基坑锚杆支护
Fig. 9.35　Anchor support for foundation excavation

⑧ Anchor bolt support (Fig.9.35) is a kind of geotechnical active reinforcement method in foundation excavation engineering. One side of the anchor bolt was driven into the soil layers and the other end was connected with supporting structure such as middle beam and top beam, etc. Pre-stress was put on the bolt at the same time. Through the tension effect of bolt, stability of foundation pit was maintained. It has many advantages including low cost, good supporting effect, easy operation, flexible use, and less construction space occupation, etc.

⑨ 内支撑支护(见图 9.21)是由竖向支护结构体系和水平内撑体系两部分构成的结构。支护结构体系常用钢筋混凝土排桩墙或地下连续墙形式,内撑体系主要包括水平支撑与斜撑(钢筋混凝土或钢材)。该支护结构适用面广,适用于各种土层和不同深度的基坑。

9.4.3　地下工程

城市化的高速发展,迫使人们开发利用地下空间,这已经成为现代城市规划和建设的重要内容之一。国际隧道协会 ITA 曾提出:"大力开发地下空间,开始人类新的穴居时代"。在地面以下土层或岩体中修建各种类型的地下建筑物或结构的工程,称为地下工程。地下工程按照用途不同,交通运输方面包括地下铁路、公路隧道、地下停车场等, 国防方面包括地下指挥所、军火库等,工业与民用方面包括地下工厂、电站、停车场、商场、人防等。图 9.36 所示为地下空间开发利用的效果图。

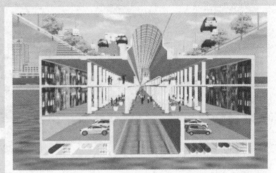

图 9.36　城市地下空间开发
Fig. 9.36　Utilization of urban underground space

⑨ Inner support (Fig.9.21) is made up of vertical support retaining structure system and horizontal support system, respectively. Vertical support structure system commonly includes row of reinforced concrete pile wall or underground continuous wall introduced in the form sections. Horizontal support system mainly includes horizontal brace and diagonal brace using reinforced concrete or steel. This support system is widely used in the foundation excavation engineering, which is applicable to different soil types and different excavation depths.

9.4.3　Underground engineering

With the rapid development of urbanization, it is urgent for people to develop and utilize the underground space, which has become an important issue in modern urban planning and construction. International tunneling association (ITA) proposed that "vigorously develop the underground space, to start a new cave era for human". Underground engineering refers to build various types of buildings or structures in the soil or rock mass below the ground. It is a series of linked subterranean spaces that may provide a defensive refuge, a place for living, working or shopping, a transit system, mausolea; wine or storage cellars; cisterns or drainage channels, or several of these. The term may also refer to a network of tunnels that connects buildings beneath street level, which may house office blocks, shopping malls, metro stations, theatres, and other attractions. These passages can usually be accessed to the public space of any of the buildings connecting to them, and sometimes have separate entries as well. Fig.9.36 shows the utilization of urban underground space.

地下工程施工中的开挖方法包括明挖法和暗挖法。明挖法可细分为基坑敞口开挖、基坑支挡开挖、地下连续墙、沉管法等;暗挖法可细分为矿山法、盾构法(见图 9.37)、掘进机(TBM)法(见图 9.38)和顶管法(见图 9.39)。

9.5 地基处理

地基处理是一门技术,用于提高地基土的工程性质。当天然地基不能满足建(构)筑物对地基的要求时,需要对天然地基进行处理形成人工地基。常用的地基处理方法有换填垫层法、预压地基法、压实与夯实地基法、复合地基法和注浆加固法等。

图 9.37 盾构机
Fig. 9.37 Shield machine

图 9.38 岩石掘进机(TBM)
Fig. 9.38 Tunneling boring machine (TBM)

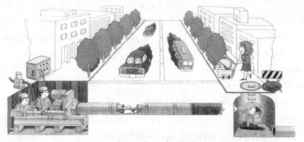

图 9.39 顶管法
Fig. 9.39 Pipe jacking method

Underground excavation construction can be divided into open excavation and concealed method, respectively. Concealed method can be subdivided into open excavation of foundation pit, retaining wall of foundation pit excavation, underground continuous wall, immersed tube method, etc. Open excavation method can be subdivided into mining method, shield method(Fig.9.37), tunneling boring machine(TBM) method (Fig.9.38), and pipe jacking method(Fig.9.39).

9.5 Ground improvement

Ground improvement is a technique aiming at improving the engineering properties of the soil mass. When the natural foundation can not meet the requirement of foundation construction, artificial foundation is needed to improve the natural foundation. Common foundation treatment methods include replacement layer of compacted fill method, preloaded ground method, compacted and tamped ground method, composite foundation method, and grouting reinforcement method, etc.

地基处理主要采用基础工程措施和岩土加固措施。某些工程不改变地基的工程性质，而只采取基础工程措施；有的工程需同时对地基的土和岩石加固，以改善其工程性质。选定适当的基础形式，不需改变地基的工程性质就可满足要求的地基称为天然地基。反之，加固后的地基称为人工地基。地基处理工程的设计和施工质量直接关系到建筑物的安全，如处理不当，常会发生工程事故，且事后补救大多比较困难。因此，对地基处理要求实行严格的质量控制和验收制度，以确保工程质量与安全。

9.5.1　换土垫层

当建筑物基础下的持力层比较软弱，不能满足上部荷载对地基的要求时，常采用换土回填法来处理，如图 9.40 所示。施工时先将基础以下一定深度、宽度范围内的软土层挖去，然后回填强度较大的砂、石或灰土等，并夯至密实。换土回填按其材料分为砂地基、碎石地基(见图 9.41)、灰土地基等。

图 9.40　换土垫层
Fig. 9.40　Replacement layer of compacted fill

图 9.41　碎石垫层
Fig. 9.41　Replacement layer of gravels

Ground improvement mainly adopts foundation engineering measures and geotechnical reinforcement measures. For some engineering projects, foundation engineering measures are adopted without changing the properties of engineering foundation. Some projects, however, still need some reinforcement of foundation soil and rocks at the same time in order to improve its engineering properties. The former is called natural foundation with selected foundation type without changing its properties. The latter is called artificial foundation due to some reinforcement measures. The design and construction quality of foundation treatment is directly related to the safety of the structure. Engineering accidents often occurred if measures can not be handled properly and remedial measures are much more difficult to carry out after accidents. Therefore, strict quality control and inspection system should be guaranteed to ensure the engineering quality and safety.

9.5.1　Replacement layer of compacted fill

Replacement layer of compacted fill method (Fig.9.40) is often used to deal with poor ground when the bearing soil layer under building foundation is relatively weak and can not meet the load requirement of foundation. Soft soil layers were first dug out with a certain depth and width below the foundation in the construction. The sand, stone or lime and so on with relatively higher strength were filled back to replace the soft soil. According to backfill materials, it can be divided into sand foundation, sand and gravel foundation(Fig.9.41), lime foundation, etc.

9.5.2 预压地基

预压地基是指在原状土上加载,使土中水排出,以实现土的预先固结,减少建筑物地基后期沉降,提高地基承载力。按加载方法的不同,分为堆载预压(见图9.42)、真空预压(见图9.43)、降水预压3种不同的方法。

排水法是采取相应措施如砂垫层、排水井、塑料多孔排水板等,使软基表层或内部形成水平或垂直排水通道,然后在土体自重或外界荷载作用下,加速土中水分的排出,使土体固结的方法。

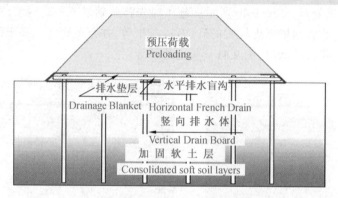

图 9.42 堆载预压
Fig. 9.42 Surcharge preloading

9.5.2 Preloaded ground

Preloaded ground method is one of the ground improvement methods. Water can be drained from the soils by applying load on the intact soil in order to make the soil pre-consolidate under the added load. It can reduce the settlement of building and improve the bearing capacity of foundation. According to the load source, preloaded ground method can be further divided into surcharge preloading method (Fig.9.42), vacuum preloading method(Fig.9.43), and dewatering preloading method, respectively.

图 9.43 真空预压
Fig. 9.43 Vacuum preloading

Drainage method aims at making the soil consolidation by taking appropriate measures such as sand blanket, drainage well, and porous plastic drainage board, etc. to form drainage channels. Water in the soil layers can be discharged quickly through the horizontal or vertical drainage channel in soft ground surface or interior under its self weight or the added preload.

9.5.3　压实和夯实地基

压实法和夯实法均是有效的地基处理方法,分别如图 9.44 和图 9.45 所示。压实法是指采用重型机械将地基土压实的方法。夯实法是指用几十吨重的夯锤,从几十米高处自由落下,进行强力夯实的地基处理方法。夯锤一般重 10~40 t,落距为 6~40 m,处理深度可达 10~20 m。采用夯实法要注意振动对邻近建筑物的影响。

9.5.4　复合地基

随着地基处理技术的发展,复合地基技术得到愈来愈多的应用。复合地基是指天然地基在地基处理过程中部分土体得到增强或被置换,或在天然地基中设置加筋材料形成的地基。加固区是由基体(天然地基土体)和增强体两部分组成的人工地基。复合地基中增强体和基体共同承担荷载。

图 9.44　压实法
Fig. 9.44　Compacted method

图 9.45　夯实法
Fig. 9.45　Tamped ground method

9.5.3　Compacted and tamped ground

Compacted ground method and tamped ground method are two effective ground treatment methods, as shown in Figure 9.44 and Figure 9.45, respectively. The compacted method refers to use heavy machinery such as road roller to compact the soil foundation. The tamped method refers to improve the soil foundation by the rammer with dozens of tons falling to the ground from dozens of meters high. Rammer weight varies from 10 t to 40 t and distance of fall varies from 6 m to 40 m. The improvement depth can be up to 10~20 m. Attention should be paid to the vibration on adjacent buildings.

9.5.4　Composite foundation

Composite foundation method has been used widely in ground improvement with the development of the foundation treatment technology. A part of soil was enhanced or replaced in the process of the ground treatment or reinforced material was set in natural foundation. The reinforced area is composed of reinforced part and natural soil part, which form so-called artificial one. The strengthen body together with natural soil part shares the load.

　　复合地基一般分为水泥土搅拌桩复合地基(见图9.46)、高压喷射注浆桩复合地基、砂桩地基、振冲桩复合地基(见图9.47)、土和灰土挤密桩复合地基、水泥粉煤灰碎石桩复合地基、夯实水泥土桩复合地基等。

9.5.5　注浆加固

　　注浆加固法是通过钻孔或其他设施将浆液压送到地基孔隙或缝隙中,改善地基强度或防渗性能的工程措施,主要有固结灌浆、帷幕灌浆、接触灌浆、化学灌浆、高压喷射灌浆法等。

　　(1) 固结灌浆

　　即为改善基岩的力学性能而进行的灌浆操作,可以减少基础的变形和不均匀沉降,改善基础工作条件,降低基础开挖深度。该法的特点是灌浆面积较大、深度较浅、压力较小。

图 9.46　水泥土搅拌桩地基
Fig. 9.46　Cement-soil mixing pile foundation

图 9.47　振冲桩机
Fig. 9.47　Vibration machine

Composite foundation is generally divided into cement-soil mixing pile composite foundation (Fig.9.46), high pressure jet grouting pile composite foundation, sand pile foundation, vibration pile composite foundation(Fig.9.47), soil and lime compaction pile composite foundation, cement fly-ash gravel pile composite foundation and rammed soil-cement pile composite foundation, etc.

9.5.5　Grouting reinforcement

Grouting reinforcement aims at improving the strength of foundation and its anti-seepage performance by drilling boreholes or other facilities to make grouting slurry been pressed into pores or cracks of the foundation. It can be mainly divided into different methods such as consolidation grouting method, curtain grouting method, contact grouting method, chemical grouting method, and high-pressure jet grouting method, etc.

　　(1) Consolidation grouting

Consolidation grouting can improve the mechanical properties of bedrock and reduce the deformation and uneven settlement. It can also improve the working conditions of foundation and reduce the depth of foundation excavation. The grouting area is large and the depth is shallow with small grouting pressure.

（2）帷幕灌浆

即在基础内,平行于建筑物的轴线,钻一排或几排孔,用压力灌浆法将浆液灌入岩石的缝隙中去,形成一道防渗帷幕,截断基础渗流。其特点是深度较深、压力较大。

（3）接触灌浆

即在建筑物和岩石接触面之间进行灌浆,以加强二者之间的结合程度和基础的整体性,提高抗滑稳定性,同时也增进岩石固结与防渗性能。

（4）化学灌浆

化学灌浆是以一种高分子有机化合物为主要材料的灌浆方法。这种浆材呈溶液状态,能灌入 0.10 mm 以下的细微管缝,浆液经过一定时间发生化学作用,可将裂缝粘合起来形成凝胶,起到堵水防渗及补强的作用。

（5）高压喷射灌浆

通过钻入土层中的灌浆管,用高压压入水泥浆液,并从钻杆下端的特殊喷嘴以高速喷射出去。在喷射的同时,钻杆以一定速度旋转,并逐渐提升;高压射流使四周一定范围内的土体结构遭受破坏,并被强制与浆液混合,凝固成具有特殊结构的圆柱体,也称旋喷桩。如采用定向喷射,可形成一段墙体,一般每个钻孔定喷后的成墙长度为 3~6 m。

（2）Curtain grouting

Curtain grouting aims at forming an impervious curtain to truncate foundation seepage by using pressure in order to make the grouting slurry pumped into the cracks. A row or several rows holes were first drilled paralleling to the axis of the building in the foundation. The grouting depth is deeper due to the drilled holes and the relative high pressure.

（3）Contact grouting

Contact grouting is performed at interface between the buildings and the rocks, which can ensure the integrity of the foundation and combination degree to improve the stability against sliding. It can also promote the rock consolidation and anti-seepage performance.

（4）Chemical grouting

It is one kind of grouting methods using high polymer organic compounds as mainly grouting material. The liquid grouting material can be injected into small pipe cracks less than 0.10 mm. The cracks can be bonded together to form gel after a certain time of chemical action, which can take effect of water plugging anti-seepage and reinforcement of foundation.

（5）High-pressure jet grouting

Cement grouting was injected through a special nozzle at a high speed into soil layers with high pressure by the grouting pipe drilling in the soil. Meanwhile, the drilling rod rotated at a certain speed and gradually elevated. Certain scope of soil structure around the high pressure jet grouting was damaged and forced to mix with the grouting. Then it was solidified into a cylinder structure and called jet grouting pile. Wall body can be formed if using directional jet. Generally the length of wall for each jet drilling is 3~6 m.

9.6 边坡工程

为满足工程需要而对自然边坡进行改造,称为边坡工程。根据边坡对工程影响的时间差别,可分为永久边坡和临时边坡两类;根据边坡与工程的关系,可分为建筑物地基边坡(必须满足稳定和有限变形要求)、建筑物邻近边坡(须满足稳定要求)和对建筑物影响较小的延伸边坡(允许有一定限度的破坏);根据边坡的材料可以分为土质边坡(见图9.48)和岩质边坡(见图9.49)。

边坡的稳定性常以安全系数或破坏概率作为评价指标。安全系数系指滑动体沿滑动面所受的抗滑力和滑动力之比。破坏概率是指安全系数小于1的出现概率。破坏概率愈小,边坡稳定程度愈高。

图9.48 土质边坡
Fig. 9.48 Soil slope

图9.49 岩质边坡
Fig. 9.49 Rock slope

9.6 Slope engineering

Slope engineering refers to reform the natural slope to meet the engineering project needs. According to the different impact time on engineering, slope can be divided into two categories, namely permanent slope and temporary slope, respectively. According to the relationship between project and slope, it can be divided into three categories such as building foundation slope(must meet the requirements of stability and finite deformation), nearby building slope(must meet the requirements of stability) and extension slope(allow certain destruction). According to the materials, slope can be divided into two categories: soil slope(Fig.9.48) and rock slope(Fig.9.49).

Safety coefficient refers to ratio of anti-sliding force and sliding force along the sliding surface. Failure probability refers to the probability of the safety coefficient less than 1. The smaller the probability, the higher the degree of slope stability is.

9.6.1 土质边坡稳定分析

土质边坡在重力或其他因素作用下向下运动,造成土体的破坏称为滑坡或土坡破坏。土坡的稳定程度一般以安全系数的大小来衡量,表示方法有多种,通常为: ① 实际的抗剪强度与维持平衡所需的抗剪强度之比;② 抗滑力矩之和与下滑力矩之和的比值; ③ 总抗滑力与总下滑力的比值等。工程实践中为选用合适的安全系数,需考虑的因素有荷载组合、建筑物按重要性的类别、抗剪强度的试验条件、计算方法的选择、施工控制的可靠程度、获得的信息完整性、工程经验、经济情况等。常用的安全系数值的范围是 1.1~9.0。由于土的复杂性,在土坡分析中所涉及的因素都是随机的,即使采用大于 1 的安全系数,也不能认为土坡的稳定性十分可靠。故对重要工程,除采用较大的安全系数外,还需进行边坡稳定的概率分析。

9.6.2 岩质边坡稳定分析

岩质边坡由岩石构成,由于坡度过大,坡脚受切或作用于坡体的力发生变化,均会使边坡失稳而破坏。岩质边坡稳定性分析以工程地质资料为基础,内容包括判断边坡破坏模式,确定计算参数,应用岩体力学和有限元等方法进行力学计算并做出稳定性评价。

9.6.1 Soil slope stability analysis

Slope stability is the analysis of soil covered slopes and its potential to undergo movement. Soil slope may move down resulting from gravity or other factors and causing soil damage known as landslide or slope damage. Slope stability is commonly measured and assessed by safety coefficient. There are many types of representation for safety coefficient, such as ① the ratio of actual shear strength and the shear strength required for balance; ② the ratio of total anti-sliding movement and total sliding moment; ③ the ratio of total anti-sliding force and total sliding force, etc. In engineering practice, suitable safety factor should be selected according to the load combination, building importance category, test conditions of shear strength, selection of calculation methods, reliability of construction control, accuracy of information, engineering experience, economy and so on. The scope of safety coefficient value commonly used varies from 1.1 to 9.0. The factors involved in the analysis of soil slope are random due to the complexity of the soils. Slope stability can not be very reliable even with the safety coefficient greater than 1. Therefore, probability analysis of slope stability should be conducted besides adopting bigger safety coefficient for the important projects.

9.6.2 Rock slope stability analysis

Rock slope is formed by rock, which can be unstable and destroyed since the slope gradient is bigger than soil slope and slope toe is sheared or the changing of external force on slope body. Safety coefficient or failure probability is often used as evaluation index for rock slope stability. Rock slope stability analysis is based on the engineering geological data including judgment of the slope failure mode, calculation of parameters, the application of the rock mechanics and finite element method to perform mechanical calculation and stability evaluation.

9.7 岩土工程发展展望

岩土工程是人类改造世界,发展生存空间,营造现代物质文明的一项重要系统工程。随着科学技术的迅猛发展,21 世纪的岩土工程必将在现有先进水平的基础上获得突破性的发展。

随着高层建筑、城市地下空间利用和交通工程的发展,岩土工程者的注意力将较多的集中于建筑工程、市政工程和交通工程建设中的岩土工程问题。土木工程功能化、城市立体化、交通高速化,以及改善综合居住环境成为现代土木工程建设的特点。人口的增长加速了城市发展,城市化的进程促进了大城市在数量和规模上的急剧发展。人们将不断拓展新的生存空间,开发地下空间,向海洋拓宽,修建跨海大桥、海底隧道和人工岛,改造沙漠,修建高速公路和高速铁路等。而展望岩土工程的发展,不能离开对现代土木工程建设发展趋势的分析。

9.7.1 岩土工程可持续性发展

近 20 年来,随着国民经济和各类土木工程的快速发展,我国岩土工程建设空前活跃;但是几乎在同期,国际岩土工程学科的领域和范围发生了明显的变化。随着人们对于资源和生态环境认识的深入,岩土工程已不限于具体"工程"的设计施工,而是扩展到环境岩土、地质灾害,以及与生态和资源相适应的岩土工程可持续发展等战略问题。

9.7 Developing prospects for GE

Geotechnical engineering is an important system engineering aiming at reconstructing the world, developing human survival space, and creating a modern material civilization. With the rapid development of science and technology, the significant breakthroughs will be achieved based on the existing advanced level in the 21st century.

Geotechnical engineers begin to focus on the engineering problems related to building engineering, municipal engineering and traffic engineering with the development of high-rise buildings, urban underground space utilization and express highway. Function of civil engineering, three-dimensional city, high speed transportation, and improvement of human settlement environment have become the hallmark of the modern civil engineering construction. Population growth accelerated the urban development and the urbanization process promoted the rapid development of the number and size for big cities. People will continue to expand a new space for survival, including developing underground space, broadening to the ocean, building sea-crossing bridges, tunnels, and artificial islands, transforming desert, and building highways and high-speed railways, etc. When it comes to looking forward to the development of geotechnical engineering, people can not ignore the analysis of the trend of modern civil engineering construction.

9.7.1 Sustainable development of GE

With the rapid development of national economy and various types of civil engineering over the past 20 years in China, geotechnical engineering has undergone an unprecedented prosperity. In the same period, however, significant changes have taken place on the subject areas and scope of the international geotechnical engineering. Along with the deep understanding of resources and the ecological environment, geotechnical engineering is no longer limited to specific "project" design and construction, but extended to geotechnical environment, geological disasters, and sustainable development, etc.

例如,大量开采深层地下水,使地面普遍下降,引发地裂缝,海水入侵及地下水盐碱化;废弃物对土地、空气与地下水的污染;大规模的水利水电工程、高速公路、高速铁路,尤其是在生态环境比较脆弱的西部高寒、高海拔和高纬度地区的工程,可能引发的次生地震、崩塌、滑坡、泥石流、融陷、大面积水土流失等;在膨胀土、湿陷性黄土、多年冻土、盐渍土地区施工作业一方面工程本身十分困难,另一方面也可能存在生态环境方面的不利影响。

环境岩土工程是岩土工程与环境科学密切结合的一门新学科。它主要应用岩土工程的观点、技术和方法为治理和保护环境服务。人类生产活动和工程活动造成许多环境公害,如采矿造成采空区坍塌,过量抽取地下水引起区域性地面沉降,工业垃圾、城市生活垃圾及其他废弃物、有毒有害废弃物污染环境,施工扰动对周围环境的影响等。另外,地震、洪水、风沙、泥石流、滑坡、地裂缝等灾害对环境造成破坏。上述环境问题的治理和预防给岩土工程师们提出了许多新的研究课题。随着城市化、工业化发展进程的加快,环境岩土工程研究将更加重要,我们应从保持良好的生态环境和保持可持续发展的高度来认识和重视环境岩土工程研究。

For example, a large number of exploitation of deep groundwater will make ground settlement, ground cracks, seawater intrusion, and groundwater salinization. Wastes will pollute the land, air and groundwater. Large-scale water conservancy and hydropower engineering, highway, high-speed railway, especially in the fragile ecological environment, cold western, high altitude and high latitude areas, could cause the secondary earthquakes, collapse, landslide, debris flow, sink, large area of soil and water loss, etc. Project in the area of expansive soil, collapsible loess, permafrost, saline soil will make its construction difficult. There are also negative impacts on ecological environment.

Environmental geotechnical engineering is a new discipline closely combined with the geotechnical engineering and environmental science. It serves for governance and environmental protection by applying geotechnical engineering views, technologies and methods. Human production activities and engineering activities may result in many environmental hazards. For example, mining will cause the collapse of mined-out area; excessive pumping groundwater will cause regional land subsidence; industrial waste, municipal solid waste and other waste, especially toxic and harmful waste will pollute the environment; the construction disturbance will influence the surrounding environment, etc. In addition, earthquake, flood, sandstorm, mud-rock flow, landslide, ground cracks and so on will also cause damage to the environment. The governance and prevention for the above environmental problems have put forward many new research topics to geotechnical engineers. With the rapid development of urbanization, industrialization process, environmental geotechnical engineering research will become more and more important. Geotechnical engineering research should be recognized and focused on from the perspective of maintaining a good ecological environment and sustainable development of environment.

9.7.2 岩土工程智能化测试技术

通过钻探取样直观地掌握岩土性状,仍将是 21 世纪岩土工程勘测的核心方法,但是钻探技术和测试技术将有长足的进步,原位试验仍是必不可少的勘测手段。随着原位试验精度的提高,不仅使钻探取样、室内土工试验的工作量减少到最低,而且岩土工程勘测的效率也将大大提高。原位试验技术的发展同样以智能化为方向,而遥感遥控和数据采集处理自动化是普遍的技术方法。

9.7.3 城市地下空间的开发利用

要还城市更开阔的空间,更多的绿色景观,更浓厚的历史与人文氛围与气息,就要更多地利用城市地下空间。21 世纪人类面临着人口增长的巨大压力,人们生活的空间是有限的,必须在保护生态环境的同时,充分开发和有效利用适宜的生存空间。一方面高层建筑仍将是解决城市人口密集区居住问题的重要手段,另一方面要充分利用既节能又不争地的地下空间。城市交通、公共设施、人居空间将大量转入地下,这也是城市防护和减灾的要求。目前,地下铁道、地下停车场、地下仓库、地下商场、越江或海底隧道等地下建筑物已经广泛地为人类服务。在电力工业方面,核电站核燃料废弃物的存放问题已经成为岩土工程的热点研究课题之一。

9.7.2 Intelligent testing technology of GE

Geotechnical properties can be obtained intuitively by sample drilling, which will also be the core method of geotechnical engineering survey in the 21st century. However, the drilling technology and testing technology will be improved greatly. In situ tests will continue to be an indispensable way. With the improvement of in situ test accuracy, expensive drilling sampling and laboratory soil tests work will not only be reduced to a minimum, but also the geotechnical engineering exploration efficiency will be greatly improved. Intelligence will be the direction of development of in-situ test while remote control and data acquisition process automation will be common technical methods.

9.7.3 Underground space development and utilization

With the changing of people's consciousness, more open space, more green landscape, more strong historical and cultural atmosphere are needed, which requires greater use of underground space. People have to face the huge pressure of population growth in the 21st century. Since the living space is limited for people, full development and effective utilization of human living space is needed based on the protection of ecological environment. So on the one hand, high-rise buildings will still be the important means to solve the problem of the high concentration of city population. On the other hand, underground space will be fully used for both saving energy and saving land, which requires more use of underground space for the cities. Urban transportation, public facilities, residential space will transfer to underground, which meets the requirements of urban protection and disaster mitigation. Currently, subways, underground malls, underground parking lots, underground malls, river-crossing or sub-sea tunnels, and so on have been widely built for serving the mankind. In the power industry, the storage problem of nuclear waste in nuclear power plant has become one of the hot research topics in geotechnical engineering.

知识拓展
Learning More

小贴士 Tips

1. 中国注册岩土工程师

注册土木工程师(岩土),简称注册岩土工程师,是指取得"中华人民共和国注册土木工程师(岩土)执业资格证书"和"中华人民共和国注册土木工程师(岩土)执业资格注册证书",从事岩土工程工作的专业技术人员。

1. Chinese registered Geotechnical Engineer

Registered Civil Engineer, shorten as the Registered Geotechnical Engineer, refers to the professional and technical person who has been involved in the geotechnical engineering after they successfully obtains both "the People's Republic of China Registered Civil Engineer (CE) qualification certificate" and "the People's Republic of China Registered Civil Engineer (GE) qualification certificate of registration".

2. 岩土工程类期刊

英国 "Geotechique",美国 "Geotechnical & Environmental Engineering, ASCE",加拿大 "Canadian Geotechnical Journal",日本 "Soils and Foundations",中国《岩土工程学报》《岩土力学》。

2. Geotechnical engineering journals

Geotechique in Britain, *Geotechnical & Environmental Engineering, ASCE* in USA, *Canadian Geotechnical Journal* in Canada, *Soils and Foundations* in Japan, *Journal of Geotechnical Engineering* and *Rock and Soil Mechanics* in China.

相关链接 Related Links

1. 岩土论坛
2. 国际岩土信息中心
3. 土木工程网

思考题 Review Questions

1. 基坑支护的方法有哪些?

What are the foundation excavation support methods?

2. 地基处理的方式有哪些?

What are the ground improvement methods?

3. 如何实现岩土工程的可持续性发展?

How to achieve sustainable development of geotechnical engineering

参考文献
References

［1］李广信. 岩土工程 50 讲:岩坛漫话[M].2 版.北京:人民交通出版社,2010.

［2］周景星,李广信,张建红,等. 基础工程[M].3 版.北京:清华大学出版社,2016.

［3］张萌.岩土工程勘察[M].北京:中国建筑工业出版社,2011.

［4］穆保岗,陶津,童小东,等. 地下结构工程[M].南京:东南大学出版社,2016.

［5］高大钊. 土力学与基础工程[M].北京:中国建筑工业出版社,1998.

［6］侴磊,徐燕,代树林,等. 边坡工程[M].北京:科学出版社,2010.

［7］刘永红. 地基处理[M].北京:科学出版社,2005.

［8］龚晓南. 21 世纪岩土工程发展展望[J].岩土工程学报,2000,22(2):238-242.

［9］钱家欢,殷宗泽. 土工原理与计算[M].2 版.北京:中国水利水电出版社,1996.

[10] Mitchell J K. Fundamentals of Soil Behavior[M].3th ed. Hoboken:John Wiley & Sons,2005.

[11] Klaus K, Alan B. Ground Improvement[M].3th ed. Boca Raton: CRC Press Inc.,2019.

[12] Braja M D. Principles of Foundation Engineering[M]. Bcston: Cengage Learning,2010.

[13] Chowdhury R, Flentje P, Bhattacharya G. Geotechnical Slope Analysis [M]. Boca Raton:
CRC Press Inc.,2009.

第10章 现代土木工程建造与管理

Chapter 10　Modern Civil Engineering Construction and Management

先导案例

Guide Case

上海中心大厦(Shanghai Tower)是上海市的一座超高层地标式摩天大楼,其设计高度超过附近的上海环球金融中心,建成后成为中国第一高楼及世界第三高楼(见图10.1)。

观光及餐饮　tourism & dining
第9区　L118-L121

酒店及精品办公　business office
第8区　L101-L115

酒店　hotel
第7区　L84-L98

办公楼层　office storey
第6区　L69-L81
　　　　L68
第5区　L53-L65
　　　　L52
第4区　L38-L49
　　　　L37
第3区　L23-L34
　　　　L22
第2区　L8-L19
　　　　L8

会议及多功能空间、零售商场
conference & shopping
第1区　L1-L5

地下室　basement
B1-B2　商业、地下通道
　　　　shopping & passageway
B3-B5　停车、机电设备
　　　　park & equipment

图 10.1　上海中心大厦
Fig. 10.1　Shanghai Tower

Shanghai Tower, China's tallest building and the third tallest in the world when completed, is a super high-rise land-marked skyscraper in Shanghai, and its design height is beyond the Shanghai World Financial Center nearby (Fig. 10.1).

　　上海中心大厦的设计建造难度非常高,依赖传统手段很难完成。而本章中将要提到的建筑信息模型(BIM)(见图 10.2)技术在该工程的实施中发挥了至关重要的作用。

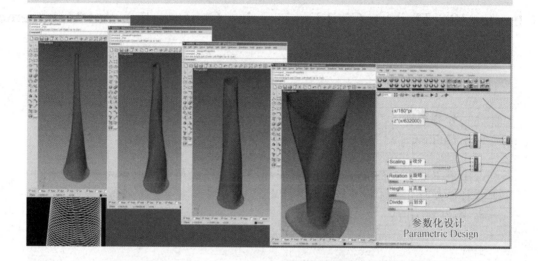

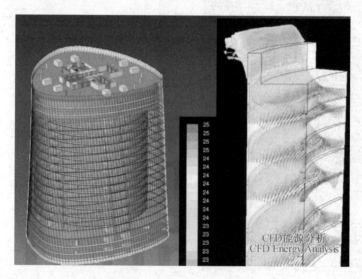

图 10.2　建筑信息模型(BIM)技术的应用 1
Fig. 10.2　Application of building information modeling (BIM) 1

　　The design and construction of the Shanghai Tower is very difficult. It's hard to accomplish only relying on traditional technology. The building information modeling（BIM,Fig.10.2）, which will be mentioned in the following parts, has played a vital role in the implementation of the project.

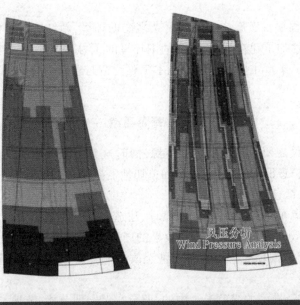

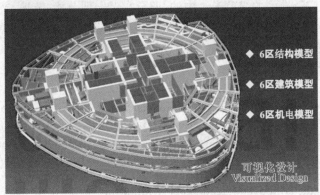

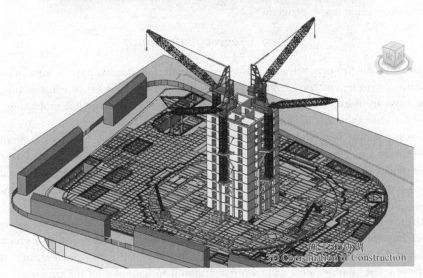

图 10.2　建筑信息模型（BIM）技术的应用 2
Fig. 10.2　Application of building information modeling（BIM）2

现代土木工程项目实施中面临着复杂多变的环境,新结构、新材料、新技术、新工艺的大量使用增加了建造难度,来自周围环境的压力对土木工程建造方法和技术提出了更高要求。本章主要介绍现代土木工程建造与管理中应用的新理论、新方法和新技术。

10.1 绿色建造

工程项目建设需要大量耗费各种资源,同时这又可能对环境、生态产生巨大影响。随着环境保护意识的加强,人类更加迫切地需要一种能够降低上述不利影响的建造模式。于是,绿色建造应运而生。

绿色建造是指在工程项目的规划、设计、建造、使用、拆除的全生命周期过程中,在提高建筑生产效率或优化建筑产品效果的同时,又能减少资源和能源消耗,减轻污染负荷,降低对周围环境的不良影响甚至改善环境质量,促进可持续发展的工程建造模式。

目前,我国广泛推广的绿色施工中的"四节一保",即节能、节地、节材、节水和保护环境,正是绿色建造的核心。绿色施工是绿色建造过程中的一个环节,即在施工各个阶段贯彻绿色建造的要求,完成工程项目建设。

The implementation of civil engineering is facing with complex and changeable circumstances. At the same time, the extensive use of new structure, new materials, new technologies and new techniques have also increased the difficulty of construction. The pressure from the surrounding environment puts forward advanced demands for methods and technologies of civil engineering construction. This chapter will introduce the new theories, new methods and new technologies which are applied in modern civil engineering construction and management.

10.1 Green construction

Construction projects require various resources, but it may have a significant impact on the environment and ecology. To reduce the adverse effects, a construction mode is more urgently needed with the increased awareness of environmental protection. Thus, the green construction come into being.

Green construction(GC) is a mode to improve the efficiency of building production or to optimize the effect of building products, to reduce the consumption of resources and the pollution load, to reduce the adverse effects of the surrounding environment, and to promote the sustainable development through the life cycle of the project in planning, design, construction, operation and removal.

At present, the "green four-saving & one-protection"(energy saving, land saving, material saving, water saving and environmental protection) is the core issue of green construction in China. Green construction is a link in the construction process, and it aims to implement the requirements of green construction for completing the projects.

　　绿色建造的核心要求如下：

　　(1) 采用绿色建材

　　绿色建材采用清洁生产技术，不用或少用天然资源和能源，大量使用工农业或城市固态废弃物生产，无毒害、无污染、无放射性，达到使用周期后可回收利用，有利于环境保护。如硅藻泥就是绿色建材的一个应用实例(见图 10.3)。

　　绿色建材最基本的功能是维护人体健康、保护环境，是保护和治理环境的有益材料。采用绿色建材要求在源头上就关注消除污染，并贯彻到生产、施工、使用及废弃物处理的全过程。

图10.3　绿色建材：硅藻泥及应用效果
Fig. 10.3　Green building materials: diatom mud and it's application

　　The core requirements of the green construction are as follows:

　　(1) Use green building materials

　　Green building materials are conducive to environmental protection. It's made by clean production technology, using no or less natural resources and energy, plenty of industrial and agricultural waste production or urban solid garbage. It has no poison, no pollution and no radioactivity. It can be recycled after the use cycle. An application of green building materials is diatom mud(Fig.10.3).

　　The basic function of green building materials is to protect the human body and environment. Green building materials are beneficial to the protection and management of the environment. It pays attention to pollution elimination from the source through the whole construction process.

（2）减少资源消耗

为了减少施工过程中材料和资源的消耗,应尽量就地取材,减少材料运输所造成的能源消耗和对环境影响。最大限度利用场地现有资源,可对建筑施工废弃物及雨水、中水尽量回收利用(见图10.4),提高资源再利用率,节约材料与资源。

图10.4　工地水资源再利用
Fig. 10.4　The water reusing in the construction site

（2）Reduce resource consumption

In order to reduce the consumption of materials and resources in the construction process, it is necessary to use local materials to reduce energy consumption and environmental impact caused by materials transportation. The waste, rainwater and reclaimed water should be reused to maximize the utilization of the existing resources(Fig. 10.4). This can improve the utilization rate of resources and save materials.

（3）清洁施工过程

施工过程中应严格遵循国家和地方的有关法规,减少对场地地形、地貌、水系、水体的破坏和对周围环境的不利影响,严格控制噪声污染、光污染以及大气污染等。在施工过程中采用清洁生产技术,制定节能措施,改进施工工艺,提高施工过程中能源利用率,节约能源,减少对大气环境的污染。清洁施工措施如图 10.5 所示。

图10.5　清洁施工
Fig. 10.5　Clean construction process

（3）Clean construction process

In the construction process, the relevant laws and regulations of the state and local should be strictly obeyed to reduce the damage to the terrain, landform, water system, water body and the adverse effects on the environment. Besides, clean production technologies and energy-saving measures should be developed to improve the efficiency of energy use, to save energy and to reduce the pollution of atmosphere through the construction process. Clean construction processes are as shown in Fig.10.5.

（4）加强文明管理

施工过程中要保护施工人员的安全与健康。要合理布置施工场地,施工期间采取有效的防毒、防污、防尘、防潮、通风等措施,加强施工安全管理和工地卫生文明管理(见图 10.6)。

图10.6　标准化示范文明工地

Fig. 10.6　The standardized demonstration of construction civilization

（4）Strengthen the management of construction civilization

In the construction process, it is necessary to protect the safety and health of the construction workers. Furthermore, it is significant to arrange the construction site reasonably, and take measures (anti-virus, anti-pollution, anti-dust, anti-moisture, ventilation) effectively. Besides, it is indispensable to strengthen the management of safety and health civilization in the construction(Fig.10.6).

10.2　虚拟建造

　　虚拟建造(virtual construction,VC)是实际建造过程在计算机上的映射。它通过计算机对建造活动中的人、物、信息及建造过程进行全面的虚拟仿真,以虚拟建造环境来模拟和预估建筑物的外观、功能及施工性能等各方面可能存在的问题,从而辅助人们提高预测和决策水平,达到项目一次性建造成功,降低成本,缩短开发周期,增强建筑产品竞争力的目的,同时增强对各参与建造过程企业的综合决策、优化与控制能力。

　　虚拟建造提供了从建筑物概念的形成、设计到施工全过程的三维可视及交互的环境,使得建造技术走出了主要依赖经验的狭小天地,发展到了全方位预报的新阶段。

　　虚拟建造的核心技术是虚拟现实。虚拟现实(Virtual Reality,VR)这一名词由美国 VPL Research 公司的奠基人、发明家 Jaron Lanier 于 1980 年代初提出,其主要内容是:实时三维图形生成技术、多传感器交互技术和高分辨显示技术。

10.2　Virtual construction

　　Virtual Construction (VC) is a reflection of the actual construction process on the computer. A comprehensive virtual simulation of people, objects, information and process of building during the activity of construction can be conducted with the aid of computer. A virtual construction environment can be used to simulate and forecast the various aspects of problems which may exist in the appearance, functions and construction performance of the structure, etc. Consequently, it will improve people's ability of prediction and decision, help them to achieve the purpose of making the project built successfully, reducing the cost, shortening the development cycle and enhancing the competitiveness of building product. At the same time, it is beneficial to enhance the enterprise abilities of comprehensive decisions, optimization and control.

　　Virtual construction provides an interactive 3D visible environment of the building's concept from formation, design to construction. It makes construction technologies get out of the narrow world relying on experience and develop a new stage of all-round forecasts.

　　The core technology of virtual construction is virtual reality (VR). VR is proposed by Jaron Lanierin (the founder of the VPL Research) in the early 1980's. Its main content includes the technology of real-time 3D graphics generation, multi-sensor interaction and high resolution display.

　　虚拟现实作为对传统的人机交互方式的深化，是一种技术领域的人机共享模式，人们能利用它仿真或创造出虚拟环境，并让人们能身历其境地进入这一环境(见图10.7)，具有沉浸性、交互性、自主性和多感知性4个重要特征。

　　在建筑领域，可用虚拟现实技术实现工业与民用建筑物虚拟仿真(见图10.8)，对一个建筑物具体的一次性仿真，可细致到水龙头、电灯开关、门把手等。还可以用它设计各种建筑产品和施工设备，这对施工工程的方案比较和优化大有裨益。

图 10.7　某校园的虚拟现实系统
Fig. 10.7　VR system in a campus

图 10.8　虚拟建造系统中的施工场景
Fig. 10.8　Scene of construction in virtual construction system

　　As the development of the traditional way of human-computer interaction, VR is a kind of computer sharing mode of technology, by which people can copy or create a virtual environment and the user will be personally on the scene (Fig.10.7). It has four important characteristics of inebriation, interaction, self-determination and multi-sensation.

　　In the construction field, VR technology can be used to realize the virtual simulation of industrial and civil buildings (Fig.10.8). A specific one-time building simulation can be as fine as to tap, light switch, door handle, etc. It also can be used to design a variety of construction products and construction equipment. It is great beneficial to the comparison and optimization of construction plans.

对于在大型项目的规划、评估上的应用前景更是令人振奋。例如,一个大型水电站的投建,工程施工的优化,大型电力设备的设计、检修和日常维护,直到操作人员的培训等等,利用虚拟现实技术实现全方位仿真,将对业主、设计单位、建设单位、施工单位以及电厂的管理和控制带来前所未有的便利。在此背景下,引出了虚拟建造的概念。

10.3　建筑信息模型

建筑信息建模(Building Information Modeling,BIM)是工程项目的物理特性与功能特性的集成的数字化模型(见图 10.9),它能够为工程项目在从最初概念设计开始的整个生命周期里的所有决策提供可靠的共享信息资源。实现 BIM 的前提是在工程项目生命周期的各个阶段不同的项目参与方通过在 BIM 建模过程中插入、提取、更新及修改信息,以支持和反映出各参与方的职责。

图 10.9　BIM 的概念
Fig. 10.9　Concept of BIM

It has an exciting prospect in the planning and evaluation of large projects. For example, the VR technology has been applied to a full range of simulation in a large hydro-power station project, such as construction and the construction optimization, large-scale power equipment's design, repair and routine maintenance, the operation training, etc. Unprecedented convenience will be brought to the owner, the designer, the constructor, the builder as well as management and control of the power plant. Under this background, the concept of virtual construction will be introduced.

10.3　Building information modeling

Building information modeling (BIM) is a process involving the generation and management of digital representations of physical and functional characteristics of a facility (Fig.10.9). It can be used to support decision-making on a facility from earliest conceptual stages to design and construction, covering its operational life and eventual demolition. The precondition of BIM is that different project participants at all stages of the project life-cycle insert, extract, update and modify the information in the modeling process of BIM to support and reflect their responsibilities involved.

传统的建筑设计很大程度上依赖于二维图纸(平面图、立面图、剖面图等)。3D模型增加了高度,实现了空间化表达。BIM 则将这一模型扩展到了 3D 甚至更多的维度(nD),不同的维度对应于不同的功能。例如,时间作为第 4 维,费用作为第 5 维(见图 10.10),BIM 也就不仅包含了几何信息,还包含了建筑空间关系、光照分析、地理信息以及建筑构件的数量和属性信息。

3D-BIM应用:建筑建模
3D-BIM application: model constructing

4D-BIM应用:进度管理与施工模拟
4D-BIM application: schedule management and construction simulation

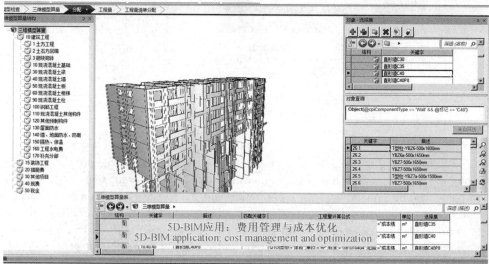

5D-BIM应用:费用管理与成本优化
5D-BIM application: cost management and optimization

图 10.10　BIM 的应用
Fig. 10.10　Application of BIM

Traditional building design largely relies upon two-dimensional drawings (plans, elevations, sections, etc.). BIM extends to 3D. It augments the primary three spatial dimensions (width, height and depth) by the fourth dimension of time and the fifth dimension of cost (Fig.10.10). BIM therefore covers more than just geometry. It also covers spatial relationships, light analysis, geographic information, and quantities and properties of building components.

　　4D 的 BIM 模型是在 3D 模型的基础上增加了进度管理功能，可以研究项目的可施工性、进度安排、进度优化、精益化施工等，提高经济性与时效性。

　　5D 的 BIM 模型是在 4D 模型的基础上又增加了费用管理功能，可以实现全寿命期的费用分析与优化。

　　BIM 理念就是设想在实体结构建设之前构建一个虚拟结构，以减少不确定性，提高安全性，解决相关问题，并模拟和分析其潜在影响。每一个项目的分包商在项目建造之前都能在模型中输入关键信息，以有可能在施工场地外进行相关系统的预制和预组装。施工现场的损耗可以实现最小化，生产出的制品也可以按照准时制(just-in-time，JIT)输送，避免了在现场的堆积。BIM 通过允许每个团队在项目建设期间在BIM 模型中调价或者访问信息的方式来架起设计单位、建设单位、施工单位之间的沟通的桥梁，防止信息的丢失(见图 10.11)。这有益于设施的业主或运营方。

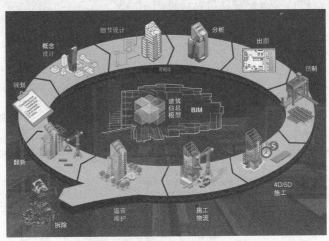

图 10.11　BIM 的全生命期应用
Fig. 10.11　BIM Life-cycling Application

　　4D BIM adds the function of schedule management on the basis of 3D. It can research the projects' constructability, scheduling arrangement, process' optimization and refining construction, which improve the efficiency and timeliness.

　　5D BIM adds the function of cost management on the basis of 4D. The analysis and optimizing of life cycle cost can be done with the aid of 5D BIM.

　　The BIM concept envisages virtual construction of a facility prior to its actual physical construction in order to reduce uncertainty, improve safety, work out problems, and simulate and analyze potential impacts. Sub-contractors from every trade can input critical information into the model before beginning construction, with opportunities to pre-fabricate or pre-assemble some systems off-site. Waste can be minimized on-site and products can be delivered on a just-in-time basis rather than being stock-piled on-site. BIM can bridge the information loss associated with handing a project from design team, to construction team and to building owner/operator by allowing each group to add and refer to all information they acquire during the period of contribution to the BIM model (Fig.10.11). This can yield benefits to the facility owner or operator.

10.4 工程全寿命期管理

工程建设活动跟工厂产品生产具有一定的相似性。工程建设和运营过程相当于工厂厂房和设备,输入土地、资金、原材料、劳动力等,经过建设和运营,输出产品和服务、产生废弃物和噪声等(见图10.12)。

建设工程项目全寿命周期是指建设工程项目从开始策划到使用报废为止所经历的各个阶段全过程。通常根据全寿命周期经济分析和管理的特点,将建设工程项目寿命周期划分为五个阶段:策划决策阶段、设计阶段、招投标阶段、施工阶段和运营维护阶段,其中设计阶段、施工阶段和运营维护阶段统称为实施阶段。

工程全寿命期管理是以工程的前期策划、规划、设计、施工和运营维护、拆除为对象的管理过程。按照工程寿命期阶段划分,工程全寿命期管理可分为开发管理(DM)、项目管理(PM)和物业管理(FM)(见图10.13)。

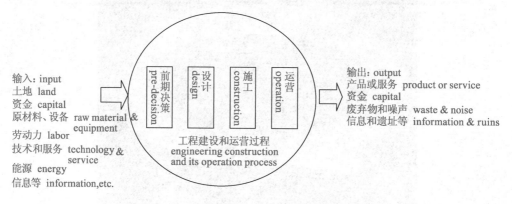

图 10.12　工程建设活动
Fig. 10.12　Engineering construction activities

10.4 Whole life cycle management

Engineering construction activities have some similarities with the factory production. Engineering construction and its operation process are equivalent to the plants and equipments. Through construction and operation, the input of land, capital, raw materials, labor, etc. become products and services, waste and noise, etc.(Fig.10.12)

Construction project life cycle refers to the whole process from the beginning to scrapping. According to the characteristics of economic analysis and management of the whole life cycle, the construction project life cycle can be divided into five stages: decision-making stage, design stage, bidding stage, construction stage and operation maintenance stage, in which the design stage, construction stage and operation maintenance phases are collectively referred to the implementation stage.

The whole life cycle management is a management process based on the planning, design, construction, operation and maintenance of the project. It includes development management(DM), project management(PM) and facility management(FM)(Fig.10.13).

工程任一阶段的参与方(投资方、开发方、设计方、施工方及供货方等)都要立足于工程的全寿命期,以全寿命期的整体最优为目标,不仅注重建设期,更注重工程的运行阶段,实现工程全寿命期的集成化管理(见图 10.14)。

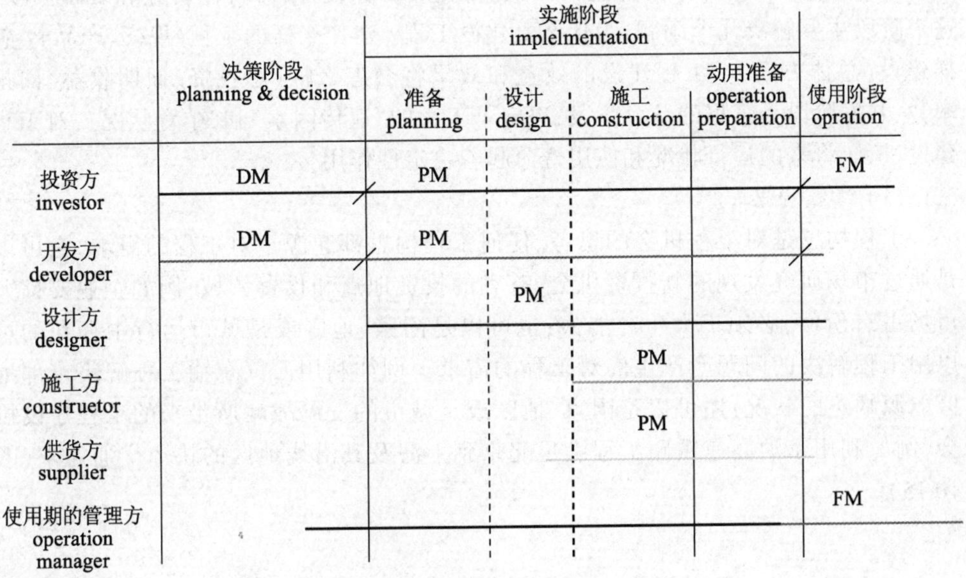

	决策阶段 planning & decision	实施阶段 implelmentation				使用阶段 opration
		准备 planning	设计 design	施工 construction	动用准备 operation preparation	
投资方 investor	DM	PM				FM
开发方 developer	DM	PM				
设计方 designer			PM			
施工方 construetor				PM		
供货方 supplier				PM		
使用期的管理方 operation manager						FM

图 10.13　工程全寿命期以及参与方
Fig. 10.13　Project life cycle and its participators

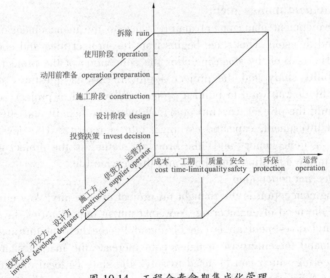

图 10.14　工程全寿命期集成化管理
Fig. 10.14　Integrated management of project life cycle

Participants involved in any phase of the project (investors, developers, designers, constructors and suppliers, etc.) should set the overall optimum of the whole life cycle as the goal and pay more attention to the operational phase to realize the integrated management based on the whole life of the project(Fig.10.14).

10.4.1　工程开发管理

工程开发管理主要指工程的前期策划和决策阶段的管理。这个阶段从工程构思到批准立项为止,其工作内容包括工程的构思、目标设计、可行性研究和工程立项。这个阶段主要解决几个问题:为什么要建设工程? 建什么样的工程(规模、产品)? 怎样建设(总体方案)? 工程建设的效益和效果将会怎么样(总投资、预期收益、回报率)? 开发阶段是工程的孕育阶段,决定了工程的"遗传因素"和"孕育状况",对工程建设过程、将来的运行状况和使用寿命起着决定性作用。

(1) 工程构思

工程构思是对工程机会的思考。任何工程构思都起源于对工程的需求。它可以是通过市场研究发现的新投资机会,有利的投资地点和投资领域,例如企业要扩大市场占有份额,必须扩大生产能力;也可以是国家、地区或是城市运行中存在的可以用工程解决的问题所产生的对工程的需求,例如利用西气东输工程解决东部地区能源缺乏的状况;还可以是国家、地区或是城市的发展战略所带来的工程建设机会,例如利用京沪高速铁路工程满足北京到上海及其沿线地区的快速交通需求(图10.15)。

10.4.1　Development management

Project development management mainly refers to the management of the project's pre-planning and decision. This stage begins from the project idea and ends at the project approval. The task of the stage includes the conception of the project, the goal design, the feasibility study and the project approval. Several questions should be answered here. Why do you want to build a project? What kind of project(scale, product) is it? How to build the project (overall plan)? What is the benefit and effect of the construction (total investment, expected earnings, rate of return)? The development stage determines the "genetic factor" and "the breeding status" of the project and also plays a decisive role in the operating conditions and the service life.

(1) Engineering conception

Engineering conception is the thought on project opportunity. Any engineering idea is derived from the need of engineering. New investment opportunities, favorable investment location and investment areas can be found by the help of the market research, such as the demand for enterprise expansion to increase the market share, or the demand for engineering which can be used to solve the state or local operation problems. For example, the West-east Gas Pipeline Transport projects are used to solve the energy shortage in the eastern region. In the meanwhile, the development strategy brings opportunities to the state and local operation. Another example is Beijing-Shanghai High-speed Railway project (Fig.10.15) which meets the requirements of the region's rapid transit.

（2）提出项目建议书

项目建议书是对工程构思情况和问题、环境条件、工程总体目标、工程范围界限和总实施方案的说明和细化,同时提出需要进一步研究的各个细节和指标,作为后续可行性研究、技术设计和计划的依据。它已将项目目标转变成具体的实在的项目任务。

工程总实施方案包括功能定位和各部分的功能分解,总的产品方案,工程总体的建设方案,工程总布局,工程建设总的阶段划分,融资方案,设计、实施、运行方面的总体方案等。

（3）可行性研究

可行性研究是对工程实施方案进行全面的技术经济论证,看能否实现工程总目标。现代工程的可行性研究通常包括产品的市场研究,市场定位和销售预测;按照生产规模分析工程建成后的运行要求;按照生产规模和运行情况确定工程的建设规模

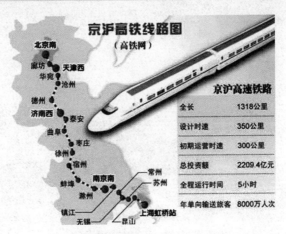

图 10.15　京沪高速铁路工程
Fig. 10.15　Beijing-Shanghai High-Speed Railway

（2）Project proposal

The project proposal is a detailed description of the engineering design, environmental condition, overall objectives, scope of the project and the implementation plan. At the same time, we need to put forward the details of further researches and the indicators, and make them as the basis for the feasibility study, technical design and plan. Thus the project objectives have been transformed into a concrete project task.

The total project implementation plan may include the functional definition and decomposition, the total product plan, the overall plan of the project construction, the total engineering layout, the phase division of the total project, the total financing plan, and the overall scheme involving design, implementation, operation and so on.

（3）Feasibility study

Feasibility study is a comprehensive technical and economic demonstration of engineering implementation plan which is used to decide whether the project can achieve the total goal. The modern engineering feasibility study usually includes the market research, market definition and market forecast; deciding the construction scale and plan

和计划;投资估算和资金筹措;工程经济效益、环境效益和社会效益分析;工程的评价和决策等。大型工程由于其影响面广,工程的评价和决策常常需要在全社会进行广泛的讨论。

10.4.2 工程项目管理

工程项目管理是指工程勘察设计、招投标以及施工阶段的管理。该阶段是将工程构思计划落到实际的重要阶段。工程的进度、质量、成本以及 HSE(职业、健康、安全)等方面的管理也主要体现在该阶段。

(1) 工程勘察设计

工程勘察是指采用专业技术方法对工程所在地的工程地质情况、水文地质情况进行调查,对工程场地进行测量(见图 10.16)。工程勘察工作是设计和施工的基础,对工程的规划、设计、施工方案、现场平面布置等有重要的影响。

图 10.16 工程勘察现场
Fig. 10.16 Engineering survey

based on the operation scale and its requirements; the investment estimation and financing; analysis of economic benefit, environmental impact and social influence; project evaluation and decision; etc. Due to its wide impact, large engineering projects should be widely discussed in the whole society.

10.4.2 Project Management

Engineering project management is the management in engineering investigation design, bidding and construction stage. It is an important stage in which the engineering design and planning will come true. In addition, the management of progress, quality, cost and HSE(occupation, health and safety) will be mainly realized in this stage.

(1) Engineering survey and design

Engineering survey is to investigate the engineering geological conditions and hydro geological conditions of the engineering site by using professional technical methods (Fig.10.16). Engineering survey is the basis of design and construction. It has important influence on the project planning, design, construction planning and the layout of the site.

工程设计是按照工程规划对工程的功能区(单体建筑)和专业要素进行详细的定义和说明,最后通过设计文件,如规范、图纸、模型,对拟建工程的各个专业要素进行详细描述。设计是由设计单位的专业人员完成。对一般工程,设计分为两个阶段:初步设计和施工图设计。对技术上比较复杂的工业工程,分为三个阶段设计,增加技术设计过程。

(2) 工程施工

工程施工过程从现场开工到工程的竣工、验收交付为止。在这个阶段,工程的实体通过施工过程逐渐形成。工程施工单位、供应商、项目管理(咨询、监理)公司、设计单位按照合同规定完成各自的工程任务,并通力合作,按照实施计划将工程的设计经过施工过程一步步形成符合要求的工程。这个阶段是工程管理最为活跃的阶段,资源的投入量最大,工作的专业性强,管理的难度也最大、最复杂。

10.4.3　工程运营管理

一个新的工程投入运行后直到它的设计寿命结束,最后被拆除,它的内在质量、功能和价值有一个变化过程。通常在运行阶段有如下工作:

(1) 运行过程的维护管理。确保工程安全、稳定、低成本、高效率运行,并保障人们的健康,节约能源、保护环境。

Engineering design is a detailed definition and description of the functional area (monomer building) and professional elements of the project, according to the project plan. Based on the design documents such as specifications, drawings, models, the various professional elements of the proposed project are described in detail. Design is accomplished by the professional staff of the design unit. The design stage of general engineering is divided into two parts: the preliminary design and the construction drawing design. A technological design process is essential to a complex industrial engineering.

(2) Engineering construction

Engineering construction process is from the beginning to the completion of the construction and the acceptance check of delivery. The project is gradually formed in the construction stage. The participants, such as the construction units, suppliers, project management(consulting, supervision) companies and design units, should complete their project tasks in accordance with the provisions of the contract to meet the requirements of the engineering design process. This stage is the most active phase in project management, with largest investment in resources and more professional works. The management is also the most complex and difficult.

10.4.3　Facility management

Every new project will be put into operation until the end of its design life, and finally be removed. Its intrinsic quality, function and value have a change process. Usually, there are following work in the running phase:

(1) Maintenance management in the course of operation should ensure the safety, stability, low cost, high efficiency, guarantee of people's health, energy saving and environmental protection.

（2）工程项目的后评价。在工程运行一个阶段后，要对工程建设的目标、实施过程、运行效益、作用、影响进行系统的客观的总结、分析和评价。它是与工程前期的可行性研究工作相对应的。

（3）对工程的扩建、更新改造、资本的运作管理等。此项工作原来不作为工程项目寿命期的一部分，但现在运行和维护管理已作为工程项目管理的延伸，无论是业主还是承包商都十分注重这项工作。

工程经过它的寿命期过程，完成了它的使命，最终要被拆除。有史以来，任何工程都会结束，最终还回到一块平地。可能要进行下一个工程的实施，进入一个新的循环阶段。

对于工程来说，工程寿命期结束是个里程碑事件，而不能作为一个阶段。一般工程遗址的拆除和处理是由下一个工程的投资者和业主承担的。不作为前一个工程寿命期的工作任务。但以一个工程对社会和历史承担的责任来说，应该考虑到工程寿命期结束后下一个工程的方便性，应能够方便地、低成本地处理本工程的遗留问题。

(2) Post-evaluation of engineering projects. After the project operation, it is necessary to evaluate the engineering project objectives, implementation process, operation effectiveness and influence. It is relative to the early feasibility study of the project.

(3) Manage the expansion, renewal, reformation and the operation of capital, etc. The operation and maintenance management has now been an extension of the project management. Either the owner or the contractor should highly value this work.

After its life span, the project will complete its mission, and will eventually be removed. In human history, any project will take an end, and eventually return to the ground. And there will be another project in a new cycle phase.

For engineering projects, the end of the project life cycle is a milestone, but not as a stage. The demolition and disposal of general engineering sites are taken by the investors and owners of the next project. It is not part of the previous project. But with the social and historical responsibility, the convenience and low cost to deal with the previous project should be taken into account.

知识拓展
Learning More

请关注以下微信公众号：
1. 施工技术
2. 筑龙 BIM 门户
3. 中国 BIM 发展联盟

思考题　Review Questions

1. 查阅相关资料后思考：无线传感网络、数据挖掘、计算机视觉等现代高新技术能够给土木工程建造和管理方式带来哪些变化？

Think about the changes in construction and related management made by wireless sensor networks (WSN), data mining (DM), computer vision (CV) and so on.

2. 除了传统的施工设备，你还能想到哪些新设备有助于提高施工效率？例如无人机能用在施工与管理的哪些环节？

Besides the traditional equipments, try to find some new ones which can improve the efficiency of construction. For example, what can be done in construction and related management activities with the help of unmanned aerial vehicle (UAV)?

参考文献
References

［1］李久林,魏来,王勇,等. 智慧建造理论与实践[M].北京:中国建筑工业出版社,2015.

［2］成虎,宁延. 工程全生命期管理[M].北京:中国建筑工业出版社,2018.

［3］葛清. BIM 第一维度:项目不同阶段的 BIM 应用[M].北京:中国建筑工业出版社,2013.

［4］张晓宁,盛建忠,吴旭,等. 绿色施工综合技术及应用[M].南京:东南大学出版社,2014.

［5］周勇,姜绍杰,郭红领. 建筑工程虚拟施工技术与实践[M].北京:中国建筑工业出版社,2013.